AF809768

BEST SERMONS

THE MACMILLAN COMPANY
NEW YORK • BOSTON • CHICAGO
DALLAS • ATLANTA • SAN FRANCISCO

MACMILLAN AND CO., LIMITED
LONDON • BOMBAY • CALCUTTA
MADRAS • MELBOURNE

THE MACMILLAN COMPANY
OF CANADA, LIMITED
TORONTO

EDITED BY G. PAUL BUTLER

BEST SERMONS

1951-1952 EDITION

NEW YORK
1952
THE MACMILLAN COMPANY

COPYRIGHT, 1952, BY
G. PAUL BUTLER

All rights reserved—no part of this book may be reproduced in any form without permission in writing from the publisher, except by a reviewer who wishes to quote brief passages in connection with a review written for inclusion in magazine or newspaper.

First Printing

PRINTED IN THE UNITED STATES OF AMERICA

To

Erica

THE EDITOR'S APOLOGIA

The 52 sermons in this volume have been selected from 6,985 submitted to the editor for consideration.

A sincere attempt has been made to choose the sermons that represent the best sermonic efforts of the great preachers of our day, as well as to discover and recognize unusual excellence in the sermons submitted by younger men who may be the great preachers of tomorrow.

All sermons have been selected for their homiletic value and their spiritual message for our time.

THE INCLUSION OF ANY SERMON IN THE VOLUME DOES NOT MEAN THAT THE AUTHOR OF THAT SERMON APPROVES OR AGREES WITH THE CONTENTS OF THE OTHER SERMONS OR WITH ANY SERMONS IN THE VOLUME. Each preacher has been allowed to speak his own words in accordance with the faith of his own church or denomination. No sermon criticizing any other religion by word or inference has knowingly been included.

Best Sermons is intended as an anthology of current preaching, not a book of theology.

It is the editor's hope that clergymen of all faiths may find inspiration for finer preaching through reading these sermons, and that students in our theological seminaries and professors of homiletics will discover in these sermons a basis for inspiring and informative reading and discussion of current preaching by men of various faiths.

CONTENTS

* Sermons by members of the Advisory Committee were selected on the sole responsibility of the Editor and were contributed at his special request.

Faith

Love

National and International

The New Year

Advisory Committee

for PROTESTANT *Sermons* *

DR. PAUL E. SCHERER

DR. JOSEPH R. SIZOO

DR. RALPH W. SOCKMAN **

---

for CATHOLIC *Sermons*

THE VERY REVEREND IGNATIUS SMITH, O.P.

THE REVEREND JOHN F. CRONIN, S.S.

---

for JEWISH *Sermons*

DR. DAVID DE SOLA POOL (Orthodox)

DR. ISRAEL GOLDSTEIN (Conservative)

DR. ISRAEL BETTAN (Reform)

General European Adviser and
Adviser for Switzerland

DR. ADOLPH KELLER

Adviser for France

DR. CLAYTON E. WILLIAMS

* Dr. Lynn Harold Hough kindly read part of the sermons for this fifth volume. The editor gratefully records his appreciation for Dr. Hough's scholarly advice.

** Dr. Sockman was so busy with his three sermons every Sunday and his own next book, that he was unable to read all of the sermons for this volume.

FOREWORD

ON READING OTHER MEN'S SERMONS

Rev. Willard L. Sperry, D.D., Litt.D.
Dean, Harvard Divinity School
Cambridge, Massachusetts

MR. BUTLER has honored me by asking me to write a foreword to this latest volume in his series of *Best Sermons.* I am glad to accept his gracious invitation.

It is a very dangerous thing for a preacher to spend too much time reading other men's sermons. He may easily fall into the lazy habit of making those sermons his main, if not his only, source material. A preacher ought to preach his own sermons, not some one's else. He ought to live homiletically a first-hand, not a second-hand life. In so far as the layman is concerned he may read these pages without scruple and with profit. They were meant for him in the first instance.

Therefore, in so far as the readers of *Best Sermons* are themselves ministers and preachers, a word of caution and then of guidance is necessary. Such a reader ought not find his own next Sunday's sermon here, no matter how he may attempt to recast and thus disguise it. Yet he ought to get professional encouragement from the entirely proper knowledge that in our complicated and difficult task of speaking in behalf of religion to a world that is largely secularized he is not a lonely prophet, but is a member of a goodly company of fellow craftsmen.

His main professional interest in a volume such as this will be of two sorts: (a) a better understanding of the mind of our time, its theological perplexities and its moral dilemmas, and (b) the best strategic means of reaching that mind by the spoken and now the printed word. There is little use in thundering conventional platitudes from a pulpit, whether theological platitudes or moral platitudes. These platitudes may stroke the fur of the unco-guid the right way, and thus make them purr in their pews. But there is no surer way to rebuff the youth of our time—on whom after all the future depends—than to try to evangelize them by truisms that are merely repeated, rather than re-interpreted. A platitude, according to Coleridge, is not a lie; it is rather an important truth that has for the moment lost its power of truthfulness and lies "bed-ridden in the dormitory of the soul among despised and exploded errors." In short, these sermons were not preached in a void. The preachers understood their hearers and spoke to what the Quakers call their "condition."

In so far as these sermons are good they do not propose theological and ethical novelties. They are witness to the enduring, even the eternal, verities of the Christian religion. But they are honest attempts to revitalize and reinterpret in a contemporary vernacular, the truths without which the Christian religion cannot survive. They should be read, studied and used as adventures in the translation of an ancient classic into a living vernacular. Every preacher has to be, or ought to be, his own translator. This volume is of value in that it is an instance of the way in which fellow translators have set about their task.

I would stress, in particular, the following elements of strength and merit in these pages:

1. Mr. Butler has not used this volume as an occasion for propaganda in behalf of any particular theological point of view. The result is neither a plea for orthodoxy, or neo-orthodoxy, for conservatism or liberalism.

2. Though his public is in the main presumably Protestant, the editor has been ecumenical-plus. He has welcomed Catholic and Jewish contributors, French, English, Scotch and other sermons, as well as American messages. Meetings for church unity are usually unsuccessful in achieving agreement as to forms of doctrinal theology and more particularly of ecclesiastical polity, but they often discover a deeply felt unity when saying their prayers. The same is true of sermons, as in this case. This book is not a collection of sermons preached by a single rather stylized and highly individualized preacher within some high sectarian fences. It is a "catholic" volume, and thus in tune with the times.

3. These pages furnish a clue to the mind of the decently intelligent American citizen of this present time. They give us a clear idea as to what his theological and ethical problems actually are. In so far forth they are a guide to us in helping us understand the mind to which we too are supposedly speaking.

4. Hence we find here valuable suggestions as to the best homiletical means of reaching that mind. The absence of rhetoric, of sentimentality, of anything like "purple passages" and the constant use of direct and sober prose fits the serious mood of the times.

5. Mr. Butler has not restricted himself to sermons by ministers with an established, nation-wide reputation. A number of his contributors are men who are not so well known. Their contributions to the volume assure us of the wealth of good preaching which is going on in our churches and thus reassure us as to the influence of the Protestant pulpit in our time.

WILLARD L. SPERRY

Cambridge,
Massachusetts

PREFACE

THE IMPORTANCE OF THE PULPIT TODAY

THE VERY REVEREND IGNATIUS SMITH, O.P., PH.D.
Dean, The School of Philosophy, and Director of the Preachers Institute, Catholic University of America, Washington, D.C.

WHEN SAINT PAUL urged Timothy to "preach the word . . . in season, out of season, . . . with all longsuffering and doctrine," he left the Church a great commission which added force, if any had been needed, to Christ's own command to his disciples, "go thou and preach the Kingdom of God." There are some today who have expressed doubt of the value of preaching. These men can rest their doubts, for preaching has had a place of tremendous power over the minds and hearts of men in bringing them to God from the days of Saint Peter and Saint Paul down through Clement, St. Augustine, John of Antioch, Chrysostom, Gregory the Great, Bernard, Thomas Aquinas, Peter Canisius, Bossuet, Bourdaloue, Fénelon, Lacordaire, to Ravignan, and the men of our own day.

Great preaching is as necessary today as it was in the time of St. Francis of Assisi, St. Dominic, or Ignatius Loyola. The pulpit must still bring the message of salvation, of faith and life, of moral living, convert sinners, comfort those who are living a Christian life at great personal sacrifice. Speculative truth and a code of daily conduct go hand in hand with the commandments and every revealed mystery. The wise preacher today will not merely say "do" or "don't," for modern audiences demand reasons and the pulpit should garner those reasons from theology, philosophy, science, Scripture, and history.

Millions of people today are living in the midst of indifference to religion, among people who think faith is superstition. Preaching is the most likely way to reach these people. The successful pulpit will combine teaching with inspiration, will instruct the laity that they too may help to bring the world to Christ, and will keep alive the authority of Christ and the Church in the life of man. The modern sermon must be simple in thought, construction, language; its divisions must be clear cut with every paragraph a steady and logical march to a conclusion. Short sentences, simple and dignified words, timely subjects, presented with conviction will convince hearers.

We know that our truth is divinely revealed, but we must preach it with faith, courtesy, courage, persuasion. It is impractical, imprudent and impudent to insult others and offend them with a bigoted sense of our rightness. In our preaching against sin and evil we can always be uncompromising but also polite and gentlemanly. Sin and error are to be crusaded against but sinners and unbelievers must be loved and that love must be shown by sincere concern for their welfare. We must let people know that our preaching comes from God, not merely from the mind of the preacher.

The true preacher will be pious, gentle and strong at the same time. He will be well trained, as well trained as he can possibly be, for this is a day that demands great things from the man of God. The preacher can realize with Saint Paul that he is always preaching for God's cause and in God's presence. The laity have an uncanny instinct in detecting humility and piety in their preachers and more quickly heed priests with such qualities. In writing of the practical techniques of preaching, one must attach great importance to enthusiasm in the preacher, enthusiasm for the importance of the work of preaching, enthusiasm for the word of God and enthusiasm for the welfare, temporal and eternal, of the people. No preacher worth hearing will ever leave the impression that he is hurrying through a sermon as something that he is anxious to have done; rather, he should always leave people with the feeling that he had much more to say, that it is worth coming back to hear him further.

Like an actor, the preacher must watch his voice, care for it, develop it for the pulpit. Believing, breathing, and thinking are assets every preacher can use. Even more, love must displace the hatreds and bigotries, trust must displace the suspicions that divide men and nations. Indifference to God has made about seventy million Americans Godless and when God is ignored or scorned the rights of men will not be respected. Preaching has always been important in the Church of Christ; it is still a major work of the Church and God grant us the determination to work and preach as one for Him and those for whom Christ died.

Washington, D. C.
February 15, 1952

INTRODUCTION

PREACHING: A.D. 1950–2000

WE HAVE heard much about mid-century preaching—as if something startling had taken place as we passed the year 1950 and started through the last fifty years before A.D. 2000. After reading 6,985 sermons for this volume of Best Sermons, I find that preaching is basically the same as it has been for a long time, a dynamic public presentation of faith. But there is this heartening difference—the general quality of the sermons received for Volume V has been higher than those sent for any previous book in the series. Ministers everywhere seem to be more fully aware of the need for great preaching. They are rising to that need.

They have spoken of the need of a spiritual awakening. They have denounced secularism and materialism. They have called upon people to turn to God. Gerald Kennedy asks, "What is the unique function of the Christian pulpit?" and answers by saying the pulpit must proclaim "a unique, divine saving Word." [1] Elton Trueblood says, "Our world is more shaken than it has ever been in our lifetime," [2] while Nels Ferré writes, "The greatest challenge [today] is to breathe vitality into our spiritual life." [3]

Ministers are preaching on almost every topic, in all homiletic styles—they are tackling the pressing problems of the day and are trying to solve them by the application of the Gospel in daily life. Clarence Tucker Craig sees the Church as "a fellowship of faith," [4] while Dean Charles W. Gilkey commented in a recent symposium, "It is the age-long task of the Christian preacher to bear . . . witness [of the mighty works of God] in his own tongue to his own contemporaries." [5] From experience at the Riverside Church, Dr. McCracken writes, "As soon as religion becomes a live issue it poses questions" [6] which ministers must try to answer. Out of a lifetime of preaching Dr. James Gordon Gilkey is convinced that "most people who attend church services . . . hope they will receive help" [7] with the problems of life, and Gustaf Aulén, the distinguished Swedish theologian, adds, "I am persuaded that no form of Christian teaching has any future . . . except such as can keep steadily in view the reality of evil . . . and go to meet it with a battle song of triumph." [8]

1 Kennedy, *The Lion and the Lamb* (Abingdon-Cokesbury), p. 7.
2 Trueblood, *The Life We Live* (Harper), p. 15.
3 Ferré, *Strengthening the Spiritual Life* (Harper), p. 13.
4 Craig, *The One Church*, p. 145.
5 Gilkey, "Preaching," p. 220, in *Protestant Thought in the Twentieth Century*, edited by Arnold S. Nash (Macmillan).
6 McCracken, *Questions People Ask* (Harper), p. 7. (A fine collection of Dr. McCracken's sermons.)
7 Gilkey, *Here Is Help for You* (Macmillan), p. ix.
8 Aulén, *Christus Victor*, translated by A. G. Hebert (Macmillan), p. 159.

It is good to be able to report that the sermons sent to me for this volume show great faith in preaching, in God, Christ, and salvation. In spite of all the tensions and disturbances of today, the ministry has optimism and courage. Once men like the great Jonathan Edwards [9] held sinners over hellfire to scare sin out of them; today the pulpit presents Christ and his redeeming love for men to accept without force or fear. Dr. Joseph R. Sizoo states without hesitation: "This is the age of preaching. The sermon is coming back into its own." [10]

Modern sermons attempt to tell the old, old story in new words, with fresh appeal, to stimulate men to faith, to inspire them to Christian living. Preachers are seeking to catch what Thomas Gray called "thoughts that breathe, and words that burn," with forceful diction, mental energy, imagination, persuasiveness and conviction. The half-dozen very best sermons of every year have the feel of genius as though the preacher had touched the divine lyre that made him preach with words that sing. These are the inspired sermons that open men's minds and hearts to God in the midst of a troubled world.

Truly, sermons are highways to God. Every word, gesture, thought, expression, explanation, exposition, is the attempt of a consecrated man of God to make the ways of God plain to men, to open men's understanding and love so that even one born blind of soul or heart can see. Father Thomas V. Liske says the preacher must communicate "truth effectively every time he speaks." [11]

But like the great highways and streets of our land, a sermon can become cluttered with too much traffic, too many stop signals, too many detours or digressions to get anywhere. With the millions of automobiles on the highways of America in 1952, states are constructing great parkways, throughways, leading directly to one's destination at a penny a mile. A sermon may have many illustrations as open windows or doorways to make ideas clear; but these must not lead up blind alleys or be time-wasting anecdotes.

Every sermon should be carefully constructed on the best homiletic methods, with main theme, divisions, conclusion—but all should move easily and surely as one unit from first word to last, with every non-essential carefully omitted. Anything that gets in the way of the spiritual purpose of a sermon should be ruthlessly cut off, for we are dealing with souls, which never come on a bargain counter.

Preaching is the attention-getter for the church, the only religious education many people ever receive, and the only contact they ever have with God. Therefore, a sermon must grip the minds of people at once, must have suspense, holding power, and a conclusion worth listening for; it must be delivered with enthusiasm and conviction. Great sermons are born in the souls of men of God, out of the needs of their fellow men, and

[9] Jonathan Edwards (1703–1758).
[10] Sizoo, *Preaching Unashamed* (Abingdon-Cokesbury), p. 13.
[11] Liske, *Effective Preaching* (Macmillan), p. 2.

come from God, who still uses the minds of men to help other men. A great sermon will stimulate, inspire, open new doors—like a great book or a work of art—if it is successful; and, like a book or a play, it should be exciting, should suggest more than it says.

Sermons are made of faith and dreams, of men's hopes and deep despair, of God's love and man's fears, of trust and failure. Religious dramas are excellent; radio is far-reaching; television is quick, powerful; but nothing can ever take the place of a live pulpit with a living preacher in it.

Hundreds of books have been written on homiletics or the art and method of preparing and delivering homilies (the ancient name for sermons). Augustine began his long written discussion of preaching about A.D. 430 in *De Doctrina Christiana,* Francis Bonaventura made a long advance in his *Art of Preaching* about 1200 and Jean Claude wrote his famous *Essay on the Sermon* about 1660. From these works down to the treatises by Broadus, Quayle, McDowell, great pulpiteers have reviewed the hows and whys, the principles and techniques, the preparation and delivery of sermons. No minister can afford to miss these works on preaching.

But young ministers should also study, *read,* or *hear* the *sermons* of the pulpit masters—Augustine, Gregory, Aquinas, Paulinus, Abélard, Knox, Calvin, Wesley, de Ravignan, Lacordaire, Montsabre, Fosdick, Sockman, Gannon, Niebuhr, for style, content, phrasing, motivation, art, language, diction. It is here, I hope, that *Best Sermons* affords a distinct service to seminary classes, individual students, active pastors and preachers by bringing together in one volume year by year fifty-two of the "best" examples of modern preaching of different types, on various subjects, from a half-dozen foreign countries and from different faiths as models to be read and studied and analyzed, and in which sermon ideas, illustrations, and sermonic materials may be found. Dean Sperry wisely reminds us that a man must never become a slave to other men's sermons, but while he is growing his own style and acquiring his homiletic training he will need to see what other ministers do. Every minister must develop his own style of preaching and sermonizing, yet it will do him good to hear other men preach, and to read some of the best sermons delivered every year.

Normally ministers preach the Gospel to bring sinners to repentance, comfort church members, strengthen active Christians; but when necessary men of God have risen in every century to thunder like angry prophets to denounce sin in high places, corruption in government, dishonesty in business, wickedness in home life, national life, personal life. At such times the minister must speak with a message of fire in words that burn.

There is much that a "preacher" needs—education, social life, knowledge of men, philosophy, wide reading, recreation, a hobby, time to study, and determination to spend time inviolate on his sermon preparation. Every age has produced its own great preachers. Tomorrow will bring its great men in the pulpit if we keep our training in theological seminaries strong and our self-discipline after seminary and university days on a high plane.

Albert Schweitzer said one of the greatest things of modern times in his "Epilogue" when he wrote that we must believe in "the conquest of the spirit of the world by the spirit of God." But "the miracle must happen in us before it can happen in the world." [12]

G. Paul Butler

New York and Fairlee Haven,
February 15, 1952

[12] Schweitzer, "Epilogue" to *The Theology of Albert Schweitzer,* by E. N. Mozley (Macmillan), p. 116.

ACKNOWLEDGMENTS

AFTER reading 31,838 sermons submitted for consideration for the five volumes of BEST SERMONS, and especially after reading the 6,985 for this fifth volume, I am more encouraged than ever, about the high quality of preaching being done in our day. My first thanks go to the thousands of ministers who graciously allowed me to read their sermons and select the fifty-two Protestant, Catholic and Jewish sermons included.

So excellent and so universal has preaching become in our day that this may indeed, as Dr. Sizoo has said, "Come to be known as the age of preaching." It is a real satisfaction to have Lowell Ditzen, Russell Stroup, Lowell Atkinson, Frederick Bruce Speakman, and other newcomers in the volumes. These younger men have something to say, and say it well.

As a matter of principle, I have made an effort not to include men whose sermons have been in my previous volumes in order to give new ministers, priests, and rabbis representation here. But—it would be an injustice to the "preacher" and the reader to omit James Stewart, Ralph Sockman, Paul Scherer, Gerald Kennedy, Father Gannon, Dr. Sizoo, Bishop Sheen, Dr. McCracken, Père Riquet, Father Martindale, Dr. Louis Finkelstein, and other great contemporary preachers of The Word simply because they have been in my earlier volumes. I am happy, therefore, to repeat some of these men and find from my correspondence and contacts over the country that readers of BEST SERMONS hope to find the great minds of the modern pulpit here. In thousands of letters and statements, ministers have urged me always to include several of these men.

I have spent my great effort—and the major portion of my time and editorial work and reading—in an attempt to give new, young ministers of promise a hearing. It would have been a relatively simple matter to ask the twenty-five leading preachers of the world to give me a sermon for inclusion in the volumes year by year. But that is not what I have done. Rather, I have sent 22,500 invitations to clergymen for this volume alone, and received in reply 6,985 sermons, all of which I have read with care, and have had the final selections read and checked by my advisory committees. This has meant that an average of nearly 135 sermons have been read for each of the 52 included.

For five volumes now the men on my advisory committee have helped me with their knowledge of preaching and have worked with me faithfully; I am glad to record my appreciation to them as individuals and as a threefold committee: Dr. Paul Scherer, Dr. Joseph R. Sizoo, and Dr. Ralph Sockman advised with the selection of Protestant sermons. Dr. Sockman was in the midst of his own three sermons a week, a special new book and important addresses, so that he could not read all of the sermons chosen this time as he has for the previous volumes. Dr. Lynn Harold Hough gave me hours of his time in the midst of preparation for his lectures on the "Five Great Humanists," to read some of the sermons by new men in this volume. Clayton Williams of the American Church in Paris searched for the outstanding Catholic and Protestant sermons in France and assisted with their translation. Dr. Adolph Keller keeps an eye on Switzerland for me. Father Ignatius Smith's advice on Catholic preaching was invaluable; Father John Cronin was most helpful in advising on Catholic sermons. My Jewish advisors, Dr. Israel Goldstein, Dr. David de Sola Pool, and Dr. Israel Bettan, have again been helpful in choosing the Jewish sermons for inclusion.

In writing the "Foreword" on preaching Dean Sperry has drawn on his wide experience in training preachers at Harvard, and Father Smith added an interesting Catholic touch in his "Preface." Yet both are universal enough so that Catholic and Protestant clergy alike can read and apply them to their own preaching. I had hoped to have a statement on Jewish preaching by Dr. Pool, but his attendance at conferences in France and Spain made his paper too late for inclusion in this volume.

Publishers have kindly granted permissions to reprint copyrighted quotations used in the body of various sermons selected for inclusion in this volume and it is a privilege to express my appreciation to:

Doubleday and Company, Inc., for permission to reprint lines from "The Recessional," by Rudyard Kipling and "Man Making" by Edwin Markham from *The Gates of Paradise and Other Poems.*

Charles Scribner's Sons, for permission to quote from Henry Sloane Coffin's book, *The Meaning of the Cross.*

E. P. Dutton & Co., for permission to quote from C. E. M. Joad, *The Return to Philosophy.*

The Macmillan Company, for permission to reprint the selection from "Foreign Missions in Battle Array," by Vachel Lindsay, from *Collected Poems;* a "London Idyll," by John Presland, from *Poems of London;* and *Letters to Young Churches,* by J. B. Phillips.

Harcourt, Brace and Co., for permission to use the excerpt from T. S. Eliot's poem, "Little Gidding"; a quotation from Lewis Mumford's *Faith for Living;* a quotation from Stringfellow Barr, *The Pilgrimage of Western Man;* and the poem by Dorothy L. Sayers, "The Zeal of Thy House."

The Pulpit, for permission to reprint the sermons of Bishop Gerald Kennedy and Dr. Walter Roberts.

Harper and Brothers, for permission to quote from *The Best of Studdert-Kennedy,* and for portions of biographical notes on clergymen whose sermons appeared in previous volumes.

Houghton Mifflin, for the courtesy of quoting "The Eternal Goodness" from the *Works of Whittier.*

The Methodist Advance for Christ and His Church, for permission to reprint "Endless Line of Splendor," by Bishop Oxnam.

Dr. Eric M. North for permission to reprint four lines of "Where Cross the Crowded Ways of Life."

Pilgrim Press, Boston, for permission to reprint ten lines of John Oxenham's poem "The Ways," from *"Gentlemen—the King."*

Fleming H. Revell Co., for the quotation from Henri Bergson in *You and Yours* by Guy L. Morrill.

Pacific Coast Publishing Co., for the poem by Hermann Hagedorn, "The Bomb That Fell on America."

Willett, Clark & Co., for the poem "Build Me a House," by Thomas Curtis Clark, from *Home Roads and Far Horizons.*

Edgar A. Guest, for permission to reprint six lines from "We Who Are Americans."

Bishop G. Ashton Oldham, for permission to reprint "America First."

In the preparation of the manuscript so many people helped that it is impossible to mention all by name—denominational leaders, bishops, scholars, councils of churches, ministerial associations—yet I hope they will know I have not forgotten them. In seventy-one cities groups defied the heat of the summer of 1951 to meet with me. Bishop Tippett called twenty-five leading ministers to confer with me in San Francisco; Bishop Kennedy even when on a world tour left plans for conferences in Seattle and Portland; leaders in Los Angeles, Phoenix, El Paso, San Antonio, Denver, Omaha and on across the country extended every courtesy to me in my wide sermon search.

Guy Brown, J. P. R. Budlong, and the staff of the Macmillan Company have taken such a personal interest in BEST SERMONS that appreciation must be expressed to them here.

It is the editor's conviction that anyone who will read these 52 sermons with an open mind will find faith if he does not already believe!

G. P. B.

"And, as ye go, preach."

Power to Work Wonders

REVEREND RALPH W. SOCKMAN, PH.D., D.D., LL.D.[1]
Minister, Christ Church, Methodist, New York City, and Preacher on the National Radio Pulpit

Robert Browning is often referred to as the "poets' poet." Ralph Sockman is in the same sense the "preachers' preacher," yet he speaks the language of the multitude with an unexcelled beauty and power in his words. To hear him preach is a spiritual event.

Recently selected by New York University as the new director of the Hall of Fame for Great Americans, Dr. Sockman is one of the great preachers of our day. At ten o'clock every Sunday morning he has a nation-wide radio audience on the National Radio Pulpit of the National Broadcasting Corporation and the National Council of the Churches of Christ in the United States of America. At eleven and five each Sunday he preaches to his own congregation at Christ Church, where many New Yorkers go week after week for a spiritual message and visitors from all over the country fill all the remaining seats.

Dr. Sockman was born in Mount Vernon, Ohio, graduated from Ohio Wesleyan University, took graduate work at Union Theological Seminary and holds his Ph.D. from Columbia University. The honorary doctorate has been conferred upon him by Ohio Wesleyan, New York University, Wesleyan University, Dickinson College, Rollins College, Washington and Jefferson College, Florida Southern College, and Northwestern University.

While he was still a student, he became associate minister of the Madison Avenue Methodist Church, and has been minister of this church since 1917. He served with the Army Y.M.C.A. during 1918 and from 1927 to 1929 was president of the Greater New York Federation of Churches. In 1933 the present new church building was erected at the corner of Park Avenue and Sixtieth Street and the name was changed to Christ Church, Methodist. Strangers who attend the church soon find that New York has a warm heart and that the minister of this church has warm friendship to give them all.

In 1941 he was the Lyman Beecher Lecturer on preaching at Yale University, is on the Harvard University Board of Preachers, and he gave the Fondren lectures in 1943. He is a director of Union Theological Seminary in New York, and of New York University, a trustee of Drew Theological

[1] Sermons by the members of the Advisory Committee were contributed at the request of the editor and are included on his responsibility.

Seminary, of Ohio Wesleyan University, also Syracuse University, Goucher College, and Santiago College in Chile. He has served on the Board of Foreign Missions of the Methodist Church, is president of the Church Peace Union, and is one of the most far-sighted churchmen today.

He is widely known for his excellent books, including Suburbs of Christianity, The Unemployed Carpenter, Paradoxes of Jesus, The Highway of God, Date with Destiny, The Lord's Prayer, Now to Live, Live for Tomorrow, Recoveries in Religion, Morals of Tomorrow, Men of the Mysteries, *and* The Higher Happiness.

During 1947 to 1949 he was visiting professor of homiletics at the Divinity School of Yale University, giving the brilliance of his homiletic insight and his creative impulse to the men who come from all over America and other lands to perfect their preaching technique. He lectures on preaching at Union Theological Seminary, New York. "Power to Work Wonders" was delivered on Sunday, January 21, 1951, on the National Radio Pulpit of the National Broadcasting Company, and was also preached to his own congregation at Christ Church. He catches facets of the wonder of life and faith, the miracle of reconciliation, and shows how the Gospel can help men in our world today. The sermon upholds the reputation of this great preacher who never fails to capture the minds and hearts of his hearers.

Sermon One

WE ARE told repeatedly that the Russians have been restrained from attacking us by the fact that we are superior in atomic power. Perhaps that is true. How much the Soviet leaders are held back by this fear, we have no means of knowing.

But I submit this morning that there is another power which the Kremlin cannot match or even appreciate. It is a force which, in the long run, so far surpasses atomic energy that the nation which possesses the most of it will eventually win. I refer to what John's gospel had in mind when it said of Christ in the prologue that "as many as received him, to them gave he power to become the sons of God."

Perhaps some of you are saying, "Oh, what a let-down. I want to hear something practical and realistic, and now the speaker is going to talk about some vague thing called the power of God." Well, if that is the way you feel, you are right in line with the communists, for that is what they think. They laugh in scorn at any power religion can release.

The other night I engaged in a televised debate with a communist leader, Dr. Katz-Suchy, the head of the Polish delegation to the United Nations.

Like all communist propagandists, he would have us believe that Karl Marx started a great movement which will free the world from the shackles forged by so-called Christian civilization.

Now the fact is that Christ came as the liberator, and communism is a throw-back to a pre-Christian enslavement of the individual. In the times before Christ, the individual person felt himself in the grip of two unbreakable forces—the force of fate and the power of the government over him.

Into this static world Christ came with the liberating idea that man could break through the bonds of fate and government and remake the things that are into the likeness of things that ought to be. Christ came to give men the power to set themselves free. It is a tragic shame that Christianity has allowed communism to pose before the world as the great liberator; for Marxism makes man a mere economic unit subject to the iron laws from which Christ freed us. It is high time that we who call ourselves Christian leave the defensive attitude into which we have allowed communism to drive us and reassert Christ as the true liberator for the people of the world.

But let us begin with ourselves. What can the gospel promise mean to us when it says: "As many as received him, to them gave he power to become the sons of God."

First of all, when we receive *Christ into our minds,* he gives us the power to think as sons of God. Have you ever reached the point of confusion and distraction that you said, "I just can't think."

When our minds reach that state, what can Christ do for them? Last summer I had an experience which may serve as a suggestion. In the city of Rio de Janeiro the traffic confusion makes this New York traffic seem almost orderly and I can think of nothing worse with which to compare it! The din, the speed, the snarls defy description. But when you ascend from the honking whirl to the summit of Mount Corcovado and stand beneath the great statue of the Christus, you look down on the most thrillingly beautiful panorama of city and harbor which this world affords. Up to that height the sound of the traffic does not reach. And from that height one can look with a calm perspective which brings peace to the soul. There you can think.

So when our minds are so distracted and our thoughts become muddied and muddled, we can look up to the Christ and rise into a high mindedness where little things and petty puzzlements do not reach us.

And Christ helps us not only to think, but to think straight. He lengthens the outlook and also shows the next step toward the desired goal. It has been said that the basic question in living is not what to do, but what to do next. There is always a next thing to do. And Christ gives us a sense of priorities, just as he did in the home of Mary and Martha, when he found the latter flustered and said to her: "Martha, Martha, thou art careful and troubled about many things; but one thing is needful and Mary hath chosen that good part."

Christ gives us those luminous moments when we see through things to their meanings. Professor Bliss Perry, late of Harvard, writes that when a student at Williams College he once complained to his father, a professor there, about the time wasted in attendance at college chapel services. His father replied: "If you are turning a grindstone every moment is precious; but if you are doing a man's work, the inspired moments are precious." When we turn our thoughts to Christ, we get inspired moments. There are occasions you know, when we must take time off in order to take eternity in.

When we receive Christ into our thoughts, he gives us the power to think and to think straight and also think positively. Dr. Emmet Fox gives what he calls the "Golden Key" to prayer. It is this: "Stop thinking about the difficulty whatever it is and think about God instead." Don't look back over your shoulder to see how things are coming; just let the trouble stay behind and think about God. "God is Love," "God is Wisdom." Personally I see the value of his principle, but the words "Love" and "Wisdom" are a bit too abstract. Christ, however, makes God vivid to me. When I think of the historic Jesus, walking the roads of Palestine, clear of eye, calm of voice, compassionate of heart; when I think of his courage as he faced his critics and of his undimmed radiance as he went to his cross—well, I find my nerves quieting, my mind clearing, my spirits rising. And I'll make this proposition to any one of you that tomorrow in the midst of any perplexity—I don't care what it is—if you will take three minutes to think about Christ you will come back to your problem with a clarity of insight which will work wonders. As the Scriptures say, "There is a One able to do exceeding abundantly above all that we ask or think."

Yes, and not only will Christ enable you to think straight and positively but also creatively. These minds of ours can surprise us with their powers of creative thinking, once we get them out of their grooves. In our popular search for peace of mind, we are making a mistake in seeking that peace so largely by adjustment to things as they are. We are in danger of mistaking ease of mind for peace of mind. I have a chair in my home to which I am perfectly adjusted. It fits me so well that I can relax comfortably in it. But I cannot write sermons in that chair. Perhaps you think my sermons would be more pleasant if I did, but I can't! I cannot get my mind creatively at work when my body is too comfortably relaxed. The Spanish philosopher Unamuno had a real point when he asked God to keep us from peace and give us greatness. It was what Jesus must have had in mind when he said, "I came not to send peace, but a sword." Of course, the Prince of Peace did not mean that he came to bring material swords, but he did come to set men striving for a better world with a soldier's fervor and force of spirit.

Christ gives us patience and fortitude to endure the things that cannot be changed. But he also came to give us courage to challenge the things which should be changed—and wisdom to know the difference. The follower of Christ hears what Kipling's explorer heard, an "everlasting whisper" calling:

"Something hidden go and find it, go and look beyond the ranges, something lost behind the ranges. Lost and waiting for you. Go."

One day Jesus was walking along the Jordan River. Two men followed him. When Jesus turned and saw them, he asked: "What seek ye?" They replied, "Master, where dwellest thou?" Jesus answered, "Come and see." They came and saw where he dwelt. One of the two was named Andrew. Something about the Master so impressed him that Andrew went and called his brother Simon Peter. Thus two humble fishermen whose thoughts had been limited to catching and peddling fish became fishers of men, and the names of Saint Andrew and Saint Peter are enshrined in almost every community around the world. When Christ is received into the mind, he gives the power to think straight, positively and creatively. In mind they become sons of God, with power to work wonders.

But to receive Christ into the mind is not enough. We must also receive him into the *heart*. Then he gives the power to feel like sons of God.

So often we hear the old proverb, "As a man thinketh in his heart, so is he." I used to think it meant simply that thoughts shape character. But it means more. The old wisdom writer is warning against being deceived by those persons who invite us to their tables and say nice things to us when their hearts are not really in their words. It is what a man feels in his heart that counts.

My religion is not vital until my heart is in it. Christ is not a real power in my life until he gets into my heart, for the heart is the seat of the affections.

Christ works wonders with our affections. He enlarges them. If we just followed our own inclination and moods, most of us would not learn to like very many people. We might find our circle of affection growing smaller with the years. Dr. Henry C. Link tells of a young woman who found herself becoming lonelier as the years passed. The number of her friends was shrinking. Looking into her case history, Dr. Link found that she had never exerted herself to be agreeable to people who did not appeal to her. If company came to call on the family, and she did not think she liked them, she would remain in her room. Thus never putting herself out to make friends, she found herself being more and more left out.

Christ leads us to put ourselves out in order to take friends in. He inspires us to see good in persons we never noticed or liked. He quickens our imagination to feel how others feel. And in these days of tension between nations, imagination is even more important than information. Cold facts make cold wars, and unless the present trend is halted, will lead to hot wars.

Moreover, Christ-like imagination is needed in order to find out the truth about other lands. You see, we demand an almost impossible service of our press and radio. We expect them to give us the news of the whole world every day. In order to do this, our newspapers and radios have to generalize from what a few leaders say. If the White House speaks, Europe says that is America talking. If Downing Street speaks, we say that is England talking.

When the Kremlin says something, we say that is Russia speaking. Well, America is more than the White House, England is more than Downing Street, and Russia is more than the Kremlin. And if we are ever to develop understanding and brotherhood among the peoples of the world, we must have imagination enough to see into the faces, the families, the human values of the people in Asia, in Russia, in Europe.

Here is where Christ helps when he gets into our hearts. When we look up to God the Father of all men and pray to him through Jesus Christ Our Lord, our imaginations are quickened to see how life looks to men under other flags, other colors of skin, other creeds of faith.

Christ in the heart not only helps to control and enlarge our affections, but he also gives power over our motives. There is a lot of stupid talk to the effect that human nature can't be changed. How often we hear that man is a fighting animal, always has been and always will be, and therefore wars are inevitable. Of course, there is an instinct of pugnacity in human nature. But, as Professor Gordon Allport of Harvard tells us, motives can be completely altered and thus our instinctive drives can be redirected. General William Booth, for instance, of the Salvation Army was as brave a soldier and as hardy a fighter as any man who ever wore the uniform of a nation's army. But his fighting spirit was directed against sin and poverty and suffering.

What if the two most powerful nations of the world, America and Russia, which now seem to be drawing ever nearer to mortal combat, could change their motives and enter into a genuine rivalry of service, each trying to outdo the other in helping backward peoples to fight poverty and hunger and disease! America had such a motive when she began the Marshall Plan, but the Russians misunderstood and misrepresented our intentions. And to us the Soviet Union seems to have been steadily pursuing a selfish policy of greed and grab. And so our missionary motive which started the Marshall Plan has been changed into a program of defense with the largest peace-time budget in our history.

Of course, our nation has to be militarily strong to match the power of Russia. But must we match the motives of Russia? Under God, no! If we are a Christian nation, let us take our motives from the cross rather than from the Kremlin. Let us keep clear in our own minds that the billions we are to spend in Europe are to help keep Europeans free and not to use Europe as a mere outpost for our own defense. Let us keep clear in our own minds that what we do in Korea is for the good of the Koreans and not merely for the sake of ourselves.

Let us give the two million tons of grain which India so desperately needs and let us do it because the starving people of India are our brothers in the family of God.

If our motives are sincerely Christian in the desire to serve, the world will eventually discover it. The underprivileged people of the Orient will some day wake up to know the difference between Christian power out to serve and a communist power out to exploit. And remember, ultimate victory in the Orient lies not with the nation which wins the immediate battles in Korea but with the nation that wins the hearts of the Oriental people. God grant that America may be the victorious nation because she remains true to the "Son of Man who came not to be ministered unto but to minister."

Sometimes, I confess, the world's fog of confusion engulfs me in despair. And then at other times the thought sweeps over me that this is the greatest, the most thrilling hour of opportunity in all history. Why do I feel this? This is the reason. For over a century America has been sending missionaries over the world to proclaim Christ as Lord and Savior. Now we are confronted by a powerful government which professes to be godless. Now the world at last has a chance to see the difference between a nation which declares "In God we trust," and one which officially denies God. If America can meet this test by keeping her motives truly Christian, if she can show herself the nation that ministers rather than seeks to be ministered unto, then I am confident the minds and hearts of the world's hungry hopeless people will be open to receive the Christ who gives power to become the Sons of God.

Back in 1943 when the war was passing its turning point, a leader in Europe said: "Without the miracle of American production, the United Nations could not possibly have hoped to win." The leader who said that was Joseph Stalin.

I believe America can still repeat the miracle of material production. But I pray that she will have the power of the spirit to work still greater wonders through Jesus Christ Our Lord.

PRAYER: *Eternal and ever-loving God, who didst come all the way to the Cross to save us, help us to draw nigh to Thee in grateful remembrance of Thine unmerited mercy. Thou hast given us the good earth and we have soiled it with the blood of our brothers. O God, forgive us. Thou hast given us these bodies to be the temples of Thy Holy Spirit and we have used them for selfish indulgence. O God, forgive us. Take these lives of ours and use them in Thy service. Give to the church a new vision and renewed unity! Grant to the United Nations the wisdom and courage so to act that it shall not lower its standards nor lose the confidence of the people who are praying as never before for peace. Give us grace to come through the trials and confusion of the present hour more than conquerors through Him that loved us, Jesus Christ Our Lord. Amen.*

Standing Up to Life

Reverend Joseph R. Sizoo, D.D., Litt.D., S.T.D., LL.D.[1]
A Minister of the Reformed Church and President of New Brunswick Theological Seminary, New Brunswick, New Jersey

Dr. Sizoo is one of the great preachers of our day. His sermons have a historical perspective which makes his religious interpretations doubly valuable for men and women in our troubled times.

Born in the Netherlands, he was educated at Hope College, Columbia University, and New Brunswick Theological Seminary. Hope College and Rutgers University conferred the honorary D.D. upon him; Hastings College conferred the Litt.D.; Columbia University the S.T.D.; George Washington University the LL.D., in recognition of his brilliant preaching and church and national leadership.

He has been a minister of the Dutch Reformed Church all his life except for twelve years when he had a distinguished ministry at the famous New York Avenue Presbyterian Church in Washington, D.C. It was here that diplomats, senators, businessmen and people from all walks of life crowded his church to hear his brilliant preaching.

The Collegiate Church of St. Nicholas called him to New York, where he won recognition and great respect for the substantial character and spiritual quality of his sermons. This church was the oldest Protestant congregation in New York City with a continuous ministry and was originally organized at the Battery as "the Church in the Fort." Theodore Roosevelt attended St. Nicholas Church when he was a boy.

In 1942 he was president of the General Synod of the Reformed Church in America, served a term as president of the Greater New York Federation of Churches, as vice-president of the Protestant Council, and as chaplain of the Twelfth Regiment of the New York Guard. His books have been widely accepted. Three of them, Make Life Worth Living, Not Alone, On Guard, *and* Preaching Unashamed *have had a great influence. During the war he visited army and navy bases and spoke to the enlisted men and chaplains.*

He is more and more in demand as a preacher in various important cities of America and in important churches from coast to coast. This sermon is a grand study of people who do not quit when life and things get hard. Peter stood at the door and "continued knocking." It is a sermon of the grandeur

[1] Sermons by members of the Advisory Committee were contributed at the special request of the editor and are included on his responsibility.

of soul that God gives a man who follows him through adversity and heart-break and suffering. The editor is proud to present this sermon to the readers of Best Sermons.

Sermon Two

TEXT: But Peter continued knocking. ACTS 12:16

THIS IS THE story of a man who tried and failed, but refused to give up. He was one of those rare individuals who addressed himself to a specific task. He struggled hard to reach his goal, but when he arrived the door shut in his face. Through circumstances beyond his control and which he could never foresee, his mission failed. What he set his heart upon faded away like an old soldier. But when the door closed he refused to give up and was determined to see it through.

That lesson is stretched across this story of Peter's imprisonment. It was early in his ministry of witness to the word. With great zeal and utmost devotion he gave himself to preaching the good news of Christ crucified and risen again. He went up and down among his countrymen bearing witness to the power of that Gospel. Soon he found himself enmeshed by hostile forces and in the end he preached himself into prison. Prayers were made for him continually and one night an angel came; the chains fell off his hands, the guard fell back in a stupor, and the doors of the prison opened. There is here an interesting aside. The angel said to him, "Put on thy sandals." It seemed strange that the angel which could break his chains and open prison doors could not also put on his sandals, because time was of the essence. But he was called back with the admonition, "Put on thy sandals," as if to say, "God will always do for you what you cannot do for yourself; but God will never do for you what you can do for yourself." You can almost see him go out into the night, hiding in the thick shadows, to the one place where he knew he would be received and where he would be safe. It was one of those moments when every split second counted. When he came to the door a maid saw him, shook her head and walked away. You read, "She opened not the door." Peter did not fly into a rage or become hysterical. There is no evidence anywhere of panic or fear. The very next sentence reads: "But Peter continued knocking." That was Peter's finest hour.

All that has meaning and significance for us. Let no one suppose this book is antiquated. H. G. Wells once remarked that we should either have a new Bible or that it should be brought up to date because it has lost touch with our world. Well, every page of this book has its relevance for human life in this day and every day. So it is here. There are times when we address ourselves to some high adventure, undertake some great resolve or give ourselves to some holy cause. We hold nothing back. We give ourselves to it in great devotion, utterly unconcerned about ourselves. But in spite of everything that we do, nothing comes of it. The harder we try, the more do we become entangled by defeat and despair. One day you knock at the gate of ambition until your knuckles are raw, but no one opens the door and nothing comes of it. You go out in search for health; you obey all the laws and regulations to achieve it, but the door remains closed and nothing happens. Or again, you knock at the gate of happiness, for it is the one thing you want above all other things. You hear voices and laughter in the house of mirth, but no one opens the door and nothing happens. How often we meet that in life! Robert Browning once said that the reason why men fail is because they are unwilling to pay the price. That is only a half truth. Columbus wrote in his diary that he could have accomplished so much more had those upon whom he counted played their part. You cannot evade experiences like that. All of us have our closed doors.

In hours like that it is so easy to walk out on the vision or turn your back on some high commitment. Many settle back in some grim frustration, shrug their shoulders and say, "What is the use?" Others will pin a crape on their arm and feel sorry for themselves. There are those who are seized with panic and become hysterical; others are overwhelmed by cynicism and become embittered. Here is a teacher who with great devotion gives herself to her little group; but the children are listless and indifferent. It is so easy for the teacher to say, "Why should I care if they do not care?" Here is a businessman who holds standards of high integrity and honor. He turns neither to the right nor to the left; but all around him are those with lower standards of integrity who seem to advance and profit by it. It is so easy for this businessman to say, "Why should I hold my banner high? Why not live by their code and get rich?" Here is a politician who still believes that his call to office is by the mandate of the people. He believes that public office is a public trust. He determines that he shall seek the highest interest of those whom he serves. But there are those less scrupulous in office who give people what they want and thereby get themselves reëlected. It is so easy for him to say, "Why should I burn myself up? Why not give people what they want and so get reëlected?"

The person is not normal who is not sometimes staggered by these closed doors, whose soul is not chilled by cruel and undeserving reversals. It is a cheap thing to run up the white flag and say, "It is only a figment of the imagination." You cannot dodge hours like that. They come to all alike, old and young, rich and poor. They haunt like a vise and follow like a shadow. There are times when we must cry with the poet:

Leave me not weary and dying,
Lost and in pain
Like a forgotten thing, lying alone
In the rain.

Whether or not you set your heart on the search for happiness or health or ambition or friendship, sooner or later we all face these closed doors. Circumstances beyond our control and which we could never foresee keep us from realizing our goals. Life has a strange way of throwing road blocks.

What do you do when things like that happen? What can you make of this experience? How can one face these dilemmas?

What has the religion of Jesus to say to people in an hour like that? By the ability to stand steadfast, life is always tested. Impetuous Peter never revealed the grandeur of his soul more than in the hour when he continued knocking. The worth of a person is laid bare in, "and having done all, to stand." Those who take counsel of fear and frustration never build a better tomorrow. Fidelity to some high purpose, whatever the odds may be, marks the value of one's inner being. Before these closed doors life is always asking, can you keep sweet in the face of bitterness?—patient in the face of frustration?—tolerant in the face of disillusionment? Can you toil without reward, and labor without returns? There is that inner voice which is always raising the inquiry, will you stand your ground?—will you see it through?—"are you able to drink the cup that I drink of and be baptized with the baptism with which I am baptized?"

It is the person who stands his ground, refuses to give up and is determined to see it through who lifts the level of life and writes his name across immortality. The high achievements which history relates belong to those who refuse to turn their backs on some high adventure. All the progress in science, all advancement in civilization and all the enrichment of our culture has come to pass through those who refused to take no for an answer. There was a time when Gideon, at the threshing floor at Gilboa, was able to muster only three hundred men to stand against an army of ten thousand; but he refused to give up. Washington at Valley Forge was in command of troops who were ill fed, ill clad, underpaid and in half mutiny; but he did not walk out on the venture. There was a time when Lincoln had against him many of the most influential leaders of the north. He said one day to a friend, "I have just finished reading on my knees the story of the Son of God in Gethsemane; I am in my Gethsemane now." But Lincoln refused to abandon what he set out to achieve. Have you ever thought how much we owe to those who refused to turn their backs on some great resolve even though the doors remained closed? What if they had walked out on the vision?

Some years ago a little boy suffered a cruel accident. For untold and wearying months his mother nursed him. The doctors came one day to say, "He will never walk again." But the mother refused to accept life's closed door and together, the boy and the mother, were determined to see it through. Some years later this boy became a young man and stood at a tape when the pistol cracked over his head and he won the mile record in the Olympics. One day a singer sang a song. The dramatic critics praised her to the skies. Some said that God in His goodness had given to this world a new voice. Then she was stricken with a vague disease and the physician said, "She will never sing again." But never is a long time. She determined to see it through until one day once again her golden voice rang through the Metropolitan Opera as people choked back their sobs of praise. How true it is: by your ability to stand fast, life is tested and the grandeur of one's soul is laid bare in the determination to see it through.

That was the glory of Jesus. There has never walked this earth one who gave himself so completely to the needs of men. He dragged the sorrows of this generation across his soul. He identified himself with all in distress. He was the most compassionate person who ever lived. He went about helping and healing and doing good. But the harder he tried, the more was he entangled by a hostile generation. But he did not walk out on the adventure. He stayed with it to the end of the end. No more magnificent sentence was ever written about him than this: "Having loved His own which were in the world, He loved them unto the end."

Something of that we must recover for our time. Much is said of the hopelessness of our present world. If it is true that on the continent one finds much of nihilism, then it is also true that we in our land have more than our share of cynicism. Many will tell you goodwill is a wonderful thing, but it won't work. They have convinced themselves that hunger will always gnaw, disease will always ravage. It is so easy to shrug the shoulders, walk out on the adventure and say, "It is no use." Many today are disposed to abandon the high adventure of freedom, justice and goodwill. That is the great peril of our time. We desperately need men and women who will face these dilemmas without fear and see them through until the barriers break down. Believe me, the hope of the world is in the hands of those who will not take counsel of despair.

You say, "This is all very well, but how can one live with that faith and where shall one find it? What are the inner resources upon which it rests?" Let me go back to the story. Peter saw it through because he never lost faith in human nature. He knew that those in that home were capable of something better. In spite of all appearances and no matter how disillusioning the hour, he was convinced they were still his friends, would come through, open the door and welcome him. He never lost faith in human nature. That is still true today. It is to be questioned if one can long face these experiences without it. Human nature is very disappointing at times. We fall apart rather quickly. People at their best are very brittle and you can never tell where or when it will occur. Sometimes the most selfless people give evidences of cruel selfishness. The most compassionate show signs of callousness and the most saintly stagger across the threshold of indiscretion. But in spite of everything man is still a child of God, made in His image, fashioned after His likeness. He is capable of far better things than He sometimes reveals. You may deface the image of God, but you can never efface it. There are deep reserves of decency and grandeur in every human life which only need releasing to come to the surface. Jesus never lost faith in people. He took a tax-collector and made him the writer of the first gospel. He took a cynical skeptic and made him the preacher of a universal, eternal love. Sooner or later, if you are willing to wait long enough, that way of life is always vindicated.

All this is equally true on the world level. Here we are seeking honestly and sincerely to live at peace with all mankind. We want to share with the world our goodwill and understanding. We want no one's land; we want no one's possessions. The only thing we seek is that men shall live together to bring in a better tomorrow for all God's children. We regret that sometimes this motivation seems clouded, for as a nation we blunder along rather badly. We have a way of contradicting ourselves. Sometimes we blow hot and sometimes we blow cold. Indeed, we confess that at times we are a paradox to ourselves. And yet, deep down underneath everything and in spite of our frailty, we want to live and let live and help live.

Today we are accused of cruel self-motivation. Our desires are wilfully and cruelly and falsely misinterpreted. What shall we do? Above all things we need today a patience to stand before this closed door. The keepers of the iron curtain may be ruthless, but behind that frontier of steel there are people much like ourselves with the same hopes, the same desires, the same aspirations. They, too, want to live in dignity and in freedom and in goodwill. One day when this insanity ends and the angels of their better selves have the opportunity to express themselves, these iron curtains will be torn apart, the door will be opened and we will be welcomed in friendship. It may be difficult to believe that, but it is infinitely more difficult to believe the only alternative.

> When today is ending He will come some day,
> Just as of old, just in the same old way,
> Naught shall be changed. The sunlight still shall fall
> With lengthening shadows on the floor and wall.
> The little tasks all finished, I shall wait for Him
> And shall not wait in vain.
> For He shall come, shall place upon my brow
> The old sweet kiss and He shall say, "Thou,
> Thou who hast waited I have come at last."
> The hideous dream of war is past, is past.
> O my beloved, let thy weeping cease
> For once more men are brothers, there is peace.

Again Peter saw life through because he never lost faith in God. He put himself in the keeping of God and left himself there. He did not ask too many questions about life; he simply lived it. He was convinced that we live in a moral universe and that history belongs to God. He knew that he was held by a love that would not let him go. He could say with Job, "Yea, though he slay me yet will I trust him." Many times he must have quoted the assurance of Paul, "I know in whom I have believed and am persuaded that He is able to keep that which I have committed unto Him against that day." Peter was so convinced of this truth that one of his last messages to the early Church was, "Who hath begotten us unto a lively hope; to an inheritance incorruptible that fadeth not away."

When Cecil Roberts, gifted contemporary British novelist, came to this country during the war he told some of us that as he was driving his car from his home to the port of embarkation he saw along a rural road a freshly dug grave that had a stone. He wondered what it was and why it was there. So he retraced his steps to the nearby village to inquire about it. They directed him to a little thatched-roof cottage in which there lived an elderly woman. She told him that her pet marmoset had been killed and she asked the authorities if she might bury it where it was killed. Because she was an old woman they gave their consent. Then Cecil Roberts went back to this little grave, brushed away the leaves that had gathered and saw written on the stone, "There is not darkness enough in all the world to put out the light of one small candle."

So I say to you when life tumbles in and the doors close; when the lamps go out and the lights begin to flicker; when hope no longer sees a star and love no longer sees the rustling of the leaves; when it is touch and go; remember there is the love of God to enhearten; the hand of God to hold; the strength of God to sustain and the light of God to guide. So may one stand steadfast before life's closed doors unafraid.

CHRIST

The Mind of Christ

Reverend J. Arthur Lewis

Minister of Collins Street Baptist Church, Melbourne, Australia

The Reverend J. Arthur Lewis is known as the outstanding Baptist preacher of Melbourne today. His sermons attract large crowds Sunday morning and evening and he is also highly rated as a Bible teacher. He is respected among his brother ministers as a scholarly and courageous leader. His city church has about two thousand members and conducts an active social and religious program.

"The Mind of Christ" is a fine example of a sermon which applies the teachings of Christ to the Christian life of today. The editor is happy to present this outstanding message by a clergyman "from down under."

Sermon Three

TEXT: Treat one another with the same spirit as you experience in Christ Jesus. PHILIPPIANS 2: 5 (Dr. Moffatt's Translation)

DO YOU EVER sit down and wonder what is wrong with the world? Do you ever ask yourself why it is that Christians seem to have so little influence, why they seem to achieve so little, for all their numbers, in putting the world right? To each of those two questions there is ultimately but one answer. It is this: we lack the mind of Christ.

There are some subsidiary reasons for the plight of man today, which we must not overlook. One is that there are so few Christians in the world. There are not enough of us. For all our numbers, the amount of Christian leaven in the world is all too small for the mass it has to permeate and change. Another reason is that even those who are Christians are not Christian enough. But to keep to the main point: what is wrong with the world today is that it lacks the mind of Christ; and there can be no hope for the world until it possesses, or is possessed by, the mind of Christ. Let me tell you why that is so.

First of all—what is this "mind of Christ," for which I am making such large claims? The Apostle Paul tells us what the mind of Christ is. "Being in the form of God He—Christ—thought it not robbery to be equal with God: but made Himself of no reputation, and took upon Him the form of a servant, and was made in the likeness of men; and being found in fashion as a man, He humbled Himself, and became obedient unto death, even the death of the Cross." And therefore, says the Apostle, "Look not every man on his own things, but every man also on the things of others. Let this mind be in you, which was also in Christ Jesus." Dr. Moffatt translates that last sentence thus: "Treat one another with the same spirit as you experience in Christ Jesus."

Now let us see what we have here. Remember, we are trying to elicit what is the mind of Christ. We find that our Lord lays no obligation upon us that is not first found in Himself. Nay, more: there is nothing He tells us either to do or to be that is not already found in God. We are to forgive one another because God forgives us; we are to be merciful because God is merciful; to be holy because He is holy; to love our enemies because, when we were His enemies, God loved us. To have the mind of Christ, then, is to deal with others as God in Christ has dealt with us. We are to put into life that spirit which is the ground and source of our own redemption. The very essence of that spirit is to give rather than to get; to seek the good of others before our own; to leave the self-centered life for a life of service; to be willing to suffer that others might gain, rather than gain while others suffer. That is what we find in God—happily for us. That is what we find perfectly expressed in Christ, Who is God manifest in the flesh. That is the mind of Christ.

But if the mind of Christ is thus rooted in the very nature of God; if all this is laid on you and me just because God is all this; then certain gripping and even staggering consequences follow. Let me indicate one or two. First of all, all that I have just been saying comes home to you and me with all the force of a Divine imperative. Its validity lies in the fact that it is Reality. It therefore becomes binding on our hearts and consciences because it is Divine. This is no mere human speculation, some piece of philosophy garnered from the wisdom of the world. This is Divine truth, because it is rooted in the very nature of the Divine. The second consequence is this: either God Himself is making a mistake as to the basic principles of true living—which, of course, is absurd; or we are! And as a third consequence: since God cannot be mistaken, here in Christ's mind, in His life, His ethic, His Cross, we have demonstrated before our very eyes the abiding principles on which alone true life can be built: not self-assertion, but self-sacrifice; not getting by hook or by crook, but giving to the limit of our capacity; not competition, but suffering; not grabbing, but yielding even to the uttermost, in order that others might live. That is what we find in God—happily for us. And we are to "Treat one another with the same spirit as we experience in Christ Jesus."

I said a moment ago that we are making a mistake as to the basic principles of true living. The big problem of life, now as always, is the problem of living together. Wherever you turn your eyes, you find chaos today. There is chaos in twice ten-thousand homes, for which the divorce courts and the law courts provide ample evidence. There is chaos in our industrial life, as our ceaseless disputes make so painfully evident. There is chaos in our national life, and chaos in our international life. Why is this? Because men haven't yet learned the art of living together.

Moreover—and I cannot make this too clear—to remove the material hindrances to true living is one thing, but to impart to men the secret of true living is quite another! The various political gospels of our day are all blind to that distinction. If you could at one stroke abolish all slums, insure for all men an adequate remuneration, house them all decently, and educate and clothe them all nicely—even that emancipation would not of itself teach men what Life really is, nor would it teach them how to live it. "There is no political alchemy by which you can get golden conduct out of leaden instincts." Only Jesus Christ can do that. It is only He Who can finally make men understand that a man's life consisteth not in the abundance of the things that he possesses. The workers of the world are winning for themselves today all sorts of freedoms and privileges. No one would begrudge them that. They should have had them long ago, and they ought to have yet more of them. Contrasted with conditions in the labor world of even fifty years ago, the workers of the world are living today in a comparative paradise.

But what is the good of all these gains if men still don't know how to live together, or if they confuse "Life" with money or property or meaningless activity, or with the antics and morals of certain film stars, who spill their foul filth and deadly devilishness on to the screens of our moving picture theatres? That kind of thing isn't life. It is the stark denial of it, the absolute antithesis of it. It is the mind of the devil, not the mind of Christ.

The Christian Church, then, holds and teaches that to seek for the reproduction of Christ's mind in the mind of the community is the greatest and noblest aim that we can cherish. It *is* possible for human life to be lived on a basis of mutual cooperation. It *is* possible for all work to be so done that it reflects the glory of the Carpenter of Nazareth. It *is* possible for all business organization to become an expression of service and good-will. It *is* possible for wealth to be so distributed that it shall make life rich for all, and not merely for the favored few. It *is* possible for all government to be administered with respect for personality, and for all laws to be the embodiment of Christ's mind and will. It *is* possible for the whole fabric of life to be inspired by the Holy Spirit of God. You ask me how? I reply: "Let this mind be in you, which was also in Christ Jesus. Treat one another with the same spirit as you experience in Christ Jesus."

To you who are Christians I make one simple plea. It is that we who are known as Christians should have faith in our Lord's way, that we should stand ever loyal to His central principles, and strive to root deeper and deeper in ourselves the conviction that only by a closer conformity to His Spirit can men achieve freedom, life, and peace in any realm. Christians are to be like Christ in their whole mental and spiritual nature and outlook. To be a Christian is to have the mind of Christ.

To you who are not Christians I also make one simple plea, but let me preface that plea with an affirmation. I affirm my faith in Jesus Christ as the wisdom of God; eternal, abiding, and authoritative for all time and in every sphere. I know very well what you, as a non-Christian, have been thinking while you have been reading these words. You have dismissed me and the things I have been saying as too impractical and remote from the world of things as you know them. The very idea of taking the things I have been talking about into the work-a-day world, the very idea of applying the principles and the mind of Christ to the political and industrial and commercial problems of our time, is too laughable for words. "Soft-headed sentimentalists" is our latest nickname—coined by a Minister of the Crown in the Federal Parliament! I know the motto that stands on your desk: "Business is business. Each for himself, and the devil take the hindmost."

But, my hard-headed and intensely practical business friend, you who reckon that your head is screwed on tight, the world as you have run it is a howling chaos. The world of industry is full of snarling suspicion, the family ideal is crumbling, Europe is a baffling problem, to say nothing of China and Russia and India. Is this where the practicality of your hard-headed business men has led us? Be honest with yourself: haven't you led us all "down the garden path"? You politicians and statesmen, is this the best you can do for us—two world wars each of which was to be "the war to end all war," and now a third world war already a common topic of conversation, and preparations against the evil day already being put in hand all over the world? Is this where your philosophy of life leads us to? Aren't you proud of your hard-headed practicality, your sheer common sense, your mighty achievements whereby tens of millions lie moldering in earth's battlefields, and tens of millions of others still suffer the tortures of the damned?

Now for my plea to you who are not Christians. It is that in the name of God and of humanity you should stand aside for a while, that you should surrender this ridiculous notion of your own marvellous ability and competence to run this world, and that you should at least and at last try the way of Jesus. "Let this mind be in you, which was also in Christ Jesus. Treat one another with the same spirit as you experience in Christ."

When we are ready to do that, we shall have put our feet at last on the road that leads to the Kingdom of God.

PRAYER

Almighty God, Who art the Father of all men, most heartily do we pray that Thou wilt both lead and keep Thy children in the way of peace. Teach us to put away all bitterness and misunderstanding, both in the Church and in the State; teach us to forgive, as we have been forgiven; teach us to seek not our own good, but each other's good; that we, with all the brethren of the Son of Man, may draw together as one family of mankind, and dwell evermore in the fellowship of that Prince of Peace, Who liveth and reigneth with Thee in the unity of the Holy Spirit, world without end.—Amen.

The Peril of Conformity

Reverend Robert James McCracken, D.D., S.T.D., D.H.L.
Minister, The Riverside Church
New York City

Dr. McCracken has had a very busy life. Born on March 28, 1904, in Motherwell, Scotland, in a home that had inherited the great traditions of Scotch Presbyterianism, he attended school in Motherwell, and having been spiritually awakened in the Motherwell Baptist Church, became a member of that church in 1918. He attended Glasgow University, from which he graduated in arts with an M.A. degree in 1925 and with a B.D. in theology in 1928. Soon after he became minister of the Marshall Street Baptist Church in Edinburgh.

After four years in Edinburgh, Dr. McCracken was called to be minister of the Dennistoun Baptist Church in Glasgow, and in 1933 was appointed lecturer in systematic theology in the Baptist Theological College of Scotland. For the next four years he carried responsibilities both as preacher and teacher. In 1937 McMaster University at Hamilton, Ontario, invited him to become associate professor of Christian theology and philosophy of religion. Then followed a year of special study at Cambridge University, after which Dr. McCracken returned to Canada. From the outset he associated himself with ecumenical interests and activities. All the Protestant communions in Canada have sought his spiritual leadership. During the last three years that he was in Canada he was on one occasion or another daily preacher at the Toronto Conference of the United Church, daily preacher at the General Council of the United Church of Canada, lecturer on the Pitcairn-Crabbe Foundation at Western Theological Seminary in Pittsburgh, lecturer before the McGill Theological Alumni in Montreal and before the Knox College Alumni in Toronto. For five years he was the guest preacher during the summer at Yorkminster Church in Toronto. He had a term of notable service as president of the Baptist Convention of Ontario and Quebec.

As preacher, pastor, teacher and administrator, he had a wide and useful experience before becoming the successor to Dr. Harry Emerson Fosdick, in New York, where, on October 2, 1946, he was installed as minister of the Riverside Church. He has proven himself a worthy successor to Dr. Fosdick in this important pulpit.

In recognition of his work, he received the honorary degree of Doctor of Divinity from McMaster University in May, 1946; from Bucknell University, 1947; the University of Glasgow, 1949; Colgate University, 1950; the degree of Doctor of Sacred Theology from Columbia University, June, 1950; the degree of Doctor of Humane Letters from Bates College, June, 1951. Since 1949 he has been Lecturer in Practical Theology, Union Theological Seminary, New York, and is frequently the speaker on the Sunday Radio Chapel and Faith in Our Time broadcasts. The editor of Best Sermons *is proud to present this sermon of Dr. McCracken's. As he discusses "The Perils of Conformity," he opens a whole new, fresh train of thought; his use of Voltaire, his emphasis upon the need for real convictions and spiritual resources show him as a growing preacher of power and insight in our own time.*

Sermon Four

TEXT: Be not conformed to this world: but be ye transformed by the renewing of your mind. ROMANS 12:2.

VOLTAIRE once remarked that every man must be either the hammer or the anvil. One would like to think that the alternatives need not be quite so extreme. Actually they are not. Voltaire said what he did to drive home the point that mankind tends, generally speaking, to fall into two classes, one small and one large—those who make history and those who are made by it, those who are creators of fact and those who are creatures of circumstance, those who put color into their environment and those who take their color from their environment. To which class do we belong? Is society molding us more than we are molding society? Are we conforming to the world and its ways or are we being transformed by the renewing of our mind?

In the early settling days this country was noted for its rugged individualism. There came to its shores men who had resolved to break with the Old World and set up here a new pattern of life. This nation came out of the throes of a revolution and the heroes of those days—Washington, Jefferson, Franklin, Paine—were all nonconformists. The doctrine of the rights of man, so precious to them, made a virtue of self-reliance and encouraged the habit of self-assertion. Later on, Thoreau wrote: "I desire that there be as many different persons in the world as possible; I would have each one be very careful to find out and pursue his own way." And Emerson counselled his hearers and readers: "Every institution is but the lengthened shadow of an individual." "Build your own world." "Envy is ignorance." "Imitation is suicide." No doubt this individualism had its drawbacks and weaknesses. It produced people who attached to themselves and their opinions an importance for which there was not always adequate warrant. It introduced the cult of popular personalities, especially successful money-making personalities, and concentrated attention on men and their doings more than on movements and their significance. But it had sound and wholesome features. Men should know their worth, learn to stand on their own feet, discover their native talents and make full use of them. They should be creators of fact, not creatures of circumstance. "Whoso would be a man must be a nonconformist."

These are things that need to be said today. If individualism was the prevailing feature of American life in the nineteenth century standardization is the prevailing feature now. We more and more incline to wear the same kind of clothes, eat the same kind of food, read the same kind of books, see the same kind of movies, listen to the same kind of radio programs. The uniformity extends not only to dress and manners but to speech and thought; indeed speech and thought are so conditioned as to be for millions practically automatic; individuality of expression and independence of judgment are in danger of becoming lost arts. We have to reckon in America with a mass mind whose opinions are molded by the press, the films, the radio, the book clubs, and whose tastes, standards, ambitions and responses tend to be stereotyped.

For illustration, take the case of advertisements. It would be difficult to exaggerate the part they play in creating a mass mind. In newspapers and magazines, on the radio, goods of every description—beer, cigarettes, automobiles, jewelry—are urged on us as absolutely indispensable. Advertisers are out to catch the eye and the ear. To promote sales any means seem permissible—persistent reiteration, ridiculous exaggeration, sex stimulation. The appeal is directed to vanity—Don't you want to be a Man of Distinction?—to cupidity—Why not take advantage of such a bargain?—to envy—Why should you do without a commodity owned by your neighbor? It is obvious that little thought is given to the effect of the advertisements on character. In some instances not only truth but taste and decency are at a discount. The advertisers have something to sell and they mean to sell it. The profit motive overrides ethical considerations. I have no hesitation in saying that this is one of the most pernicious and vulgarizing influences operating at the present time on American manners and morals.

I draw attention to the matter because the imitative instinct is a powerful instinct. We are all to some extent suggestible. A man has only to stand still on the street and stare upwards with an expression of rapt interest on his face and a crowd will quickly gather around him staring upwards too. It is not for nothing that we have a proverb, one of the most familiar of proverbs, which says that in Rome we must do as Rome does. I was amused the other day when I came across this sentence in a novel. "He wanted passionately to be indistinguishable on the surface from other men; he wore his moustache like a club tie—it was the highest common factor." We may smile at that but we should not pretend to be superior about it. Most of us are afraid of anything that savors of eccentricity. We dress, we speak, we act, we think like the people in our class and circle. Group habits are, of course, an essential element in all group life and many of them involve no moral issue. But conformity to social pressure and convention as soon as it becomes servile and slavish is for character a dangerous thing. It robs us of individuality and independence. It means that we do things not because we want particularly to do them but because others do them and they are expected of us. It means that we listen to what others have to say before we speak, and when we speak it is to echo the sentiments we have just heard expressed. We don't have robust convictions; we have impressions, theories and opinions; and even then before we air them we are careful to find out in what direction the wind is blowing. We don't put color into our environment; we take our color from it.

Is there any temptation stronger than this one—to do in Rome as Rome does? In his autobiography Dr. Joseph Fort Newton maintains that there are four things, and only four things, that a person can do with his life. First, he can run away from it, or try to, as Jonah did. Second, he can run along with life, hunt with the pack, think with the herd, in obedience to the phrase, "Everybody's doing it, why shouldn't I?", which may be the very reason why he should not do it, if he has any standards or any character. Third, he can take hold of life with singleness of purpose, and submitting himself to discipline, can run it to some end. Fourth, he can give himself up utterly in self-surrender to a Cause or a Person and let the Cause or the Person run his life. If lines of division were drawn through this congregation after that fourfold pattern would we not find the majority group consisting of those who are running along with life, adapting themselves chameleon-like to their environment, taking on the complexion of their surroundings, molded by society, not its molders?

I am thinking especially of young people. You should be on the way to mental and moral independence. You are at the stage where you should be building your own world, standing on your own feet, acquiring your own preferences, forming your own judgments, learning if need be to carry yourself in the face of opposition. Over the gateway of University College in Aberdeen are the words: "They say. What do they say? Let them say." Admirable words for those who are learning their own worth, though it is one thing to pay one's respects to them and another to put them into practice. You are aware of all sorts of subtle but strong group pressures. You feel that you have to satisfy the requirements of sociability. You dislike being different from the other members of your circle. The last thing you want is to be taken for a kill-joy or a poor sport or to convey the impression that yours is a holier-than-thou attitude. It is your wish to be congenial and acceptable in company. You dread going friendless. Some of you have no inclination to drink hard liquor or adopt a free and easy attitude in the matter of sex but you fear that abstinence on your part will be construed as a criticism and a condemnation of those who do have the inclination.

It is a difficult situation. It is the kind of situation, however, that tests your stamina. It shows of what mettle you are made and whether you have principles as well as preferences, convictions as well as opinions. To say No when everybody else is saying Yes calls for the highest and rarest form of moral courage. It requires special grace to say No and keep the affection and respect of those who are saying Yes. Nicolay and Hay in a passage devoted to an analysis of Lincoln's political popularity have left this sentence on record: "He was everybody's friend and yet he used no liquor or tobacco." The thing can be done. Robert Louis Stevenson never set himself up as a preacher or moralist but he once made a remark that all of us can with advantage turn over in our minds: "To do anything because others do it, and not because the thing is good, or kind, or honest, in its own right, is to resign all moral control and captaincy upon yourself and go post-haste to the devil with the greater number."

Natural history tells us that the animals that became vertebrates, that acquired a spinal column and a backbone, were the nonconformists. The rest continued as jellyfish or became clams—a static species. Human history tells the same story. No man ever became a truly great man by running along with life, accommodating himself to his environment, yielding to social pressures. One of the lessons taught in chapter after chapter of Arnold Toynbee's *Study of History* is that those civilizations which survive and endure do so because they have leaders who step out into the wind and accept the challenge of the storm. I think of what his biographer describes as one of the most masterly speeches ever delivered by the late William Temple. It was from first to last a plea for nonconformity, an appeal to the members of the Christian Church not to take their character from their environment but to put character into it. "Privileges which are abused are forfeit. Throw off what hampers your service, even though it be venerable with the history of ages or consecrated by dear familiarity. . . . Come out from your safety and comfort; come out from your habits and conventions. Listen for the voice of the wind as it sweeps over the world and stand where you may be caught in its onward rush. . . . Here is your task. Will you perform it? Or will you stay as you are, to flicker out, a lamp that gives no light, unmourned and even unnoticed?"

Those who have done most for the world and have lifted the level of life for the ages have been the nonconformists. They have been men of fixed purpose and staunch principle. They have had standards and kept them, no matter what the cost. They have been ready to face criticism, ridicule, ostracism, impoverishment. They have not run away from life. They have not run along with life. They have taken hold of it firmly and have organized it around a Cause or a Person. In the motion picture version of

The Life of Emile Zola there was a stirring scene. Discharged from the publishing firm of Monsieur Larue, Zola turns upon his former employer and says: "While you, Monsieur Larue, continue to grow fatter and richer publishing your nauseating confectionery, I shall become a mole, digging here, rooting there, stirring up the whole rotten mess where life is hard, raw and ugly. You will not like the smell of my books, Monsieur Larue. Neither will the public prosecutor. But when the stench is strong enough, maybe something will be done about it." There was a man who scorned to compromise with principle, who would not prostitute his gifts or sell his soul for cash, who saw with eagle eye the evils of his day and had no mind to tolerate or hold his peace about them. There was a man who was resolved to be not a creature of circumstance but a creator of fact.

And why? Because his life was organized around a central purpose. Because he had found himself and knew what he wanted to do with his time and his gifts. The reason why so many people are at the mercy of circumstance is that they have neither discovered a faith by which to live nor a cause to serve. There is a story about a man who asked the question, "What is the devil?" and before anybody could reply he answered his own question. "The devil," he said, "is not a huge monster with horns and a harpoon tail and a wicked glitter in his eye. No, the devil is just taking the line of least resistance. It is inertia. It is doing nothing." The definition might not satisfy a theologian but it is not without its point. What happens to those who take the line of least resistance? Standing on the bank of a stream we watch the current carrying objects along. We see chips of wood caught into a side eddy, revolving for a while, and then escaping, only to be caught in another eddy, and finally finding their way into a stagnant pool where they will probably remain till they rot. There are men and women like those chips of wood. They go with the stream. They run along with life. They take their color from their company. They reflect the standards and practices of their environment. You cannot calculate what shape their life will take or in what direction it will move. Everything depends on the people they meet, the things that happen to them, the current of life round about them. What they need, if they are to drive rather than drift, is a faith by which to live and a cause for which to work.

In all this I have been thinking of the three Hebrew youths, Shadrach, Meshach and Abednego. Do you remember their story? Nebuchadnezzar was a dictator. He set up a great golden image on the plains of Dura and commanded all his subjects to prostrate themselves before it. The people flocked out to do his bidding, but not Shadrach, Meshach and Abednego. From them came words of moral idealism and spiritual defiance which to this day act like a tonic. "O Nebuchadnezzar, we are careful to answer thee in this matter. . . . We will not serve thy gods nor worship the golden image." They were nonconformists. There was iron in their blood. They would neither bow nor bend. Why? Because their lives were organized around a steadying conviction and their strength and stamina were derived from God. I repeat that last phrase. Their strength and stamina were derived from God. If you want, not to take your character from your environment but to put character into it, you will need to draw on spiritual resources greater than your own. And you may. You can be strengthened by God in your inner life if you will open it daily to His influence. It was of this Paul was thinking when he said: "Be not conformed to this world but be ye transformed by the renewing of your mind."

THE CHRISTIAN LIFE

The Love Without Demand

Reverend Bernard Iddings Bell, S.T.D., D.D., Litt.D., LL.D.

A Priest of the Protestant Episcopal Church; Consultant on Education and Religion to the Diocese of Chicago; Representative of the Episcopal Church at the University of Chicago

In "The Love Without Demand," Dr. Bell shows a different side to his preaching. Most people know him best for his university sermons, talks or addresses and articles on education. Here he talks of love, divine and unselfish, in a world of selfishness and war. For several years he has been Consultant on Education to the Protestant Episcopal Bishop of Chicago and representative of the Episcopal Church on the campus of the University of Chicago.

He was born in Dayton, Ohio, and educated at the University of Chicago and Western Theological Seminary, Chicago. His interests have always been chiefly in religious education and in the relationship between religion and politics.

Before the first World War he was dean of the Episcopal Cathedral in Fond du Lac, Wisconsin, and during that war was aide to the senior chaplain at Great Lakes Naval Training Station. The war over, he became warden of St. Stephen's College, a country college of Columbia University, and professor of religion in Columbia University. In 1933 he retired from academic connections and since then has devoted his time to preaching, lecturing, research and writing.

A priest of the Episcopal Church, he has written twenty-one books of which the best known are Beyond Agnosticism; The Church in Disrepute; The Altar and the World. *His most widely sold book was entitled* God Is Not Dead. *He has written twenty some articles for the* "Atlantic Monthly" *on religious topics and has contributed to* "Harper's," *the* "Criterion," *the* "New York Times Magazine," *to many church papers. He has been in residence and lectured in six of the great English schools, including Rugby and Charterhouse, and has preached in thirteen English cathedrals. He has been Stated Preacher at various times at Harvard, Yale, Princeton, Chicago, Columbia, Williams, Amherst, Wellesley, Vassar, Smith, Mount Holyoke and many other colleges.*

Sermon Five

THE CHRISTIAN religion, brethren, when it is understood, leaves us in no state of uncertainty about what the God is like who creates, sustains and judges the universe; creates, sustains, and judges you or me within it. God does not ask us to try to relate our lives to a mystery which cannot be resolved. For us men and for our salvation God emerges from the humanly incomprehensible realm called Heaven and takes on Himself manhood like our own. He did this once in Palestine, as we remind ourselves each time we say the Creed. He keeps on doing it. The Christian religion is based on the proposition that a character named Jesus is more than a man in history, a man who lived once, merely as a man, in a particular place at a particular time; that He is man but more than man; that He lives eternally, in Himself a revelation of the ultimate mystery of existence, God revealing Himself in human terms. Every time we come into His house, every time we kneel at His altar, every time we wish for God, Jesus is present. Christians seek God in Jesus. They find God in Jesus. In Jesus they touch the Infinite, the Infinite God who is no indefinite supposition but a friend.

But we cannot come to know the reality of His friendship by merely desiring it. He lays down a condition which, if left unfulfilled by us, prevents our entering upon and remaining in the divine joy. It is required that we who would be received by Jesus, the incarnate God, must desire that which Jesus, the incarnate God, commands.

What does He command? He commands that we shall love, love Him with heart and soul and mind and strength, and that because we love Him in this fashion we shall also love every man, woman, child, and chiefly those with whom our contacts are immediate and inevitable. We do not, cannot, gain from Jesus, the incarnate God, that peace and gaiety of soul without which life is barren, disappointing, *unless we love.*

What does it mean, to love? It means to place another at the center of one's life. Do I love my child? I do if my child matters to me more than I matter to myself, if I regard myself as existing to make it possible for the child to come to a good maturity. If this desire is lacking in me, if what I chiefly desire is rather that my child shall amuse me and serve me, or, even worse, that my child shall not bother me, shall not prevent my doing as I please, when I please, then I do not love my child; I love myself. Do I love my friend? I do if I am willing to be to that friend a servant of that friend's sustenance, looking on that friend with thought not of what that friend may do for me but of what I may do for that friend. Do I love my wife or husband? Only on terms of self-investment does even conjugality release me from that enslavement to myself which is the chief cause of human woe, which is indeed the essence of damnation.

In our day the word "love" is greatly abused. "I am in love," we say, and mean by this that we wish to be loved. Now being loved is a pleasant experience, but it is not a necessary one. "To be loved"—that is passive voice. I do nothing. I sit serene, complacent, somnolent, while some one loves me. To love is not a passive thing. "To love" is active voice. When I love I do something, I function, I give. I do not love in order that I may be loved back again, but for the creative joy of loving. And every time I do so love I am freed, at least a little, by the outgoing of love, from enslavement to that most intolerable of masters, myself.

My beloved, my parent, my child, my friend, my mate, may or may not love me in return, or may once have loved me in return but have ceased to love me in return. That does not fatally matter. If he or she does not love in return, I still can love. "If ye love only them that love you, what profit is that to anyone?" asks Jesus. "I say to you, Love even them that are your enemies, them that reject you, treat you with contempt. So shall you become the children of the Eternal Father." *No one can take away from me the one great treasure, the love I bear without demand of love.* The beloved may reject, hate, kill; but there is one thing the beloved who spurns my love can never do. He or she can never break my heart. When one cries out that the beloved has broken one's heart, it is not love that has caused the heartbreak. It is rather disappointed self-esteem.

This, brethren, is a hard saying, this assurance from God Himself that happiness, the fulfillment of one's destiny, is arrived at only in terms of loving without demand of love. It is so hard a saying that we keep struggling to find some way to happiness less onerous. "Let us eat, drink, and be merry," we say. We eat and we drink; but we are not merry. Or we seek to find in amusement the key to felicity; and before long we are bored. Or we cry out for applause; but soon we are more and more aware that those who flatter us are apt to be fools, or knaves with an ulterior motive in their adulation, or both. Or we crave meaning in terms of exercise of power; and find ourselves hated, condemned to dwell at last with our tiresome selves. Or we suppose that increase of knowledge will suffice; only to find that the more facts and processes we master, the more the meaning of things and of ourselves continues to elude us. So disappointing are these lesser avenues into significance that most of us, before too long, give up search for real happiness, for a joy which nothing, no one, neither life nor death, can take away. Instead, we compromise with life, hiding our lonesomeness even from ourselves, daring no longer to live in terms of high romance. No, nothing can take the place of love.

How very hard it is to love! How silly when people talk of its being easy to love. "One does not need prayer and sacraments and all the paraphernalia of religion," we keep hearing. "All that is necessary is that we love one another." Of course! Of course! To love one another, without demand of love, to do it domestically and in business and in education and in politics and in affairs international and in the Church, that would indeed be enough —but so to love is a thing beyond our powers. Loving without demand is the most difficult thing in all the world.

Yet without such loving, there is small hope for society or for you, for me. Without such loving, divorces multiply, homes are broken up, children are tortured by divided loyalties. Without such loving, society becomes a bedlam of contending individuals and mutually predatory classes. Without such loving, industry insists, even in the face of world emergency, that it will not carry on if it gain for the owners less than normal profits; and labor demands ever more pay for less work; and farmers will not raise man's food except they be unjustly subsidized; and so we all pursue the way to inflation and a common ruin.

Without the loving which makes no demand on love, as some of us who work for God in universities know and keep on insisting, education is almost certain to produce knowledge without wisdom, to implement evil as well as good, to augment cupidity, to render men the more dangerous to one another.

Unless we learn to love without demand of love, war is a perpetual inevitability, war, that monster whom we have equipped with claws of unbelievable deadliness. Without such loving, a pawn-nation like Korea is fought over until in one year three million of its people, mostly noncombatants—the old, mothers and expectant mothers, babies and little children—innocent of offense except that they live in that unhappy land, are "liquidated," as the current phrase puts it, while seventy percent of all Korean buildings—homes, factories, temples to the gods—are obliterated. So it will be, or worse, in every land, including ours, when modern war is waged therein. Without such love as makes no demand on love, man's long career on earth approaches an inglorious end.

Loving without demand of love, then, is no impractical sentimentality, not a matter for pretty speech in pulpit or elsewhere. Far from it. Loving without demand of love is the only means to prevent speedy and complete disaster to the whole world. We must love or we shall die, hopeless, frustrated, defeated.

We *must* love without demand of love. But we *cannot* love that way.

We are not up to it. This is man's tragedy, the tragedy called Sin. For what is Sin? It is the refusal to love. And the wages of Sin is death.

But the Christian religion does not leave us hopeless. Jesus, incarnate God, not only commands this incredibly difficult loving. He also promises divine assistance in loving, and for centuries now has been bestowing it on those with wit to accept it from Him. He reveals to us that God is Himself the perfect Lover. We who are such incompetent lovers, we who are afraid to love without demand of love, when we approach God in the sacrament, in prayer, find ourselves touched by Him, are empowered to dare the impossibly necessary. We touch the Divine Lover. His symbol is the crucifix whereon the Christ God is forever rejected but forever undefeated. He comes into the world that He may love the world into effectiveness and joy; and the world knows Him not. Yet He keeps on loving. He comes to His own, to you and me who say that we are His; and we draw back; we receive Him not. Yet He goes on loving us. Finally His love breaks down our cowardice. He gives us power to become the sons of God, free to live, free to dare, free to care, free to suffer vicariously for the unloving. And so He can make of us coredeemers with Him of a sick world. Loving with no demand of love remains as hard as ever, sometimes heartbreakingly hard; but it is no longer a thing impossible.

Unless we are trying, striving, agonizing, aided by the King of heaven, to love as He would have us love, as He Himself loves, religion is only a cheap amalgam of sentimentality and superstition. God's promised assistance, God's growing comradeship, God's peace—these are meaningless phrases unless we know what loving means, its necessity, its pain, its glory.

Hear the word of the Apostle, praying for you and me and all the brethren as once he prayed for the Ephesians:

(May) Christ dwell in your hearts by faith, that you, being rooted and grounded in love, may be able to comprehend with all saints what is the breadth, and length, and depth, and height, and to know the love of Christ which passes knowledge, that you may be filled with all the fullness of God (Ephesians 3:17–19). Amen.

The Everlasting Mercy

Reverend Willard L. Sperry, D.D., Litt.D.
Congregational Minister and Dean of the Chapel, Harvard Divinity School, Cambridge, Massachusetts

Dean Sperry is respected for his insight into the problems of men of our day, for his spiritual guidance of Harvard students, and for the excellence of his preaching. As dean of Harvard Divinity School and professor of practical theology he has exercised a profound influence on the theological training of hundreds of ministers in important churches all over the country. As chairman of the board of preachers to the University (college chaplain) he has drawn to Harvard many of the world's outstanding ministers for special sermons and courses of religious lectures for the last twenty years.

The distinguished theologian was born in Peabody, Massachusetts, in 1882, was a Rhodes scholar at Oxford (first-class honors in theology), studied at Yale, and has received the doctorate from Yale, Amherst, Brown, Williams, Harvard, and Boston. He served in the pastorate in Fall River and Boston, joined the faculty of Andover Theological Seminary, and has been dean of Harvard Divinity School since 1922. For four years he was also dean of the National Council on Religion in Higher Education (1927–31), has given many famous lecture series, including the Upton Lectures at Manchester College, Oxford, the Hibbert Lectures, Essex Hall Lectures, London, and the Lyman Beecher Lectures at Yale. He is a Fellow of the American Academy of Arts and Sciences and is known for several significant books, The Discipline of Liberty, Reality in Worship, The Paradox of Religion, What You Owe Your Child, Wordsworth's Anti-Climax, What We Mean by Religion, Summer Yesterdays in Maine, Rebuilding our World *and* Jesus Then and Now.

Dean Sperry is accustomed to distinguish between words written to be read and written to be spoken. His sermons deny themselves the leisure and literary elaboration which is found in his books. They reveal a certain bluntness and brevity in sentence style, which he thinks suited to the spoken word, and to the school and college groups to which he habitually speaks.

"The Everlasting Mercy" was preached in the chapel at Harvard on Easter Day, April 9, 1950. Dean Sperry has spoken to college and university students for so long that he is able to phrase his messages to exactly catch the interest and imagination of young men and women who are searching for the in-

tellectual and spiritual answers to the problems of today and tomorrow. It is a distinct privilege to present Dean Sperry again to the readers of Best Sermons.

Sermon Six

TEXT: We trusted that it had been he which should have redeemed Israel. LUKE 24:21

THOSE of you who were here two weeks ago heard Reinhold Niebuhr who was the preacher for the day. His sermon ended with a striking sentence, "Only the infinite mercy of God can meet the infinite pathos of human life." Again those of you who were here last Sunday, Palm Sunday, heard Canon Hodgson of Christ Church, Oxford, preaching on the text, "When I became a man I put away childish things." [1]

These two sermons are the premise, and a fitting premise, for our meditation on this Easter Day.

Most of us have heard Niebuhr often and many of us have wrestled with the elusive pages of his books. But if you will remember that last sentence of his last sermon to us you will have the pass key to his whole conception of Christianity. The sentence had not been written out beforehand and read to us. It was struck off in this pulpit, like a spark on the anvil of his own mind. "Only the infinite mercy of God can meet the infinite pathos of human life." No one can understand Christianity who does not feel the infinite pathos of human life and no one can have faith in Christianity who does not believe in the infinite mercy of God.

The four gospels are the record of the brief public ministry of Jesus of Nazareth. That ministry probably lasted only a few months—less than a single year. A sense of imperative and imperious destiny is felt throughout the record. What is more the note of doom is sounded from the first. When the infant Christ is presented at the temple the aged Simeon says to Mary the mother, "Yea, a sword shall pierce through thy own soul also." This note is sounded more and more unequivocally as the story goes on. "It cannot be that a prophet perish out of Jerusalem. . . . The things that are written must be fulfilled in me. . . . I lay down my life for the brethren. . . . I go my way to him that sent me."

And one cannot read the Gospels to their end without feeling that the stories of Easter come too soon, too abruptly, almost too easily. We are not given time enough to reorient our minds. There is an instinct in all of us

[1] Canon Hodgson's Sermon is also included in this volume.

which requires a happy ending to every story, however sad and tragic. The very fact that we instinctively require such an ending is itself significant. But the happy ending cannot be had too cheaply. It cannot be a mechanical and manufactured happy ending like that at the close of the book of Job. It must be felt to be inevitable and true, like the last stories in the gospels.

Of all the mysterious stories of the resurrection of Christ none is more moving than that of the walk to Emmaus, which was read as our second lesson. Coleridge once said, "I know that the Bible is true because it finds me." However one may construe the letter of the Emmaus story, it finds us religiously. And it finds us because in the first instance it does not deny or ignore the infinite pathos of human life. "We trusted that it had been he who should have redeemed Israel." Those words were true of the disappointed and disillusioned men to whom they were first attributed and they are as true of our generation as they were of that first generation.

In the face of the worldwide suffering and tragedy of our time we are ashamed to make the little tale of our own fortunes or misfortunes the measure of Christianity. The thought that, one by one, we still matter to God may reassure us, but somehow we do not matter to ourselves as much as once we may have done in our childishness. What weighs more heavily upon us is the sombre dread that the whole cause of our religion may already have gone down in ruins, or be in the last stages of some such collapse in history.

> Loud mockers in the roaring street
> Say Christ is crucified again
> Twice pierced the gospel-bearing feet,
> Twice broken His great heart in vain.

It is at this point that the infinite mercy of the Emmaus story meets and matches its infinite pathos. The day was far spent, "and he went in to tarry with them, and as he sat at meat with them he took bread and blessed it, and brake, and gave to them. And their eyes were opened and he was known of them in breaking of bread."

The story of Christianity was from the first a story of intimate and humble things. What was true in Bethlehem was still true in Emmaus. Christians have always looked forward to the day when the Kingdoms of this world shall become the kingdoms of our Lord and Saviour Jesus Christ. That day did not come in the lifetime of the first disciples nor has it yet come in our time. But just as those two disciples at Emmaus suddenly knew that the cause had not failed and that Israel might yet be redeemed, so we may be reassured that Christ is not twice dead in our own time, and dead in vain. Only one must know where to look for the religion which bears his name, and in what terms to expect it.

We live here a culturally mature and sheltered community. Most of its accepted moral standards and many of its spiritual ideals derive from historic Christianity. These are so much a part of our second nature that we do not always identify them with religion as such. Even here, those of us who know the community intimately, can cite man after man whose first and constant concern is to try to live a Christian life. The restraints, the imperatives, the high invitations of Christianity matter to them more than anything else.

As for the wider world, we often forget how stubborn and wayward a thing human nature still is. William Ernest Henley once described Piccadilly at night as "a tidal race of lust from shore to shore," where one felt only

> The lewd perennial overmastering spell
> That keeps the rolling universe ensphered,
> And life and all for which life lives too long,
> Wanton and wondrous and forever well.

And if one turns from the streets of the city to the sober halls of government, one realizes how far, as yet, the selfless genius of Christianity is from making a conquest of the imperious will-to-power which all men inherit by nature and to which so many of our near great men fall unconscious victims. The Christian religion is not a religion which celebrates power so construed. Century after century it has had to reckon with the fact that all power corrupts and absolute power corrupts absolutely. The siege which Christianity lays against all such corruptions is a protracted siege. It is a long long way from primeval chaos and the yelp of the beast to the City of God.

But if you will turn away from the more obvious and equivocal signs of our times and look closely and intimately at countless human lives today you will have no cause to despair of our religion or to say that Christ twice dead is dead indeed. Indeed, it was the very author of those lines of verse who went on to say of what he saw and heard in the roaring streets of the world,

> I hear, and to myself I smile
> For Christ walks with me all the while.

There are hundreds of persons right about us and thousands upon hundreds of thousands in the wider world who bear the burden and heat of the day, who overcome temptations, who face and survive the tribulation that is in the world, and who wipe away the tears from their eyes, sustained through all the vicissitudes of human life by their faith in Christ and their determination to follow him. Only you must get behind the forbidding façade of much of the life of these times and look more closely and more deeply than one often looks to know these things. What was true then is true now, Christ is still best known to us in the breaking of the bread.

And then a word about Canon Hodgson's sermon last Sunday. He bade us put away childish things, and that means putting away the first instinctive judgment of our immaturity that what happens to any one of us is the most important thing in the world and that Christianity is a magical device for getting good things only to happen to us. The Canon bade us become men, and becoming men means taking a wider, a more generous and mature view of our place and part in the world.

St. John's Gospel tells us that we do not have to wait until the hereafter for our immortality. We can enter into eternal life here and now. Most of us have had hints of what such an experience may be. Thus, a wartime book of letters written back to his home by a sensitive soldier, who seems to have foreseen the probability of his own death says, "If anything should happen to me, don't be too sad. We have had some experiences together that will always last. So, the first time we all heard the Brahms Requiem was just such a bit of eternity."

If you did not hear the Bach *St. John's Passion* yesterday or the day before, you probably have heard someone who did hear it. Thus, a very unchurched and anti-clerical listener said, "What better way could one find to spend Good Friday." By common consent perhaps the most moving part of the whole austere transaction was that of the single voice repeating the words of The Christ, "My kingdom is not of this world." One must be very sophisticated or cynical to be immune and indifferent to that appeal from our troubled time to the peace of eternity.

There may be, there probably are a good many childish Christians left in the world, who think of their religion as a kind of foolproof magic which they can make work in their own behalf. But there are more Christians than ever before who are growing up.

We might have taken this half hour to discuss again the conventional doctrine of our immortality. Perhaps that should be the only theme for this day. But I am inclined for the moment to mention another aspect of the matter which is constantly in my own mind. We are all involved in an age of drastic changes, taking place so fast that we cannot even grasp the facts as they hurry by. At one moment we feel impotent and helpless to do anything to direct the facts, in another mood we are grateful that we do not have to bear great responsibilities and make the grave decisions.

But what one feels about these times is the sheer political and economic opportunism of much that is said and done. The action is exciting, but much of it seems ephemeral. It is as though the curtain were down on the stage of history, the action takes place in a narrow space in front of the curtain, there is no depth of background.

Ten years ago William Allan Neilson was speaking to the freshman class here in Harvard on the day of their matriculation. He stressed only one thing, "I beg of you above all else to take long views, for only long views will suffice you in the kind of world in which you are to live."

Christianity has always taken long views. But it goes even farther than that. There is a classic phrase coming down from the distant past which bids the Christian try to see life and to live *sub specie aeternitatis,* at the level of eternity. That mysterious challenge goes beyond anything we can understand, yet we have intimations of it. In his Second Inaugural Address, and in a time not unlike our own, Lincoln took long views of our national life. Indeed, those serene words have in them the awareness of the deep background of eternity. He said at the time that he did not think they would be immediately popular, but he thought they would last longer than anything else he had ever said.

It is not enough for a Christian to try to live an opportunist life in a perplexing and baffling world. He is under bonds to try to live eternal life under the eye of God and in the midst of time. Easter is, in days like these, a reminder and a pledge that for those of us who profess and call ourselves Christian there is an eternal realm and that what we think and say and do, even in our humblest capacity, must be ready to meet not merely the judgment bar of history, but the last judgment of the eternal order. The smoke and stir of this dim spot that men call earth is not the whole account of the final account of the life of any one of us. Heaven and earth, said Jesus, shall pass away, but my words shall not pass away. Let us, therefore, be freshly mindful of that eternal life into which he bade us enter, even here and now.

THE CHRISTIAN LIFE

Can Americans Continue Living Together?

THE VERY REVEREND VINCENT J. FLYNN, PH.D.

President, College of St. Thomas and St. Thomas Military Academy (Roman Catholic), St. Paul, Minnesota

Father Flynn is one of the younger priests and college presidents of the Catholic Church, who has three great passions, the literature of the Renaissance, the education of young men, and religion. He was born in Avoca, Minnesota, September 11, 1901, studied at the College of St. Thomas, St. Paul Seminary, the University of Minnesota and the University of Chicago. He spent 1934 and 1935 abroad for research and discovered a manuscript giving new light on Anglo-Italian relations during the Renaissance, was awarded a Guggenheim Fellowship to pursue further research during the summer of 1942, and published a new edition of William Lily's Short Introduction to Grammar (1567) *in 1945.*

He was ordained to the priesthood of the Catholic Church in 1927 and has spent most of his mature years teaching English at St. Thomas College. Since January, 1944, he has been president of the college. During the summers, 1939–41, he was visiting lecturer at the Catholic University of America, is Catholic cochairman of the North Central region of the National Conference of Christians and Jews, a member of the Modern Language Association, and a member of the Catholic Historical Association. In 1942 he edited Prose Readings, *an anthology for Catholic colleges. His travels abroad have taken him to England, France, Italy, Germany and Belgium.*

This address was given before the Association of American Colleges as the presidential address of the year. Father Flynn is one of the few Catholics ever to be elected president of the Association of American Colleges. In his address he shows his understanding of the American way of life, the needs of education today, and what true freedom of worship means for men of all faiths. His plea for peace and harmony and charity among all Americans is worthy of being painted in letters of gold in many public auditoriums and in the hearts of millions of our people. His language and word choice offer an excellent example to younger men of the power of the right phrasing in any sermon or address.

Sermon Seven

IT IS a pleasure and an honor to speak to you as president of the Association of American Colleges. I am proud of the Association; I am proud to have been its president during the past year.

First of all, I am proud of being president of the Association for a personal reason. Quite naturally it is gratifying to me to feel that I have the esteem and friendship of my colleagues. It is naturally a matter of some satisfaction, that, after having been on the Board for three years, my fellow college presidents would ask me to be their president. But apart from this personal satisfaction, which I, being human, have no desire to suppress, there is another consideration involved, of far greater importance.

I take pleasure in being a member of this organization because to me it seems to furnish an example of the unity and harmony so badly lacking in our nation, and in the world at large. In our deliberations, there are, of course, differences of opinion, but I know of no similar organization where there is less friction and more fundamental goodwill than our own; and I think it a tribute to the fair-mindedness of the membership of the Association, which is predominantly Protestant, that they should have chosen for their president a priest of the Catholic Church.

Man is by nature one. That is the teaching of any theology that I know of, and sociology, as well; that is to say, human beings are essentially the same. This does not mean, of course, that all are alike in every respect; in fact, it is part of the mystery and fascination of life that no two human beings are alike in all respects. But when we say that all men were created equal, we mean that men in their very essence are the same, regardless of race, color or creed. Christianity teaches that we were all created by the same Father, all redeemed by His Son and that we all have a common destiny.

Until recent years, this concept of the oneness of man, although admitted speculatively, had not been brought into the realm of practical political thinking. A little over a century ago, Tennyson, dipping into the future, saw a vision of "the Parliament of man, the Federation of the world." The first faint realization of his dream was, of course, the League of Nations. Today, with the extraordinarily rapid development of transportation and communication, as well as the discovery of new means of destruction, there is a more urgent clamor than ever before for world unification of some sort. We have our United Nations organization, struggling bravely to maintain its existence, in the face of enormous difficulties; we have the World Federalists, who wish to go a step further; because of our recognition of the mutual interdependence of nations, we in the United States are trying to help Europe get back on its feet. It is becoming increasingly clear that the problems of one nation are the problems of all; that, in fact, we are our brothers' keepers.

To achieve world unity is, of course, a colossal task. The difficulties are all too obvious. But this is no excuse for abandoning the attempt.

If world unity is a good thing, obviously unity within a nation is a good thing, and, likewise, within the separate political and social units of the nation—the various states or territories, the counties, the cities, the families. Anything which sets a man's heart against his brother is wrong; anything that binds him to his brother in charity is good; and because I believe that the Association of American Colleges falls into the second category, I think that it is good.

Now it would be, indeed, an extremely optimistic person who would expect perfect harmony among all men for any length of time. But that does not lessen, in my opinion, the moral obligation of mankind to strive for this goal. And there is always the possibility of a certain amount of success! Even the most pessimistic would admit that.

If we are to have any success in achieving unity and harmony among men, we shall need first to discover what it is that divides and separates them. To me it seems that the answer lies in the very nature of man as he exists today. Briefly, man, as we know him, is an imperfect creature both in his intellect and in his will. Men disagree because in many matters the truth is by no means immediately evident; its discovery is often a long, tedious process. Men differ in intellectual powers, as well as in the cultivation of those powers. Only God is all-wise; only He comprehends truth in its fullness. And men disagree also because of an exaggerated selfishness in their hearts. Most of us, according to Aristotle, follow our senses rather than our reason in competing for "wealth, honors, and bodily pleasures." It is the exceptional man, says Aristotle, who is truly a lover of self, in desiring that he should always "act justly, temperately, or in accordance with any of the other virtues," and in general in trying "to secure for himself the honorable course." And from the imperfect understanding of an issue by an imperfect intellect, and from an inclination towards an irrational selfishness, come the deforming passions of fear, of anger, of hatred, of the desire for revenge.

These are the things, in my opinion, which divide and separate man from man.

The next question is, what kind of unity do we want, and how can we achieve it?

I think that we may as well face the fact that we are not likely soon to get the whole race, or our whole nation, or our whole city, to agree on anything. Ideally, of course, if we accept the theistic position, we should all have the same faith. But actually we live in a world filled with Buddhists, Mohammedans, Shintoists, Jews, and so forth, not to speak of the various Christian bodies. All believe in some sort of Supreme Being, all have some sort of moral code. If we are ever to have "one world," or anything like it, we must, I think, first accept the fact of diversity, and, secondly, we must tackle the problem with charity as our basic attitude. We shall need, to be sure, our experts in politics, economics and sociology, but all of our technicians will avail us nothing unless we work together in a spirit of mutual love. Let me explain what I mean.

When I use the word "charity" or "love" as existing between man and man, I mean that one has a desire for the total good of the other. Life, as someone has said, consists in a series of relationships with other persons. The relationships are not between ideas; they are between human beings. Now, it is quite possible for me to love a man while hating his ideas. According to the precept of Christ, we must therefore love even our enemies, and do good to those that hate us. We need not agree with the ideas of our enemies, or with the things that they do. But we must wish their well-being in time and in eternity, and be willing to do anything within reason to help them attain it. And the reason for this is the unity of the race—we are all children of the one Father. Until the leaven of this idea somehow permeates the mass of men, or at least their leaders, international unity will have, I fear, very little chance.

The same is true, I believe, with regard to national unity, state unity, civic unity and even family unity. You may begin at the top or the bottom of the scale, but you will find no real unity in a body of men without charity. Differences and disagreements will exist, even in the best of families. But I appeal to your experience: where love is present, agreement can usually be reached—at least in the field of action, if not in that of opinion. Where love is not present, no unity is likely. There may be uniformity, or conformity, enforced by the police. Totalitarian states try to reach unity through this method. But it is a method repugnant to all that we Americans cherish.

Since ours is a national organization, let us look for a moment at the present state of our nation, from the point of view of national unity. We find here in America persons of the most widely diverse ancestry, representing all, I daresay, of the races of men, and holding various political and religious beliefs. Whether we wish it or not, this diversity is going to continue for a long time. Can there be any harmony in this Pentecostal population?

I think that there can be. During the war, Americans exhibited a high degree of unity indeed. The political parties got together on a bi-partisan program; labor and capital buried their differences; army and navy chaplains of different faiths worked side by side, and there was comparative peace between their various co-religionists. But now that the emergency—that particular emergency, the threat of utter destruction—has passed we are demonstrating to the world that, as a people, we can quarrel among ourselves as lustily as the next.

While I deplore any unseemly disputes, of course, in the realms of politics and economics, I am not nearly so troubled by them as I am by those in the realm of religion. And here the situation is none too good. In recent years there has been a resurgence of bad feeling between Protestants and Catholics in the United States. Without attempting to fix the blame, we can all agree, I am sure, that the fact is highly regrettable. It is regrettable because, as I have already said, anything that sets brother against brother is regrettable; and it is especially regrettable because it gives scandal to the unbeliever, who, like ourselves—according to Christian teaching—has been made in the image and likeness of God. But is he likely to abandon his unbelief and join our ranks when he sees us quarreling like heathen? Woe to him by whom scandal cometh. "Which Christianity?" is the question the puzzled pagan asks when we invite him to accept the faith of Christ. By this shall all men know that you are my disciples, if you have love one for another.

Now what can we possibly do to keep the peace between persons of different religious faiths? Shall we fight it out with firearms? Few, I hope, wish that solution. Shall we send missionaries to each other? In a sense, we are always doing precisely that. By our lives (God forgive us) as well as by our words, we are trying to persuade others of the correctness of our views. But what if we cannot? Must we then accuse each other of bad faith? Hardly. The Christian believes that it is only God Himself who can scrutinize the secrets of the human heart; for that matter, any man of sense will agree that he cannot with certainty know the motives of another.

What program then shall we recommend? The program that men of good will have always followed; the program that even civilized pagans have followed, at least to a degree. We shall not go about picking quarrels about doctrinal differences, but we shall try to coöperate with each other in matters that affect the common good. I do not mean that doctrinal differences are unimportant. They are important, unless truth is unimportant. But we must accept the fact that, for the present at least, we must live in a world of many religious faiths. The question is, How can we live and work together without compromising our integrity?

One answer is that we are already doing it, both in America and abroad. For years it has been common to see Catholics and Protestants working side by side in matters of civic and community concern—the Red Cross, the community chests, school boards, city improvement committees, and so forth. Similarly we find both banded together for their common good in labor unions, in the National Association of Manufacturers, in business houses, in the various associations of professional men. For centuries, this has been true among men devoted to the fine arts, and for almost as long, I should say, among scholars.

Let me give you an instance from my own experience. Some years ago I had the privilege of spending fifteen months in various libraries in England, France, Italy, Germany, Austria, and Belgium. At Oxford, Cambridge, and the British Museum, I was treated with the same courtesy as I found at the Vatican library. And similarly in the libraries of France, Germany, Belgium and Austria, when I seldom knew whether I was speaking to a devout Catholic, a devout Protestant, a Jew, a free-thinking agnostic or a professed atheist. And when I was at the Vatican I learned that the building had recently been repaired by the Carnegie Foundation for International Peace, because the members of the Foundation had regarded the library as promoting peace—scholars from all lands use it—and that money from the same source was being provided to pay for a new card catalogue. And this fact is also instructive: when I was about to leave London for Rome, I was given a letter of introduction to an official in the Vatican library by a Jewish scholar from the University of Heidelberg, who happened to be working in the British Museum at the time, and no one, either at the Museum or at the Vatican, thought it odd. Of course I could probably have gotten into the Vatican library through other channels; but the point is that my friend in London had a friend at the Vatican, and he wished me to receive special attention! Furthermore, when I mentioned to the officials at the Vatican the names of two of my friends at the University of Chicago (where I received my degree), Professors Charles H. Beeson and B. L. Ullman, I was treated with added respect: Mr. Ullman had been for years a correspondent of the then pope, Pius XI, when the latter was librarian at Milan, and later in the Vatican; Mr. Beeson, under whom I had studied, was recognized abroad as one of the two or three leading paleographers in America.

But, it may be pointed out, our present-day religious quarrels are due, not so much to doctrinal differences, as to the feeling of one body that its interests are being threatened by another, or a fear that they may be threatened in the future. To be specific, some Protestants in the United States fear that if

Catholics ever became strong enough to do so, they would deny freedom of worship to persons of other religious faiths. And some Catholics have a paralyzing fear of certain Protestant organizations.

Now there is historical reason for fear, I admit, in both instances. For various reasons, civil liberties have not developed throughout the world as they have in the United States. But, to my mind, the important question is this: Are we in this country likely to follow exactly the same pattern in human relations as that which has been followed elsewhere?

Personally, I do not think so. I believe that most of the Protestants I know have no desire to curb my freedom of worship. And I trust that most of the Catholics I know are equally fair.

American history, I believe, supports the view that we are, in truth, a freedom-loving people, and that we are steadily making advances in the domain of human rights. Our forefathers came to this country, not to think alike in political and religious matters, but to live peaceably with their own convictions. No matter what has happened elsewhere, I believe that we in America have developed, and that we can in the future further develop, so strongly our understanding of the rights of the individual conscience that no fear in this regard need be felt by anyone. I feel that we Americans can learn to live together in peace. If we are to look at the matter realistically, the cross-currents of interest—political, economic and social—are so diverse in America that people here are not, in the main, arrayed for or against each other chiefly on religious grounds. And I doubt, too, whether, in the foreseeable future, any responsible religious body would wish to enslave another. For that would mean that the majority of Americans would be reversing their thinking completely; and it takes a long time indeed for people to make such a shift. And the mere fact that we Americans are gradually getting to know each other better, helps the situation enormously: it is usually the stranger whom we dislike and mistrust.

Another reason for optimism in this matter, I think, is the development in some quarters of a new approach to the question of religious differences. Whereas some persons have begun the discussion by declaring that error had no rights, others have been pointing out that rights inhere in persons, not in abstractions; and that all persons, even those in error in their thinking, have rights; and that, furthermore, we must be extremely careful of the rights of the individual conscience even when it is in error. Reduced to practical terms, this means that for every person his conscience is the guide which he must follow, even though he be objectively in error. For example, if a college instructor's conscience tells him—however erroneously—that he must liquidate the president of the institution, then he is morally bound to do so. Certain inconveniences, to be sure, result from this doctrine; but it is still sound, according to my theological friends. The doctrine of the rights of the erroneous conscience is, of course, recognized in the prevalent attitude towards conscientious objectors to military service; most people regard the position of the objectors as erroneous, but still respect their right to act according to their own convictions. The same doctrine underlies the legislation in some churches, among them my own, which forbids members to baptize a person against his will, even though these churches regard the person in objective error if he does not wish to be baptized.

There are, to be sure, people in this country who would impose an artificial harmony upon us by trying to make us all agree upon everything. As a step towards this idealistic condition, they would do away with the private educational institutions. They would say, for example, that it is wrong for the Lutherans, the Baptists, the Catholics and other religious denominations to have their own colleges; they hold that such a condition makes for division among our people. But if I understand my history correctly, the early settlers came to America so that they might believe as they saw fit, in matters political, economic, religious and otherwise. And since they themselves founded church colleges, I think that they would most heartily applaud our having colleges today maintained by the different denominations.

Why do I speak about these matters to college administrators? Because I think that they are relatively free from the prejudices which militate against our national unity; because I think that they have been foremost in the fight for a true unity, a unity based on understanding and sympathy, not force; because I think that, by virtue of their position, they can do more than any other body of men of like size to promote this idea of unity; because I think that they have given an admirable example, in this Association especially, of the patient forbearance and of the Christian charity which must underlie all our efforts to effect peace and harmony among Americans, and among all the peoples of the earth.

THE CHRISTIAN LIFE

On Allowing Ourselves to Be Cheated

Reverend Frederick Bruce Speakman, D.D.
Minister of Third Presbyterian Church, Pittsburgh, Pennsylvania

Frederick Bruce Speakman was born in Chandler, Oklahoma, and received his early education in the Southwest. He attended the College of Emporia, where he participated in athletics, football and boxing, and was president of the freshman class. He graduated in 1940 at the University of Oklahoma, was ranking member of Phi Beta Kappa, president of his junior class, and listed in "Who's Who" in American Colleges in 1940 and 1941. He received his Master's degree in classical languages from the University of Oklahoma in 1941 and in 1942 attended Harvard University, where he received a Master's degree in history.

He took his theological training at Princeton Seminary where he received the Bachelor's degree in theology in 1945. That year he was awarded the Newberry Fellowship in Graduate Study which is biennially awarded to the winner of a nation-wide competition among theological students. The same year he was granted the ranking theology fellowship given by Princeton Seminary on the basis of scholastic standing and a competitive thesis. He was also awarded the Hugh Davies Prize as the outstanding preacher in his graduating class.

He was appointed teaching fellow at Princeton, teaching theology for the Seminary in 1945 and 1946 while he finished his resident requirements for the Doctorate of Theology. He received the Doctor of Divinity degree at Washington and Jefferson in June, 1950. For five years he was pastor of Central Brick Presbyterian Church, East Orange, New Jersey, and came to the Third Presbyterian Church, Pittsburgh, in October of 1949. He served in 1950 and 1951 as Acting Dean of the Chapel at Pennsylvania College for Women.

Paul Scherer and Arthur Gossip are his favorite homiletical stylists. For relaxation he has a wood-working shop; he boxed in college, and at thirty-three is becoming one of the recognized younger preachers of America. His sermon, "On Allowing Ourselves to Be Cheated," is an example of his present preaching at Third Presbyterian Church, Pittsburgh.

Sermon Eight

TEXT: Wherefore do ye spend money for that which is not bread?

ISAIAH 55:2

IT WOULDN'T BE fair to say which department store it took place in, for the lady was very angry as she flounced into the elevator, whether at a salesgirl or the credit department or prices in general, we couldn't tell. But neither could we avoid overhearing her quiver-lipped remark to her companion, "It isn't the money I mind, it's that I hate so to be taken in!" And you could fairly feel the sympathy of her fellow passengers as we cleared the mezzanine and headed up, in the awkward silence which is manners in elevators. Whatever had happened we shared her anger, raged with her a bit that she'd been cheated. For we had this in common, we too so hate ever to be taken in.

Now never before to my knowledge have I done much meditating in an elevator. But this incident did evoke quite a chain of related thoughts. We all of us so hate to be cheated! Then why do we put up with it? That's what Isaiah asked the people of his day. *Why do you spend your money for that which obviously is not real bread?* Why do you sacrifice in order to acquire that which so obviously is not worth having? And add to that a figure that the late Archbishop Temple often used. This life is like a vast store window, he urged, in which pranksters have switched the price tags. So that things of great value are priced as worthless baubles, while cheap tawdry trash is priced so high! And the tragedy of it is, he adds, that we go along with the joke,—we believe the switched price tags, we let ourselves be taken in! And go past that just one more step to the claim of the American philosopher, W. P. Montague. He claims there's one simple question you and I must answer for ourselves before we'll be able to make any sense or any profit at life's strange rates of exchange. This question, *"Are the things that matter most finally at the mercy of the things that matter least?"* Doesn't that somehow pull Isaiah and Temple and you and me together in a common problem? It seems that there is a colossal cheat running close to the heart of life. But why do you and I who so hate to be taken in, why do we go on day after day, allowing ourselves to be swindled?

Let's stake our claim alongside Montague's question first. You and I haven't really signed up with Christianity if it hasn't convinced us that the things that matter most are not finally at the mercy of the things that matter least! That what is really valuable, human personality, the glories of the mind, the sacramental wonder of love, the costly dreams of the heart, the way great art can, across the years, carry its message of rare experience, the mystery of music, the vision of mystics, the heroism of sacrifice, the lift of high loyalties, the gallantry of faith, that all this is not finally to be rubbed out and become as if it never were before the ongoing remorseless, senseless processes of time and change and decay. When the wife of Palmer of Harvard died, he put it so grandly, "I would have only contempt for the idiocy of a universe that out of deference to the rearrangement of a few particles of matter, could exclude, extinguish, discard, forget, so fair a spirit as she was." It's the bedrock of religion, you see, this conviction that what we care for most, when we're at our best, is not the helpless prey of the world's brute forces. That whatever is humanly meaningful came from somewhere important, for some important reason, and is headed somewhere that matters because of some purpose that matters. That's why the Christian faith, when true to itself, has been able to say of the deep joys and real values and high achievements of human existence what the Psalmist said to God, "Lesser things shall perish but Thou shalt endure: yea, all of them shall wax old like a garment; as a vesture shalt Thou change them and they shall be changed: But Thou art the same, and Thy years shall have no end." And if that kind of God has convinced us it's that kind of life, so that what's highest in us is deepest in the universe and surest of victory, then it's exactly in that conviction we should be finding the dynamic for all the practical consequences of our lives. Then why do we so often go on *allowing ourselves to be cheated,* by sacrificing what matters most in order to purchase what matters least.

For if anything is still needed to persuade us that we are not creatures of logic, but patrons of impulse and irrationality, surely this should convince us. To look honestly over our lives and the other lives we know best, and see how few of the bargains we drive can look up and face Christ's short, blunt, elementary question, *What Profit?* Ben Franklin coined a proverb for the world when he told of his first venture as a purchaser, how he surrendered to the man behind the counter a literal fistful of coin for a silly whistle. And how, since that youthful fleecing, he had looked at friends and had thought so often, there it is all over again, that man paid too much for his whistle. And that's more than a homespun bit of wisdom. It's of a piece with a bafflingly consistent human tragedy, the way we go along with the swindle, the way we'll take the truly valuable and use it as barter to traffic in trifles.

Have you a friend who's wrecking his health? Try to analyze what he's trading his health for! Chances are it's far from a bargain. Some imagined prestige of career. Some badge of busyness. Some pathetic belief in the security of gain. Or just refusal to surrender some habit which others may take in stride but which to him is poison. And for any of these to sell health! Do you know a home that's on the precipice? Try to tag the basic trouble. How often it will be nothing of importance. Just hearts that are clogged with the lint of everyday living. Tension from the trivia of selfishness. A man and woman discovering that growing pains are not restricted to childhood. So often little more than the sum of petty and needless annoyances. And for any of these to wreck a home! Have you ever allowed a friendship, a real one, to be ruined by a triviality? So that now you've lost that relationship and in its place have gained, what? A sense of grieved pride? A suspicion of other friends? A cynicism? A permanent pout? Hardly a bargain! Or how many do you know who have sacrificed this quality we call personal integrity? The ability to come to themselves at night. This complex of basic self-respect and inner well-being and bedrock decency, have shredded that whole fabric in order to attain to social conformity! What a swindle! Social conformity, the mythical distinction of the chameleon. The being able so to blend with any background that we're noticeable only for being not noticed. The rubbing out of anything like a conviction that might cause our set to deem us odd. This colorless cousin of that will-o-the-wisp popularity, social conformity. To purchase it, what gilt-edged qualities our generation has handed in!

Oh, it's too widespread to more than sketch at random. It's simply all across the face of life. What of that colossal cheat traditionally called the immoralities? Surely nowhere else are the costs so heavy and the returns so staggeringly meagre. This gay adventure of flouting the real moral code remains gay only in best-selling fiction. It turns so quickly into a drab and far from glamorous affair. The price is crippled personality, and the net gains, well, list them! A hardening of the arteries of the soul. A heart that's prematurely gray. What were pleasures when kept in focus, now become compulsive servitudes. Something fine gone out of life not likely to return. And there it is all over again. "If there's anything I hate it's to be taken in," said our elevator acquaintance. No, you don't resent it enough, claims Isaiah. Or why would you go along with the price tags, asks the Archbishop. Yes, concludes the philosopher, why, in the light of what you believe, why do you keep so living that you place the things you care for most, fully at the mercy of what you care for least?

I submit to you that we've often done it with religion. That we've often done it with the Church! Of all the church lore concerning Pittsburgh which I've heard since becoming naturalized this strikes me as an incomparable incident. Some years ago in a church not far from us, a new minister found the people divided evenly into two warring camps. They sat across the church from each other. They sang and prayed at one another. Each applied the sermons to the other group. And for a while the minister played the game of litmus paper, turning red when with the acids and blue when with the alkalis, trying to discover the source of such a cleavage. And when he had traced the enmity to two influential women, one at last confessed the origin of the feud. It seems that at a church dinner these two ladies were jointly in charge of a dessert. And they decided on homemade doughnuts. "But," recited the aggrieved one, "my doughnuts were sugared, and hers were plain. And when dessert was served my family and I were served her plain doughnuts while she and her family helped themselves to my sugared!"

You've a choice as to whether God wept or laughed to see a church, the laboratory of His truth, with its climate of worship become a mockery, and its prime ingredient love, turned farcical, and its total impact a flat denial of the nature and purpose of faith, all because of a matter involving sugared doughnuts! But that's what so many of our troubles boil down to. *We've let ourselves be cheated* of so much that's our inheritance as Christians by placing its big results under the tyranny of trifles. The next time some flaw appears in a service of worship, the next time the feelings you've spread upon your sleeve are bruised, the next time you're tempted to look at church organizations as you might some P.-T.A., the next time you're on the verge of treating others here according to a code that doesn't belong here, the next time you're about to jeopardize your whole feeling for Christian truth because of your annoyance at some pathetically human spokesman for religion, could you remember the phrase, "sugar doughnuts," and let that warn you you're about to be cheated, you're about to exchange something important for something that simply doesn't matter!

Christianity should be doing big business with you and me in these months and years ahead. There's one tremendous gain that's potential in a time of crisis, if there can filter into our living a priorities point of view. If we can blue pencil the trivia. If the pressure of these hours will drive us straight into the arms of the things that matter most. If there's anything that should dismay a Christian today it's to find himself engaged in pettiness, trafficking in trifles. Only these really matter: *Whatever keeps us increasingly in touch with God, whatever keeps us increasingly in touch with men, whatever keeps us increasingly in touch with God's purpose for us and the destiny ahead of us.* These supremely matter! In an hour when the wisest know their wisdom is not enough, and the strongest are distrusting their own strength, and those who thought they had arrived are asking where they are, let us find new loyalty to the Christian sense of what is vital, let's settle for nothing less than the abiding treasures of life!

Think It Not Strange

REVEREND LOWELL M. ATKINSON, PH.D.
Minister, St. James Methodist Church, Elizabeth, New Jersey

When the Hackensack Methodist Church celebrated its Hundredth Anniversary, its Centennial Booklet was dedicated to the Pastor, Dr. Lowell M. Atkinson, with these words, "Dr. Atkinson has brought intellectual distinction and spiritual vigor to the pulpit ministry of our Church."

Born in Collingswood, New Jersey, he is the son of a Methodist preacher and likes to think of himself as essentially just that—a Methodist preacher. He holds four degrees, receiving his Ph.D. from Drew University in 1947. At that time he was awarded the Ezra Squirer Tipple Fellowship in Preaching and in 1949 occupied the fellowship at Oxford University. He also studied at the University of Birmingham Summer School in Shakespeare at Stratford-upon-Avon. While at Oxford, he served as pastor of the Marston Road Congregational Church.

Filled with enthusiasm by his experience of England, Dr. Atkinson wrote an article "On Visiting England" which was immediately published in "The English-Speaking World," used by the Methodist Board of Missions as coun-

sel for exchange preachers going to England, and reprinted in the American magazine "Freedom and Union." In 1951, Bishop Oxnam named Dr. Atkinson an alternate delegate to the Methodist Ecumenical Conference at Oxford.

Dr. Atkinson is pastor of the St. James Methodist Church in Elizabeth, New Jersey, and Dean of the Blair Youth Conference. His published articles include "Plutarch," in "Religion in Life" (Fall, 1948), "The Achievement of Arminius" in "Religion in Life" (Summer, 1950), "They Lived Happily Ever Afterwards" in "The Christian Advocate" (September 30, 1948), "The Wonder of the Presence of God" ("Christian World Pulpit," London, 1950), "Dean Hough in the Classroom" ("Drew Gateway," 1947), "The Awful Splendor of Answered Prayer" ("The Upper Room Pulpit," 1951). Dr. Atkinson's hobbies include salt-water fishing, swimming, boating, tennis, volley ball and handball.

This sermon considers the suffering, hardships and victorious living of the early Christians. His illustration of Ignatius of Antioch, Sir Andrew Barton, the death of Peter in Rome, and his use of adversity in present Christian living is good reading. The sermon gives promise of a great preaching career.

Sermon Nine

TEXT: "Beloved, think it not strange concerning the fiery trial which is to try you, as though some strange thing happened unto you." I PETER 4:12

CHRISTIANITY was cradled in adversity. The rude manger of the Infant Jesus was prophetic. A hard world opposed his wonderful truth with cold hostility. Callous men recoiled in bitter antagonism from his gracious message. Christ was at cross-purposes with the world's selfish ways, and the cross marks the meeting of Christ and the world. Suffering was no strange thing to Christ. It was of the essence of his gospel and it was the climax of his ministry. The cross of pain has become the world's symbol of salvation!

Suffering was no strange thing to the early Christian Church. The little letters of the New Testament were written to strengthen the hearts of Christians in the midst of peril. To men who dared to brave death came words of encouragement and hope. The early Christians did not lightly repeat the Lord's Prayer or the Apostles' Creed. At the risk of life they said these sacred words; and often by the gift of life they witnessed how much they valued the privilege, which we often take for granted in our comfortable Sunday morning worship. We may well remind ourselves that our Christian faith now lives in strength because the faithful died for it. They counted it no strange thing to care for Christ at the cost of suffering and peril and death. Their blood is the seed of the Church.

Does it not give us pause that these Christians of earlier days cheerfully accepted outer hardship if they might but enjoy inner comfort? For today, we seek outer comfort at all costs, even the cost of inner emptiness, unrest, and discontent. Ours is the tragedy of lives lighted only from without. All about are "the decent, godless people" going through the motions of respectability without the inner glory of the lighted mind and the kindled heart. One day, says T. S. Eliot, they will be gone, "their only monument the asphalt road and a thousand lost golf balls." Is this the price of fleeing pain—that we shall be remembered only by the external? There is a greater monument to those who thought it no strange thing to suffer for Christ—the glorious living monument of a Christian community that now girdles the globe and gives to earth its one best hope!

Why is it so? We do think it strange that we suffer, and we do raise the query, Why? This is natural enough, and by no means is it wrong. We fail when we go down the slippery descent of self-pity until the Why becomes a whine! When the natural question, Why, becomes the complaint, Why should this happen to me, it is time to recover the heroic note—"Think it not strange concerning the fiery trial which is to try you, as though some strange thing happened unto you."

Perhaps there is no answer to satisfy the mind, except the answer that makes the question seem no longer important—the coming of God as the Great Friend. Job's wild questions were flung out of his pain against a silent sky. He demanded an answer. But when God came, he forgot all about his questions. When the ache of the heart is eased, the puzzles of the mind seem unimportant.

And yet, we must face the questions of the mind with fearless candor. Why is it so, that men suffer? Could it be that these are the very terms of life that God has set for our maturing—that we must struggle and strain, that we must confront real danger, and run the risk of perilous defeat; that life's deeper wisdom is the fruit of pain and that sacrificial suffering is in the very structure of life.

Robert Louis Stevenson, in his *Master of Ballantrae,* tells of a man whose failing of mind was first recognized because he became incapable of facing trouble or pain. Stevenson makes this discerning comment, "In this life we must contemplate much that we cannot cure." Which is to say that we must face up to life, courageously and confidently, and accept our share of sorrow and pain with brave hearts. Perhaps there is purpose in pain, though we do not see it. It is our faith that God actually sees a pattern, though on the reverse side of the loom, we can see but the loose ends and apparently unrelated bits of the material of life.

John Wesley very wisely said, "Whatever raises the mind to God is good. You have accordingly found pain, sickness, bodily weakness to be real goods, as bringing you nearer and nearer to the fountain of all happiness and holiness." This is the New Testament note, the assurance that faith gives as to the soundness of life. For the Christian recalls that Christ also suffered, leaving us an example. It is the man of faith who can

> Welcome each rebuff
> That turns earth's smoothness rough.

because he looks upon life with the eyes of faith and sees a purpose running like a golden thread through it all.

How are we to live? It is the spirit that matters most, as we confront life's sufferings. The Christian has the spirit of joy in the hour of affliction. "Rejoice" is coupled with "suffer" throughout the New Testament. Edwin Markham has caught this spirit in his lines,

> Defeat may serve as well as victory
> To shake the soul and let the glory out!

Here is a Bible truth! "Rejoice, inasmuch as ye are partakers of Christ's sufferings . . . if ye be reproached for the name of Christ, happy are ye." "Blessed are ye when men shall persecute you." "Rejoicing that they were counted worthy to suffer shame for His name"—these are characteristic New Testament expressions.

St. Paul describes the buffeting of the "thorn in the flesh." He is not depressed and dejected because of his suffering. Climbing the heights of faith, he gains new perspective. Perhaps the thorn in the flesh was needed to shatter his pride, and let God's grace fully come into his life. "Therefore I take pleasure in infirmities, in reproaches, in persecutions, in distresses, for Christ's sake. . . ." Yes, defeat serves the Christian as well as victory to shake the soul and let the glory out. The outcome of faith is a gloriously invincible spirit. St. Paul describes his outer hardship and inner triumph—"troubled on every side, yet not distressed; perplexed, but not in despair; persecuted, but not forsaken; cast down, but not destroyed." Is there not the tang of true adventure in so heroic a life?

It was this victorious spirit that marked the march of Christendom in those early days. Ignatius of Antioch was arrested and convicted of the "crime" of believing in Christ. Chained to ten cruel guards, whom he described as "ten wild leopards," he made the tedious journey to Rome. His death march became a triumphal procession. He wrote beautiful letters of encouragement and hope to the Churches, the condemned comforting the free! And when at last he was flung to the lions in the arena at Rome, he cried with a loud voice, "The wild beasts are the road to God!"

This victorious note sounding in the moment of outward defeat has reverberated across the ages. It touched the medieval world to chivalry. Something of it is in the old ballad:

"Fight on, my men," says Sir Andrew Barton,
"I am hurt, but I am not slain;
I'll lie me down and bleed a while,
And then I'll rise and fight againe."

And we, who devote our mighty energies to being outwardly comfortable and to creating homes of total convenience, how are we to live? Can we hear this heroic note of joy in suffering, and victory in defeat? Like a trumpet, it sounds, "Rejoice, inasmuch as ye are partakers of Christ's sufferings. . . ." Count it your true joy, if through your brave witness of heroic sacrifice, Christ is well spoken of in your time!

Whom may we trust? "In God we trust." It is not only our slogan; it is our strength. For God gives a glorious immunity to the faithful. Not that He always gives deliverance from trouble; but He does give deliverance in trouble! And if victory often must be wrested precariously from the hour of defeat, yet we can trust the moral might of the Invincible God, and putting our lives in His Hands we may know that all is well. If we fear God, we need never fear anything else.

It helps greatly to remind ourselves of the companionship of the Friendly God. "The eyes of the Lord are upon the righteous, and His ears are open to their prayers. . . . And who is he that will harm you, if ye be followers of that which is good? But if ye suffer for righteousness sake, happy are ye; and be not afraid of their terror." The assurance of God's presence brings peace to the faithful, a peace that is beyond understanding, and also beyond the power of the world to hurt and to harm.

In the fourteenth-century poem, "Sir Gawain and the Green Knight," we watch a brave knight making his perilous journey through dismal wasteland to confront what seems certain death for the honor of his King. He is alone in this desolate region—"no soul but the Saviour to speak to." Yet here is a saving exception—he is not really alone in his perilous quest; the Saviour is with him as a living Friend! Here is his secret of comfort and courage, he has found the Saviour in whom to trust. So he cries out in strong confidence:

"By God," the knight said, "now
I'll neither weep nor groan,
Until God's will I bow
And make myself His own."

Once he had placed his hand between the hands of his King and made himself the King's man. Now he places his hand between the hands of God and makes himself God's man. In so valiant a service, obstacles are but opportunities and suffering ceases to be a problem and becomes an heroic privilege.

The story is told of St. Peter that when he died a martyr's death at Rome, he made no outcry because of his cruel fate, but only asked that he be crucified head-downwards. He was glad to die for his Saviour, but too humble to die in the same position on the cross. He thought it no strange thing to suffer for Christ, but rejoiced to be counted worthy to suffer in His name!

The triumphant self-sacrifice of Peter is commemorated in stone on the beautiful Milan Cathedral, a symbol of the heroic humility that builds moral splendor into the Church of Christ.

And we, who have lost the love of heroism, and consequently the enjoyment of peace, shall we not turn again to the heroic life of which the Bible tells, hazarding outward comfort if we may gain inwardly the mighty comfort of Christ? To those whom He counts worthy, Christ gives the gift of suffering—not as a strange thing—but as a badge of honor, that we wear proudly in His name. In times of stress, life glows with meaning, our daily experience is transfigured by a new spirit, and we need never walk alone again because we have Christ's heartening companionship. This new life of heroic faith is Christ's gift to you. He is offering it to you now. It may be yours, if you will have it so!

The Two Tentmakers

REVEREND RUSSELL CARTWRIGHT STROUP, D.D.
Pastor, Georgetown Presbyterian Church,
Washington, D. C.

Russell Cartwright Stroup is pastor of the historic Georgetown Presbyterian Church, founded in 1780. Born in Cleveland, Ohio, January 16, 1905, the son and grandson of ministers, he graduated from Stanford University in 1925 and has done graduate work there and at the University of Southern California and American University, as well as Drew Theological Seminary.

He has had a varied career as minister in the Southern California Conference of the Methodist Church, unsuccessful candidate for Congress, successful farmer, thirteen years pastor of the First Presbyterian Church of Lynchburg, Virginia and, since 1950, pastor of the Georgetown church. During World War II he served as chaplain with the infantry, landing with assault troops on five beaches in New Guinea and the Philippines, was cited five times for heroism and awarded the Bronze Star for rescuing wounded under enemy fire. His exploits in combat are mentioned by Christopher Cross in Soldiers of God *and by General Robert L. Eichelberger in* Our Jungle Road to Tokyo.

He has published numerous articles in national magazines, including an article in Harper's, *which was widely reviewed and reprinted in two anthologies of prose,* Living, Reading and Thinking, *and* The Modern Omnibus. *He received the Doctor of Divinity degree from Washington and Lee University in 1947 and the Freedom's Foundation award for a sermon in 1951.*

Using pagan poetry and the New Testament, this sermon is almost prose-poetry in certain passages. The quotations are taken from the Edward Fitz-Gerald translation of The Rubáiyát of Omar Khayyám in English Verse *(Houghton, Mifflin, 1888). His sermon makes a fitting emphasis upon the Christian message of hope in our time.*

Sermon Ten

TEXT: Delight yourself in God, yes, find your joy in Him.
PHILIPPIANS IV:4 (J. B. Phillips' Translation)

ONCE THERE WERE two tentmakers, Omar and Paul, who of course were not really tentmakers at all. Omar Khayyám, being interpreted, is Omar, the Tentmaker, but the man who bore the name was a mathematician, an astronomer, and a minor poet of Persia. Paul, the apostle, on one of his many journeys made tents for a living but we know him as a missionary, mystic, and servant of Jesus Christ. Nevertheless we may think of both as men who wove the fabric of philosophies which other men may choose to shelter their spirits on the pilgrimage of life.

It is somehow sad that Omar, who was a significant scientist, should be remembered solely for the quatrains he wrote which were pleasantly translated by a poet of the past century and became the vogue in Edwardian parlors. It was a vogue revived in the roaring twenties of our own century, that "aspirin age" in America when a lost generation lived in that unpleasant hell which lies "this side of paradise" and tried to make themselves believe that they were having a wonderful time.

One would have thought that the grim realities of depression and war would have disposed of anything so superficial as Omar's verse and his philosophy but apparently readers still buy *The Rubáiyát* for the book stores are stocked with several editions of it and we find the de luxe editions on the tables of some American homes where the Bible used to be. Perhaps this is not surprising in a country of which Commager could write, "The old taboos and the old integrities were dissolving, Puritanism gave way to hedonism, inhibitions to experiments, and repression to self-expression." [1] Where integrities are dissolving and Puritanism is being replaced by hedonism, the

[1] Henry Steele Commager, *The American Mind,* p. 429.

Rubáiyát might well become a theme song for little minds. It may not be surprising but it is distressing, for while anyone might read Omar for passing pleasure, to read him with serious intent is a sad reflection on the shallowness of our times.

Consider the most famous lines of them all:

> Come, fill the cup, and in the fire of Spring
> Your Winter-garment of Repentance fling;
>   The Bird of Time has but a little way
> To flutter—and the Bird is on the Wing.
>
> A Book of Verses underneath the Bough,
> A jug of wine, a loaf of Bread—and Thou
>   Beside me singing in the Wilderness—
> Oh, Wilderness were Paradise enow!

How many foolish folk have read that with a sigh and thought "What a wonderful life it would be!" Actually it would be a far cry from Paradise if they followed Omar literally but of course they don't. Their Eden is even less satisfying that his would be. A television set replaces the book of verses, and a smoke-filled room, the open air under a pleasant bough. Wine gives way to something stronger. Cakes and caviar appeal to their jaded palates more than wholesome bread. The "Thou beside me singing in the Wilderness" becomes a crowd in a cluttered room chattering aimlesly and endlessly like magpies or monkeys. All of which is a fairly factual description of the parties which pass for a good time in our unhappy land. If that seems "paradise" to you, imagine, if you can, spending eternity at a cocktail party. "Hell" would be a better word.

Still, the quatrains have a pleasing sound. It is so nice to be able to cloak our self-indulgence in a gaudy philosophy and put our shoddy desires into a pleasant rhyme. We have misread Omar, however, if we find in his verse a pretty picture of the pleasures of hedonism. His *Rubáiyát* is as tragic as hedonism itself and its lines are the lament of a prisoner who sang. It speaks of world weariness, self weariness, disillusionment and despair.

For Omar who studied the stars could not see beyond them.

> And that inverted Bowl they call the Sky
> Whereunder crawling coop'd we live and die,
>   Lift not your hands to *It* for help—for It
> As impotently moves as you or I.

Can you match that for sheer tragedy? It may be pleasant to sit under a bough with a book, a jug and a pretty girl. It is not so pleasant to find yourself crawling under an inverted bowl like an imprisoned fly. It is not so pleasant to lift impotent hands to an empty heaven. Omar's cry is the cry of a man trapped in a mechanistic universe where there is no God.

Such is an empty universe indeed. Where there is no God there can be no "men." In a world of blind chance they become pawns of fate, not persons.

But helpless Pieces of the Game He plays
Upon his Chequer-board of Nights and Days;
  Hither and thither moves, and checks, and slays
And one by one back in the Closet lays.

So there we are, pawns that are pushed about by some impersonal Master of the Show; made to play a game in which we have no choice and for which he has no purpose until the farce is finished and back in the box we go.

As man is a helpless pawn to Omar so life is a fleeting moment in the sad eternity of time; a tiny oasis in the wasteland of nothingness out of which we came and into which we must so soon return.

A Moment's Halt—a momentary taste
Of Being from the Well amid the Waste—
  And lo!—the phantom Caravan has reach'd
The Nothing it set out from—Oh, make haste!

Make haste to eat, drink and be madly merry for tomorrow you will surely die.

Alas in such a pitiless universe of puppets where there is no God, and man is only a mechanism in a mechanistic world, even a jug of wine is not for joy but for forgetfulness and all pleasure but a brief reprieve from the intolerable pain of life.

So, my Beloved, fill the Cup that clears
Today of past Regrets and future Fears;
  Tomorrow!—Why, Tomorrow I may be
Myself with Yesterday's Sev'n thousand Years.

As it was with Omar so it is with many in our time who turn to that way of life. Without hope of joy, they seek escape. Without hope of pleasure, they seek oblivion from pain. Without hope of life, they try nonetheless to turn their eyes from the spectre of death who sits, a skeleton, at their senseless feasts. They refuse to look but they know he is there even as Omar did, for his poem is a poem of death. Again and again the phantom appears. His presence is a haunting, ever-present terror and there is no real escape.

Men accept such a tragic philosophy for one of two reasons. It may be that they seek but, at the end of their search, find no God in the universe, no meaning in life, no significance in man. If so, if this is the kind of a world we live in and if these are the kind of creatures we are, animals only, then let us behave like animals. What does it matter? There are no values. There are no absolutes of right and wrong. There is no eternity, nothing but a brief moment of being, and it can make but little difference how we live it before we too "into the Dust descend . . . sans wine, sans song, sans singer and sans end!" This is the better reason for accepting such a philosophy. At least it is without hypocrisy.

On the other hand, and I think this is more often the case, there are those who, wishing to live a certain kind of irresponsible life, seek to justify themselves for the life they live by pretending to dismiss God and with Him man's responsibility. If I am a helpless pawn, then I can have no sin. It is the sense of sin which men would like to escape, believing so foolishly that this is the way of freedom. What they fail to realize is that man to be man at all must *have* a sense of sin. Where there is personality there must be responsibility. If there is any freedom for man it is the freedom to choose between good and evil and to be responsible for the choice. To lose one's sense of sin is to forfeit one's liberty. It is to deny one's humanity. Only as we are morally responsible are we men.

No, the philosophy of Omar is not a philosophy for men since it denies their personality. It is not a philosophy of life but of death. It is not a song of freemen but a dirge for slaves. It does not promise real pleasure but boredom and despair. As Joad, the English philosopher, has said, "The final criticism of the ideals of the modern world is that they fail to satisfy." [2] They do not give us pleasure but boredom, not life but death.

Consider, as Joad does, that heaven of hedonism, the Riviera, where the idle rich gather like spoiled children to be amused. Like all spoiled children nothing can amuse them for long. Before an hour is over they are bored and turn to something new. They spend an hour at sunbathing, an hour at motoring, an hour at polo, an hour at cocktails, an hour at gambling. When they are bored with gambling they eat and when they are bored with eating they dance, and when they are bored with dancing they make love and when they are too bored for anything they get drunk and are put to bed. "For my part," says Joad, "I do not find it surprising that the suicide rate among the unemployed rich is the highest of any class of the community." [3] Nor do I. Nor could anyone. For this is not life but death, this practical application of the philosophy of Omar. The tent which he would make for us is not a pavilion of pleasure but the canopy which shelters mourners around an open grave.

What a relief to turn to Paul. His is no languid song of summer but the shout of spring and life and hope. The verses I read to you from Philippians are a part of that scarlet thread of joy which brightens all the letters of Paul. "Rejoice," he cries, "and again I say, rejoice!" This is no sigh but a shout. "Rejoice and again I say, rejoice!" Rejoice is the first word of the passage, it is the climactic shout in the middle, and it is the assurance at the end. "Delight yourselves in God, yes, find your joy in Him." [4]

So Paul wrote and so he lived. He was a man persecuted and imprisoned. He was man impoverished and oppressed. He was a man tortured and sentenced to death. But through it all he raises his song of triumph, of life that is abundant, eternal and good. "Delight yourselves in God, yes, find your

[2] C. E. M. Joad, *Return to Philosophy* (E. P. Dutton & Co., New York), p. 66. Used by permission of the publisher.

[3] *Ibid.*, p. 67.

[4] J. B. Phillips, *Letters to Young Churches* (Macmillan, New York, 1947), p. 121. Used by permission of the publisher.

joy in Him. . . . Don't worry over anything whatever; tell God every detail of your needs in earnest and thankful prayer, and the peace of God, which transcends human understanding, will keep constant guard over your hearts and minds as they rest in Christ Jesus." [5]

The secret of happiness which Omar never learned, Paul knew. "I have learned to be content, whatever the circumstances may be. I know now how to live when things are difficult and I know how to live when things are prosperous. In general and in particular I have learned the secret of facing either poverty or plenty. I am ready for anything through the strength of One who lives within me." [6] This was not dumb resignation on Paul's part. It was joyous acceptance. Omar could not even find joy in pleasure. Paul could find joy in pain. Omar could not even find peace in oblivion. Paul could find peace in the arena of conflict. Omar was lonely in a crowd. Paul found companionship in solitude. All these seem paradoxes but they are paradoxes which become realities in the life of any Christian who has learned with Paul that "God will supply all that you need from His glorious resources in Christ Jesus." [7]

Omar looked beyond the universe and found nothing. Paul looked and saw God in the face of Jesus Christ. Omar looked at man and saw an animal. Paul looked at man and saw a son for whom Christ died. Omar lifted impotent hands to an impotent heaven. Paul lifted trusting hands to receive the infinite gifts of God. Omar felt himself a helpless pawn pushed about by blind chance to no purpose. Paul knew that all things work together for good to them that love God. Omar knew the littleness and loneliness of a man lost in the infinite and empty universe. Paul knew the wonder and glory of man found by the love of God. Omar looked at life and saw only death. Paul looked at death and saw eternal life.

What a tremendous difference between the two tentmakers—and the difference is Christ in whom alone is life and that life is the light of men. Paul did not labor alone when he wove his fabric of faith. He was ever a fellow worker with Jesus Christ and the faith he offers carries with it Christ's certainty of life abundant and eternal. Only Christ can offer that. You may turn to better men than Omar and still you will not receive the same. Leibman out of the treasury of Hebrew philosophy offered to hundreds of thousands "Peace of Mind" and that is something very precious, indeed, but it is not enough. Jesus said, "Peace I leave with you, my peace I give unto you." It is a glorious legacy but he offers more than that. Peace of mind is good but a joyous heart is better and he promises that in him our joy will be full, for such joy comes from a heart at peace with God, from a soul at one with the Father.

Phillips Brooks was strangely right when he said, "Christianity is not a

[5] *Ibid.*, p. 122.
[6] *Ibid.*, p. 122.
[7] *Ibid.*, p. 122.

religion of self-restraint but of self-indulgence." Only the animal in us is restrained and that in order that we may indulge ourselves as sons of God. Our place is not with the swine. There is no enduring pleasure there. Our place is in the Father's house where a feast is being spread. What is the chief end of man? You know the answer. The chief end of man is to glorify God and to *enjoy* Him forever. Christ gives us peace *with* joy. It is the purpose of God in Christ that the deepest desires of the human heart should be forever satisfied. So much happiness He has for us that human life is far too short to hold it all and so God gives to us eternity. Out of such assurance Paul wove the fabric of his faith.

May the same mind be in us. We cannot accept the philosophy of Omar. It is for men in despair and for cultures in decay. It is a philosophy far older than Omar; older than the Gospel which gave new hope to men. But it has never been held by men in the springtime of life; not in Greece, nor in Rome, nor in Persia, nor in America. It is the philosophy of disintegrating societies where men and women have gone flabby morally and spiritually and seek for themselves an oasis of indulgence in the waste land of futility. It is not for you. It is not for me.

Let us then strike the shabby tents of hedonism which would imprison our spirits and take to ourselves the tent of faith for our pilgrimage. So shall we leave the wasteland of despair or the unsatisfying oasis of indulgence and journey joyously toward ever expanding horizons of hope with peaceful minds and singing hearts. It will be for us a journey into life; life that is abundant, life that has no end.

CHRISTMAS

Star Over Bethlehem

REVEREND WALTER DALE LANGTRY
Minister, Prytania Street Presbyterian Church,
New Orleans, Louisiana

"Star Over Bethlehem" was preached in the old Southern city of New Orleans on Christmas Day, December 25, 1949. It was selected for its message of hope, its vision, its faith in a world where hope, vision, and faith are needed to give courage day by day. The Babe of Bethlehem still comes to the hearts of all who believe in him, making Christmas the greatest night of the year: the promise of all the centuries was fulfilled that night.

Born in Clayton, Missouri, the Reverend Walter Dale Langtry was the son of a Presbyterian minister, was educated in the public schools of Clayton, Westminster College, where he heard Mr. Churchill make the famous "Iron Curtain" speech, Washington University in St. Louis, and Louisville Presbyterian Seminary. He has done additional graduate work in the Divinity School of the University of Chicago and the University of Virginia.

Ordained by the Presbytery of St. Louis, June 2, 1935, he served a mission Church in St. Louis for two years, was one year student pastor for Presbyterian students at the University of Virginia, spent ten years in churches of North Mississippi, first at Sardis and Batesville, later at Batesville alone, and has been at the Prytania Street Presbyterian Church, New Orleans, since 1948.

In his interests, his reading is rather restricted to religious books and publications; he has great respect for Reinhold Niebuhr, Emil Brunner, John Mackay, Otto Piper, while his pulpit favorite is definitely James S. Stewart.

New Orleans is a great place, but after ten years there, he sees how much simpler life is in small towns, for all cities make life difficult and complicated. Sociability, civic responsibilities, recreation, hobbies, interchurch fellowship is harder in a big city than in a small one. "Star Over Bethlehem" has simplicity, a message for our day, a reminder of the Christ the world desperately needs.

Sermon Eleven

TEXT: "When they saw the star, they rejoiced with exceeding great joy." MATTHEW 2:10

THERE IS a significant fact about the magi and the star. They had seen it in the east, and had come to Jerusalem to find the King whose birth it announced. But they made their journey without its light. Only when they came to Bethlehem did they see it again. No wonder they rejoiced. They were on the right road, and the promise of the star was about to be fulfilled. How many doubts they must have had on their long journey! How foolish it must have seemed to go so far on such an uncertain mission. But when the star appeared again, they knew that they were right at last. They would see the new-born King, and their gifts might bring the message of their hope and trust in him.

It is no wonder that the star over Bethlehem has been the symbol of rejoicing ever since. The world had hoped so long for release from its bondage. The night had seemed so dark, and the dawn so far away. To some, there was no hope. The Greeks looked backward to their golden age, and expected nothing better than the present. So many of us now sigh for vanished peace, and vainly strive to capture the mood of a day that is gone. But wise men now, as then, look for a star, and when they see in it the coming of a Saviour, they rejoice. The song of joy is the song of Christmas. Pain there is, and was, in plenty. Suffering and cruelty and injustice and misfortune are all about us. But when the star shines for us, when our hopes are confirmed in Christ, then no matter how dark the path has been and no matter what dangers lie before us, we are sure that God is with us and that he will never forsake his own.

Even the Wise Men must have been surprised to find Jesus in a humble stable in Bethlehem. They were looking for a king and had gone to the palace in Jerusalem to inquire for him. But with the evidence of the star to confirm their discovery, they were sure they had found him whom they sought, and they rejoiced.

So it may be with our hopes. Oftentimes we do not know exactly what it is we seek. We want the best for ourselves and for others, but what we think is best is beyond our reach. Sometimes we reach the goal we seek, but do not find in it the joy we thought, for life is all a venture of faith, and the path we tread is an unknown way.

But of one thing we may be confident. If our hopes for the coming year and the eternity which lies beyond the years are in Christ, then those hopes will be fulfilled. They may be changed, but they will be fulfilled beyond our fondest dreams. For God is still busy, and if our hopes are not worthy of his purpose for us, he will still do more than we dare to dream. The important thing is that we see the star and set out on that journey of faith to which it calls us.

The greatest part of life is adjusting ourselves to circumstances. It is only in daydreams that we are able to control our fate. We can change the world, and we must, but many things in it we must accept. Health, aptitudes, abilities, these are not given equally to all of us. Our hopes for our community, our families and our nation may not be shared by enough people to make them realities. Our task is to build the character of people so that they may resist temptation and overcome the evil in themselves. The primary task of the Church is not to mend the manners of the community, but to proclaim the matchless gospel of the Lord Jesus Christ. When men hear that gospel and believe it, their lives will give evidence of their faith.

So even when we must accept in the world conditions of which we strongly disapprove, there is still a most welcome conclusion to our journey of faith. For when we bring men to the Lord Christ, we do not only the smaller things in improving the world in which we live, we bring men into the deathless fellowship of the redeemed. And this, after all, is the most important thing for us and for them.

The same things happen in our personal affairs. The hopes with which we began our careers are seldom fulfilled. But that does not mean that we entirely fail of success. It is a most chastening procedure to look back ten years or fifty, and remember what our dreams were: How few of our lives have turned out as we planned! How different what we are from what we thought to be! But it is God's way to show us the star at the beginning of our journey, only to send us where we least expected. The glory is that we may see the star again, and rejoice. For we learn, if we are wise, that success is in small things as well as great, that hopes can be changed and glorified, that the great end of life is to serve God in his way and not in ours.

The journey of the Wise Men was over when they found the new-born King. And so is ours. Whatever our purposes may be, to have Christ in our hearts is their true fulfillment. No success is worthwhile unless we make it with him, and we do not fail if we serve him with all our hearts.

It is commonly thought that the Wise Men were only three, although the Bible does not say. At any rate, the star was there for all to see. These men saw it because they were looking for it. They only demonstrated once again that we see the things we look for. Everybody has seen mold, but Alexander Fleming was looking for an agent to stop disease, and he discovered penicillin in the mold. Red skies at sunset may be a sign of fair weather or a thrill of beauty. It depends on whether we are looking for a forecast or the painting of God's "good night" across the western sky.

So it is with the star over Bethlehem. If we are looking for a Saviour, that star points us to him. If our lives are strangely empty and we would have them full, his star beckons us to find him. If we are burdened by our sins and cry for forgiveness and peace, we will find it in him. If we seek a purpose high enough to warrant our striving, a cause worth giving our lives to, a King whom we may serve without doubting, here is the answer to our prayer. But we must be looking, yearning, striving. Those who are satisfied with themselves, who ask no more of life than food, clothing, jewels and fame, will never see the star or find the King. For this King does not force us to give him our allegiance. We can ignore him. We can be so busy that we miss the glory of his coming.

We must be looking for the star if we would see it. More than that, we must be anxious for its appearance. The Wise Men surely were. Their great quest was nearly over, *yet* how little they knew of its ending. The new-born King was not in the palace of the Herods. Now they were headed for Bethlehem because of a single verse of prophecy, trying to find a child whose name and family they did not know. If they could only see the star again—have some faint clue that their search was not in vain! No wonder they rejoiced when the star appeared to guide them.

Those who rejoice to see the star today are those who are anxious for its appearance. I am glad that there are many who are in the Church and on the side of the right from a sense of duty. We need more people who are responsive to the call of duty, who will do things they ought simply because they ought to do them. But the thrill of faith comes out of a sense of great need. When happiness and joy and eternal life itself are at stake, how great is our rejoicing to see the star, to know the fellowship of the Redeemer, and to hear his gracious word, "Come unto me, all ye that labor and are heavy laden, and I will give you rest."

And if we would rejoice to see the star, we must be going in the right direction. The Wise Men used their best judgment, took the best advice they could get, studied the Word of God, and set out. If they had stayed in the East debating what this star could mean, they would never have seen it again. If they had stayed to discuss with Herod the political situation of the lands through which they had travelled, the star would have ceased to shine before they were on the road to Bethlehem. But they were on their way, and the star guided them.

There is no lack of guidance for those who would see the King. Every hotel room in America has its Bible. But to see the star, to rejoice that the Christ is always coming to those prepared to receive him, demands that we not only talk of him but go his way.

How do we know that God keeps His promises? What does the love of God mean to His saints? What is the blessedness of those who enter into the fellowship of the Saviour? The hearts of men are asking a thousand questions like this, and only those can answer who have seen the star over Bethlehem.

For the living Christ is not a possession to be bought, or a commodity to be exchanged. To know him we must give ourselves to him. To see the star, we must be on our way to give him our best gifts, to lay our lives at his feet.

There is no record of the Wise Men after they returned to their own country, but the star they saw over Bethlehem could never cease shining in their hearts. For it happens thus to those who see the star. It is always a light to guide their footsteps, a beacon of hope through all their darkness, an evidence of the fellowship of the Saviour which endures not only through life but through eternity.

The star performed its mission and faded into night. But to our hearts it still shines, beckoning us to faith in him who is the Lord of life. And if we rejoice in the shining of the star, if we welcome Christ into our hearts, it will shine for us through all eternity.

"When they had heard the king, they departed; and, lo, the star, which they saw in the east, went before them, till it came and stood over where the young child was. And when they saw the star, they rejoiced with exceeding great joy.

"And when they were come into the house, they saw the young child with Mary his mother, and fell down, and worshipped him: and when they had opened their treasures, they presented unto him gifts; gold, and frankincense, and myrrh."

Prayer: *O God, who by the shining of a star didst guide the Wise Men to behold Thy Son our Lord: Show us thy heavenly light, and give us grace to follow until we find Him, and, finding Him, rejoice. And grant that as they presented gold, frankincense, and myrrh, we now may bring Him the offering of a loving heart, an adoring spirit, and an obedient will, for His honor, and for Thy glory, O God Most High. Amen.*

CHRISTIAN CERTAINTY

Is There Nothing Steadfast in Our World?

Reverend Douglas Horton, Ph.D., D.D.

Minister of the General Council and Executive Secretary of the Congregational Christian Churches; a Minister of the Congregational Christian Church, New York City

Dr. Horton is an internationally known churchman and an outstanding figure in the councils of the ecumenical church. He is as well known in other communions for his spiritual leadership as he is among the more than a million lay members and ministers of the Congregational Christian Church.

He was active in the formation of the World Council of Churches, and is one of the central figures in the united efforts of Protestant Christendom. The World Council of Churches already links together churches of all the great nations of the world. Under its auspices a great work of reconstruction and relief has been organized for the places in Europe and the Orient that were ravaged by war.

Born in New York, he was educated at Princeton University; New College, Edinburgh; Mansfield College, Oxford, England; the University of Tübingen, Germany; and Hartford Seminary. He holds the honorary doctorate from Lawrence College, Chicago Theological Seminary and Princeton University. Ordained to the Congregational ministry in 1915, Dr. Horton served pastorates in Connecticut, Massachusetts and Illinois before he took his present post of denominational leadership.

Dr. Horton's parish is, indeed, the world, since he travels in Asia, Africa and the Near East to develop the overseas work of his denomination, carried on by the American Board of Foreign Missions. He has been a delegate to many significant international conferences and councils, one of the largest being the great international Missionary Council at Madras, India, in 1938. Some time ago he made a trip to Japan to enlist Japanese Christians in the task of remaking right relations between men and nations.

He is an author, editor and translator. Among his books are Taking a City, The Art of Living Today *and* Out into Life. *During the first World War, Dr. Horton served as a chaplain in the United States Navy. He is in constant demand as a speaker and preacher not only in the churches of many denominations but in schools, colleges and universities.*

"Is There Nothing Steadfast in Our World?" is a realistic and helpful answer to the question many of us are asking today. Is social life being destroyed? Is civilization toppling? Has Divine Providence deserted us? Dr. Horton shows the way for men to overcome the fears and uncertainty gripping people today; students in colleges are disturbed; most men have many selves. This Christian treatment of uncertainty in the midst of world change and upheaval will be helpful to people who are worried by the unpredictability of life.

Sermon Twelve

TEXT: The Earth is the Lord's, and the fullness thereof; the world, and they that dwell therein. For he hath founded it upon the seas, and established it upon the floods. PSALM 24: 1, 2

NO ONE can go up and down the land in these days without being made aware that our people are in a state—I shall not say of hysteria, though the word is not too strong to describe the condition of some—but at least of anxiety. For the very foundations of our social life seem to be shaken; the total establishment of civilization seems to be toppling.

In the cities, farseeing business men have been preparing themselves and their clients for extremities ranging all the way from the milder forms of price-fixing to the commandeering of their premises and personnel for the purposes of war. The microfilmers are busy transcribing the contents of a million documents for sequestering in rural districts which, it is hoped, will escape the holocaust of the A-bomb or H-bomb.

In a New England college a professor, to illustrate the effect of the uncertainty which had inundated the campus, told me about one of the most respected boys in his class who, having failed to appear for an important appointment, came to him later and confessed that all the while he was simply walking up and down in his room in a dark mist of hopelessness. "My education is going; my life work is going; things don't make sense any more; and I can't seem to take it." Recently the National Council called together representatives of its denominations to consider for a second time in one generation the hateful questions brought about by the unhealthy massing of populations in the environs of war industries and cantonments. The joints of the carefully constructed edifice we call our common life seem to be loosening, and we are bracing for the debacle.

It is clear that there is a kind of uncertainty that cannot be cured, since it inheres in the nature of human life. I do not know which is more inscrutable, the mind of Russia today or the will of God for tomorrow; and our unsurety at the present time combines our ignorance of each. Anxiety, however, is an emotional overtone which we add to uncertainty. This should yield to Christian treatment.

Most of us probably think of this world in which we find ourselves as fundamentally stable. When it is operating normally it seems to stand on strong foundations: it seems to be a good dependable establishment.

It seems capable of supporting a civilization within which a man may live in peace with his friends and neighbors, risk his capital with a fair confidence of return, marry and give in marriage, and all with the expectation that values of all sorts will persist from generation to generation. In this conception such times as ours today can be regarded only as disorderly and unsettled. Forces appear to be at work which are cracking our foundations and wrecking our establishment.

If those foundations and that establishment were something apart from our own personal lives, we might survey the unsettlement of our times with some degree of philosophical detachment, but each one of us necessarily builds up within himself a set of habits of thought and action which corresponds to the world he sees around him—feeling all its shocks and suffering with all its tensions. The world enters into us, in a way, so that we can truly say that we are a part of it and it a part of us.

Most men have many selves—that is, inner systems of habit which fit into various phases of their life in the world. Let me use an old friend of mine for illustration: As a professor, he has what may be called a professorial self. He lives in the ancient Latin world for a good many hours of each teaching day. He also leads an exemplary life as husband and father, and so has a domestic self which is made up of quite different ideas and attitudes from his professorial self.

Now and again he forgets and allows his Latin self to invade the territory of his home self: he quotes a Latin motto at table, let us say. Thereupon the younger members of his family, who attend a progressive school (there is Latin in the adjective but not in the curriculum), proceed with unblunted sarcasm to remind father where he is.

More interesting is the conflict of his church self—for he is a regular attendant at the local church—with his automobilist self. When the driver of another car cuts in front of him, he is likely to lean out of his window and treat the neighborhood to an eager picturesqueness of speech seldom encountered in church. He is only demonstrating systems of habitual reaction which he has acquired in different environments.

Such selves as I have been describing are so superficial that almost any one of them might be relinquished without too radical a wrench to one's inner life. But each of us similarly has a total reaction to his total surroundings; and this goes very deep. Our habits of thought and decision of this sort fit into the world as we conceive it to be as a tenon in a mortise. If that world, therefore, begins to shake, we shake with it. Rumblings in Korea mean quakings about our own heart. We are happy when the world outside seems to be stable, and distracted when it begins to totter.

But suppose our conception of the world is wrong! Suppose we have tied ourselves up to a misconception as to its basic nature. Or rather, suppose nothing. Consider, instead, the interpretation contained in the first two verses of the twenty-fourth Psalm: *The earth is the Lord's and the fulness thereof; the world, and they that dwell therein. For he hath founded it upon the seas, and established it upon the floods.* If these verses mean what I take them to mean, this conception of ours is false and our tie to it tragically mistaken.

There is nothing in the Psalmist's lines about God's having founded the world upon an unshakable base, not a word in them about his having established it upon changeless foundations. He actually declares that God founded it upon the seas, established it upon the floods.

The Psalmist undoubtedly had in mind the sketch of creation, impressionistic in detail, clear in main outline, which is given in the first chapter of Genesis. Darkness was upon the face of the deep—and the spirit of God moved on the face of the waters. And God said, "Let there be a firmament in the midst of the waters." "Let there be lights in the firmament." "Let dry land appear . . . living creatures . . . men." And it was so—all in the midst of the waters. Upon that kind of foundation God created the world.

The story out of Genesis is a sufficient symbol of the world we live in. Our history is a fluid thing. If Hollywood were to throw on the screen a series of photographs showing in color the outline of empires as they were at the turn of every century from the beginning of the human record until now, they would seem like nothing so much as the ever-changing crystals of a kaleidoscope. If in some similar way we could depict the empires of thought which men have set up, limning the mighty philosophies which have come and gone, from the beliefs of Neanderthal man to those of the confirmed communist today, we should see the same wash of variety, the same upheaval following upheaval.

If any man thinks that history is the parent of constancy, he had better read history. If he thinks that of itself it can provide a durable foundation, an absolute upon which to establish a life or a civilization, he had better read it again.

It is in fact because of the unpredictability of life that we build our civilization as a defence from it, and then allow our minds to accept the civilization as the reality. Consider how broadly a modern city shuts out actuality. You are not likely to know either of a birth or of a death in an apartment just across the street from you. The people on middle Fifth Avenue can hardly dream of the degrading squalor and gnawing poverty of upper Fifth Avenue.

By the selection of friends and places one can today create his own world, somewhat as a man under hallucination creates his, by the selection and rejection of ideas. Neither is the real world in its completeness. "Never morning wore to evening, but some heart did break"; but we are protected against doing anything about these heartbreaks, even in our own neighborhood—to say nothing of the multiplied millions of them in the hungry Orient—by the wall of the lives we lead, that is, our civilization and our adjustment to it.

The stability which we sometimes confuse with reality is the bulwark of defence we erect against the latter; but it serves not to defeat it but only to deceive us. The real world is not solid. The uncertainty in which we now find ourselves is actually normal. It is what one might expect of a world founded upon seas, established upon floods. A realistic Christian should be ready for it.

But one will say: "I can see the realism of it, but not the Christianity; is there nothing steadfast and steadying in human life?"

The Psalmist speaks the magnificent answer. There is another direction in which to look for the enduring besides the world. *"The earth is the Lord's and the fulness thereof; the world and they that dwell therein."* Let us turn away from the brittleness and fragility of the civilizations to God who holds them all in his hand as a very little thing. What we have to look for is that which is not the victim of circumstance, and where could we better look than to him who is the creator of all circumstance?

It is edifying exercise simply to watch in fancy while the Great Artificer forges and molds the worlds. Him you cannot see, but the work of his hands is convincingly before you, made more so by every triumph of scientific enquiry.

More years ago than are imaginable there is the earth, a shapeless thing, swung molten from the sun's terrific blaze, to cool and harden. Who was behind that? Now upon its seared and broken surface, groaning under the weight of many oceans, a cluster of atoms is nursed into life, and life into more life, and more life into different life, a filmy ooze on the surface of the deep. Who did that? One could go on with the story, but it is not necessary.

One fact stands clear: out of the chaos God did bring life; and out of life, mankind; and out of mankind, civilization. He founded these upon the seas; he established them upon the floods, it is true; but he did found, he did establish, them. They are in process of development and therefore not fixed, but they have not returned to the waters from which God called them. If they did, he could recall them. From one collapsing culture he can, if need be, create another. It is something to know this today, that the hand is not shortened of him who brought order out of disorder, who did found upon the seas and establish upon the floods the world and them that dwell therein. Let the flooded seas be stormy as they will: beside him they are impotent. "Well roars the storm to those that hear a deeper voice across the storm"—a voice that can command it. In God is all the constancy of power any soul requires for recourse.

No one who has beheld God's power can fail to desire him. Contact with him alone gives basic meaning to life. It was Emerson, when a fanatic told him the world was presently coming to an end, who replied with infinite insight, "Very well, we'll get along without it." T. S. Eliot in *Little Gidding,* less cavalierly emphasizes the craving of the soul for this God:

> If you came this way,
> Taking any route, starting from anywhere,
> At any time or at any season
> It would always be the same: you would have to put off
> Sense and notion. You are not here to verify,
> Instruct yourself, or inform curiosity
> Or carry report. You are here to kneel
> Where prayer has been valid.[1]

But beware, when you kneel before God honestly, and dare to ask with a hearing ear what his purpose is for you and his world. God's stable purpose is very different from a stable civilization: you can take life from it but not rest in it. A civilization is an institution, but this is an incarnation. It takes on flesh and blood and lays hold of you. When you get down on your knees, if you have truly heard the Christian gospel, you will find (in Father Tyrrell's words) a strange man looking at you from a cross, whose every quivering agony says to you: "This is the price I am glad beyond utterance to pay if only you will understand the measure of my love."

He looks in yearning at you and beyond you; and when you glance about to follow the direction of his gaze, you find that it is always to people in need—to the little children in East Harlem smelling the enticing scent of the black flower of early crime (knowing no other flower), to the old man trudging with his tired family over the homeless roads of Korea, to man as man so ready for a crazed suicide that he seems already to be sprawling and choking in his own blood.

You know beyond a peradventure that if you are to take God's grace seriously, you will follow the look of Christ into the places of need and give

[1] Used by permission of Harcourt, Brace & Co.

yourself to be an instrument of creative love in his hand. It is through you, in part, that he will found his world, and establish it. You will be to him what a man's hand is to a man.

Man needs a self fit for reality. But what kind of self corresponds to that reality? Not the kind, surely, that fits securely into a falsely secure civilization. It must have two sides, for it stands at the point of contact between God's eternal purpose and the concrete actuality of the world. It must be toward God childlike, teachable, trusting; toward the world loving, originative, hopeful, wise.

When two or three people of this sort are gathered together, they are a Church—or the rudiments of one. Our Protestant Fathers felt the need to place themselves over against the everlasting purpose of God in such a way as to be of maximum use in His world. They regarded the more ancient churches as being more a part of their civilization than of God, shaken when civilization was shaken.

The strength of the Puritans in early New England, like the strength of all other Protestant groups in their beginnings, lay in the passionate demand that nothing should stand between them and the living Christ, on the one hand, and, on the other, nothing between them and the needs of a lost world. They would be a priesthood of believers at work.

One description of a church meeting in early colonial days is given in John Cotton's *The Way of the Churches of Christ in New England:*

> Transactions of Church proceedings . . . wee do . . . by the generall and joynt consent of all the members of the Church [p. 94]. In respect of Christ (whose voyce only must be heard, and his rule kept) it is a *Monarchy;* in respect of the peoples power . . . it is a *Democracy* . . . [p. 100].

If every church in our land today could boast that in its gatherings Christ is king and its people ready, having heard the voice of Christ, to go to work democratically to do his will, what a Church, what a land, we should have!

With selves devoted to the things of God and the needs of his world, we should be ready for any shaking of the foundations there may be. We should indeed, in a sense, welcome it, "that those things which cannot be shaken may remain." This would convert us into the only kind of Christians this world has need of—militant Christians, that is, founded on belief which lives its belief to the utmost.

Many years ago in the city of Brooklyn, Dr. Richard Salter Storrs, of the Church of the Pilgrims, was asked by a little boy to come to his home where his sister was ill. Responding immediately he found the place they called "home" a wretched one-room basement in a tenement house. The mother had died; the father had disappeared; and the little fifteen-year-old sister had carried on for the younger children, being to them bread-earner and mother—and now she lay in what seemed to be the last stages of a fatal disease.

Dr. Storrs gave her the comfort of a pastor's words, but she kept asking,

"How will God know that I belong to Him when I come?" Finally the inspiration of the right answer struck him. He looked down at the little crucified creature, saw on the ragged blanket the shriveled and work-worn fingers that had kept the dishes washed and done the cooking and brought life to that family, and he said, "Show him your hands: He'll know you belong to Him."

We are looking for church members who live their belief—who will not be ashamed to show Christ their hands. This is the world's need. Such people are at home in a world founded upon seas, established upon floods.

THE CHURCH AND ECONOMIC LIFE

Christ in the Marketplace

THE REVEREND JOHN F. CRONIN, S.S., PH.D.[1]

Assistant Director, Department of Social Action, National Catholic Welfare Conference and a member of the Society of St. Sulpice; Director of the Institute of Catholic Social Studies at the Catholic University Summer School, Washington, D. C.

Father Cronin has written and lectured extensively in the social field. This sermon is one of several which he gave on current social problems from the Christian point of view. He is noted particularly for his textbooks on economic problems, Economics and Society *and* Economic Analysis and Problems. *He has also written pamphlets and articles on social problems; his articles appear in "The Sign," "The Commonweal," "Common Sense," "Survey Graphic," and in technical publications.*

His lectures have taken him to Chicago, Detroit, Toledo, Buffalo, and many other cities, where he has spoken on social questions and has also addressed conferences on interfaith questions. In 1939 he appeared at the Public Affairs Institute of the University of Virginia at Charlottesville.

In his writings and lectures he particularly stresses the need for national unity in these critical times and suggests programs to solve the social, economic, and political obstacles that would lead to real reform. During the recent war he did volunteer work in the fields of labor and rationing and has been permanent arbitrator to the Clothing Industry in Baltimore, Maryland. He is active in the Catholic Association for International Peace. He was one of the winners of the Papst Post-War Employment Awards, a contest in

[1] Sermons by the members of the Advisory Committee were contributed at the request of the editor and are included on his responsibility.

which most of the nation's leading economists submitted proposals for the best postwar use of resources in men and materials to continue and retain prosperity for the greatest number. His recent book, Catholic Social Action, *gives the thinking of Catholic leaders on social matters. He asks several disturbing questions about economic and social problems, the dignity of man, the need to place Christ at the center of all human life and activity. Business and social life and government all have a place in this sermon for today. This sermon stresses the place of America in the world today and the economic and Christian obligations laid on all Americans. Leadership has been thrust upon our country: what will we do with it?*

Sermon Thirteen

WE WISH TO MAKE the world over, in the image of Christ. Knowing that Christ as God came on earth to redeem us from sin and to teach us the ways of truth, our aim is nothing less than total Christianity. We seek to have the life-giving teachings of the Savior prevail everywhere, from the councils of nations to the humblest home. It is for this reason that the Church has a program of social principles and of social action. For us it is not enough that men accept the Gospel merely as individuals and in the narrow spheres of private action. It is good that farmers and industrial workers, business men and professional men believe in God and accept His teachings. But the world will not be Christian until men practice these truths in every phase of their lives.

We are often told that the purpose of economic life is the production of wealth. Textbooks put it that way: the laws of wealth production and distribution. But to the Christian, such a point of view is incomplete. The material world exists to serve man. Food, clothing, and housing are for man. The thousands of services, from teaching to medical care, are for man. Material wealth is a ladder which man uses to climb to higher things—to the riches of the mind, to treasures of art, and above all to the lasting wealth of virtue. It is just that simple. Economic life should help man to be a better Christian. It has no other purpose in the mind of God.

Such is the ideal. But in practice material things are often obstacles to man's progress towards heaven. This can happen in several ways. First, there are those who seek wealth for its own sake, who worship riches as men once bowed down before graven images. This is the sin of greed. In its worst form, it violates the first and greatest commandment of God: "I am the Lord, thy God. . . . Thou shalt not have strange gods before me" (Exodus 20:2–3).

Again, there are those who seek wealth merely as a step towards other vices. It may minister to an excessive desire for pleasure. It may pander to vanity and pride. Or it may lead man to the sin of Lucifer, the grasping for power and ever more power, until man says in his heart: I will be like God. All these evils have been occasioned by economic life, by the misuse of wealth.

But there are other dangers as well in the struggle for material things. Some do not receive enough in worldly goods to enable them to live a Christian life. A startling statement? But this is the thought of our present Holy Father, Pope Pius XII. He speaks of "social conditions which, whether one wills it or not, make a Christian life difficult or practically impossible" (Solennità della Pentecoste, June 1, 1941). Bitter poverty often drags a man down to earth, because it forces him to devote all his thought and energies to mere survival. It is the occasion of sins against the marriage bond, as families desperately refuse to have children which they cannot support. Such poverty has produced city slums, the festering breeding places of crime and vice. And, in our day, it has served as an excuse for the dictator state, Communist, Nazi, or Fascist, which enthrones a party and its ruler in the place of God Himself.

It is only natural, then, that the Church should speak out again and again for an economic system which would be a servant of man, not his master; a system which would minister to all, not to the few; a system which would be a material foundation for Christian living, and not an occasion of sin and oppression. But what does this mean in practice? How do we enthrone Christ in the marketplace? There are many ways to answer these questions, but two steps are basic. Christ must rule in the hearts of men. Christ must rule in the laws, customs, and organization of social and economic life.

For Christ to rule in the hearts of men, it is necessary for each of us to accept Him totally. Religion must be more than Sunday worship or family piety. Its principles must guide our conduct in the factory, at the mine, on the farm, in the office. Where greed has reigned, there must be Christian moderation. In place of the pursuit of power or pleasure, there should be a sense of service, a desire to benefit one's fellow man. Austere justice, tempered and warmed by Christian brotherhood, the love of neighbor second only to the love of God, should prevail in economic life.

But personal reform is not enough. Christ must also be enthroned in the very structure of society. We know that many will not practice the teachings of the Gospel. There will be evil men, selfish, greedy, desirous of absolute power. Some men will do the right thing only if restrained by law, custom, or by the form of social organization. That is why justice and charity must be more than personal virtues: they must also be the accepted rules of economic life. Only in this way can the pursuit of wealth take its proper place in God's plan.

A Christian economic society would produce and distribute goods and services in such a way that the individual, the family, and society itself would receive the greatest benefit. Such a society would be productive. It would use the best techniques of the physical and economic sciences to assure efficient and abundant production, provided only that man did not suffer in the process. Christian virtues do not substitute for economic law, any more than they would repeal the law of gravitation. Rather they direct and channel the forces governed by these laws into worthwhile ends.

With this in mind, we seek a level of production and distribution which would be a material foundation for right living. This means, above all, the ideal of a living wage—a system whereby a worker who is the head of a family could support his wife and children in decent comfort. We do not want children to be compelled to work at tasks unsuitable for their age and strength. Much less do we sanction conditions which force mothers of young children to leave the home for work in a factory or store. A family wage for heads of families is the norm we seek.

This example may serve to illustrate how social principles lead to social action. In the nineteenth century, the abuses of the Industrial Revolution led to sickening scenes of poverty in industrial Europe. To remedy these evils, men took two practical steps towards a better distribution of wealth. Some favored social legislation: forbidding some evils, regulating others which could not be abolished at once, and beginning a common program of action to provide such necessities as housing, medical care, and education. Some forms of child labor were abolished; the work of older children and of women was regulated. Gradually, as economic society improved and the social conscience of mankind was more developed, we got an extensive code of social legislation. Christian groups, acting on religious grounds, were in the forefront of the struggle for such laws.

The second step towards the attainment of justice for the worker was the foundation of unions. Labor unions had a hard struggle for existence in the nineteenth century. But they received tremendous encouragement in 1891, when Pope Leo XIII issued his famous encyclical letter "On the Condition of Labor." Here the pope explained clearly the rights and duties of workers, as defined by the moral law. And among the rights, he noted the very important right of organization. Moreover, Christian groups aided workers to fight for their just claims. In fact, in Europe many purely Christian labor unions were formed. While this particular pattern is unnecessary under our conditions, it does show the real interest of the Church in social action.

What the Church has done may seem strange to many of us who have held a limited view of the function of religion. Some feel that religion is almost exclusively a matter of worship, with perhaps some moral implications for our private lives. But this view misses the total nature of Christian teaching. God's law and God's purpose must enter into every phase of life. If, under God, you hold that man's dignity and his ability to observe the moral law can be substantially furthered by a sound economic foundation, then you are bound to take prudent and reasonable steps to secure that foundation. Social principles and social action form an inseparable team.

It would be possible to go much further in outlining the implications of the Christian dignity of man. Certainly it calls for a decent wage, as the material foundation for family life. But it also demands working conditions and a community atmosphere which respect man's dignity. Thus, the Christian idea of labor-management relations would call for mutual respect shown in the conditions of work and the rules of collective bargaining. It would frankly reject discrimination against anyone for trivial or unjust reasons. This is especially true when discrimination is based on race, religion or national origin.

Here again principles lead to action. In this country, for example, the Church has made continual efforts to bring about harmony, based on mutual respect, in labor-management relations. There have been many statements by our bishops regarding the dignity of labor and the urgent need for a real spirit of partnership between workers and employers. We have well over a hundred labor schools, conducted by church groups, for the purpose of infusing Christian principles into labor relations. In fact, the few exceptions to the custom that churchmen do not accept public office have occurred in this field. Priests have acted in public capacity as arbitrators, mediators, and even chairmen of state or federal boards having functions of this type.

Likewise the Church has been active in fighting discrimination against minority groups. We have Catholic interracial councils. There is a Catholic Committee for the Spanish-Speaking. The leaders of the Church have spoken strongly against anti-Semitism and other forms of bigotry. Our views are clear. Under God, there is no such thing as a second-class man. We cannot say that justice and charity are observed, so long as men are labeled inferior because of their color or their cultural background.

There is another conclusion from the Christian dignity of man. This is man's claim to freedom. Today this is particularly true in the political area. Against the dictator state, Communist, Nazi or Fascist, we hold the basic rights of the individual man, the family, and of free associations formed by men. In modern times, at least, these rights are best safeguarded by political democracy. But even democracy has its economic foundations. Men have their highest political freedom when they also have economic independence. It is for this reason that the Church has opposed socialism, even democratic but total socialism which rejects both revolution and the class struggle. When there is but one employer, the state, the citizen cannot be really free. If he has only one source for his daily bread, then that source can control his life.

The Church has argued for economic independence as the basis for man's freedom. Three popes—Leo XIII, Pius XI, and Pius XII—defended property rights against Marxist socialism or communism. But each of them has called for a better distribution of property, particularly among the workers. In fact, it would hardly be an exaggeration to call this the central social teaching of the present Holy Father. In more than a score of important addresses he has brought out this theme. "The dignity of the human person requires normally as a natural foundation of life, the right to use the goods of the earth. To this right corresponds the fundamental obligation to grant private ownership of property, if possible, to all." This will "prevent the worker, who is or will be a father of a family, from being condemned to an economic dependence and slavery which is irreconcilable with his rights as a person" (Christmas Broadcast, 1942).

The Church has fought in many ways to protect man's freedom by securing for him economic independence. Thus the National Catholic Rural Life Conference has urged the value of the family farm. It has encouraged families to live on a farm, either as a full-time occupation or as a supplement to industrial work. For industrial labor, our concept of a living wage includes saving so that the worker may own a home and achieve some degree of independence. Pope Pius XI favored experiments which would give the worker a greater stake in the plant where he works. Such devices as profit-sharing, worker stock ownership, and a voice in management have been encouraged as steps towards independence. The popes have condemned concentration of economic power, whether in the hands of the state or in private hands, as threatening the independence and hence the Christian dignity of the worker.

Freedom is the battle-cry today. But the Christian ideal of freedom is not anarchy. We do not believe that anyone can do what he wants, without regard to his neighbor. Imagine, if you will, how life would be in a large city if there were no traffic lights, no rules or regulations regarding the movement of tens of thousands of automobiles converging on the business section. That would be a form of individualism which ignores the rights of others. It would violate the basic virtues of justice and charity. The same is true in economic life. We want freedom, but not anarchy. Independence tempered by social responsibility is the Christian approach. We have the firm belief that people working together can do more than the same people fighting one another. This is why we favor labor-management coöperation in the factory. But we would go further and not restrict collaboration to the individual plant or firm. We believe that groups with common interests should organize to promote both their own interests and the common good of the economic community. Industries and professions should be so organized that everyone concerned would have a voice in meeting general problems.

There are two main reasons for favoring an organized approach to economic problems. The first is negative. If people do not group together to act on common responsibilities there will be a demand for government action to meet these needs. This would tend to move more power toward the center and to leave less with the individual or the smaller group. Even though the purpose is good, there would be some lessening of freedom. It is far better to have self-regulation by all persons in the industries and professions, capital and labor united for this purpose, than to have controls imposed from without.

A second reason for common action is the value of coöperation in contrast to struggle. What works well in the home and in the factory should at least be tried in broader areas. Of course, there is a legitimate place for rivalry and competition. There are fields in which interests clash and in which independent action may be desirable. But this is no reason for doing nothing or too little about common interests and problems.

As a means for translating these ideas into action, the Church has not only promoted collaboration among economic groups, but it has also urged that government agencies consult and work with voluntary groups in their respective fields. We prefer self-regulation to the greatest degree possible. But when this is not feasible, then we wish maximum participation by the affected groups in the making of decisions.

These, then, are the highlights of a program of social action based on Christian social principles. We seek an economic system which harmonizes with the dignity of man. This system would support the family. It would be a society based on partnership rather than conflict. It would be permeated by justice and charity. To bring this about the Church seeks to change the hearts of men. It seeks to educate, to persuade men into attitudes of friendliness and coöperation. But attitudes are not enough. These principles must pervade our laws, our customs, our organizations. Thus the work of Christian social action is also a work of legislation and organization.

THE CHURCH AND EVANGELISM TODAY

Endless Line of Splendor

Bishop G. Bromley Oxnam, D.D., S.T.D., Litt.D., LL.D., D.Sc., L.H.D.

Resident Bishop, The Methodist Church, New York

One of the Presidents of the World Council of Churches

Bishop Oxnam is one of the great religious statesmen of our time. A native of California, he studied at the University of Southern California, Boston University School of Theology, Harvard University, and Massachusetts Institute of Technology. He also did research in Japan, China and India.

His first pastorate was at Poplar, California, in 1916. In 1917 he founded the Church of All Nations in Los Angeles and developed there the leading social service institution on the Pacific Coast. He was elected Bishop of the Methodist Church at the 1936 General Conference and was assigned to the Omaha Area, where he served for three years. In 1939 he was assigned to the Boston Area, and in 1944, to New York. In addition to his episcopal duties, he was president of the Federal Council of the Churches of Christ in America 1944–46, and was a member of the Federal Council's Commission to study the bases of a just and durable peace. At Amsterdam, Holland, he was elected one of the Presidents of the World Council of Churches in September, 1948. He is a member of the Board of the National Council of the Churches of Christ, U.S.A.

He is secretary of the Council of Bishops, president of the Foreign Division of the Board of Missions and Church Extension, vice-chairman of the Methodist Commission on Chaplains and of the General Commission on Army and Navy Chaplains, and has been serving the National War Labor Board as a special mediator in labor disputes.

Bishop Oxnam has been closely identified with education. During his pastorate at the Church of All Nations, he was professor of social ethics at the University of Southern California from 1919 to 1923. In 1927 he became a member of the faculty at Boston University School of Theology as professor of practical theology and the city church, and was president of DePauw University from 1928 to 1936.

His work has been recognized with honorary doctorates by twelve institutions: The D.D. by the College of the Pacific and Wesleyan University; the S.T.D. by Yale University; the important Litt.D. by Boston University and Northeastern University; the LL.D. by Ohio Wesleyan, Wabash College, the University of Southern California, Allegheny College, and Dickinson College; the D.Sc. by Rose Polytechnic Institute, and the L.H.D. by De Pauw University. He was a member of the President's Committee on Higher Education, 1946–48.

He is a member of many learned societies and was decorated by the late King George II of Greece with the Cross of the Royal Order of Phoenix. He is a dynamic preacher, a forceful leader, a good executive. He administers 1,285 Methodist churches in the New York area.

The Bishop's travels have led him into most of the countries of Europe and Asia. He was a member of the American Delegation to Russia in 1926, of the Japanese Education Commission of the International Missionary Council in 1932, and a delegate to the World Conferences at Edinburgh and Oxford in 1937. Among his written works are The Mexican in Los Angeles, Social Principles of Jesus, Russian Impressions, Youth and the New America, The Ethical Ideals of Jesus in a Changing World, Behold Thy Mother, Preaching in a Revolutionary Age, Labor and Tomorrow's World, The Stimulus of Christ, The Church and Contemporary Change.

In 1940 he gave the Enoch Pond Lectures at Bangor Seminary, the Merrick Lectures at Ohio Wesleyan University in 1941, lectures at Florida

Southern College School of Religion in 1941, the Fondren Lectures at Southern Methodist University in 1944, the Lyman Beecher Lectures on Preaching at Yale University in 1944, and the Earl Lectures at the Pacific School of Religion in 1945, Alden-Tuthill Lectures, University of Chicago, 1948; Hoover Lectures, University of Chicago, 1949; Ezra Squirer Tipple Lectures, Drew Theological Seminary, 1949.

"Endless Line of Splendor" was given at the District Superintendents' Conference at Indianapolis, Indiana, on October 2, 1950, on the inauguration of the emphasis on Our Church in Advance for Christ and His Church. Before fifteen hundred district superintendents, bishops, church executives, pastors, and laymen, Bishop Oxnam delivered this great religious address, tracing "the endless line of splendor that moves up and on" from Antioch to today. Readers will readily recognize the Bishop's use of Vachel Lindsay's immortal missionary poem and Dr. Halford E. Luccock's book, Endless Line of Splendor. *While the message was mainly for the forty thousand Methodist pastors and their eight million members, clergy and members of other denominations will see the application of the Gospel command to go into all nations and make disciples for Christ.*

Sermon Fourteen

NO REPORTER for the Roman paper *Acta Urbana* would have seen a story in the announcement that "The disciples were called Christians first at Antioch." His attention would have been directed to the household of the Imperial Legate or to the scandals associated with the famous gardens of Daphne. Of what possible interest to Roman readers were the little handful of Jews and Greeks who talked about one Christos, and were thus nicknamed "Christians"? They met in a side street, sang hymns, shared in a strange supper, and listened to a man named Barnabas who had been sent up from Jerusalem to inquire into the work of this community that had grown up following the dispersal of the leaders after Stephen's death.

Antioch was one of the great cities of its day. More than three centuries before Paul and Barnabas preached there, a conqueror had seen in that site by the River Orontes an ideal location for a capital. Just as Washington was planned by Major L'Enfant, so Antioch was designed by Xenarius. Antioch, with broad, colonnaded streets intersecting at the center, was the Greek substitute for the Semitic capital of Syria, Damascus; and with its navigable river and nearby port of Seleucia of Pieria, it became a maritime city. It was called Antioch, the Beautiful. The pleasure-garden of Daphne was located five miles from the city. It was ten miles in circumference, a garden with the Sanctuary of Apollo, groves of laurel and cypress, flashing fountains, colonnades, halls, and baths.

Antioch was a city content to seek satisfaction in pleasure, with little thought of the morrow, less consideration of eternity. How could the Roman reporter know that these "Christians" were the first missionary church of a world religion, the first of "an endless line of splendor," a minority with a mission, charged with winning the world in the name of a World Savior?

I have chosen the theme "Endless Line of Splendor" because no phrase of poetry expresses with truer apprehension the essential story of the Church than that of Vachel Lindsay. I have chosen it too because it is the title of Halford Luccock's brilliantly written history of Methodism which will be read by hundreds of thousands of our people as we enter the second phase of The Advance for Christ and His Church and center attention upon Our Church.

> An endless line of splendor,
> These troops with heaven for home,
>
> . . . .
>
> This is our faith tremendous—
> Our wild hope, who shall scorn—
> That in the name of Jesus,
> The world shall be reborn! [1]

"An endless line of splendor" !

A peasant girl, with a song on her lips. *"My soul doth magnify the Lord, and my spirit hath rejoiced in God my Saviour . . . He hath . . . exalted them of low degree: He hath filled the hungry . . ."*

A prostitute! And there were with him at the foot of the cross Mary, his mother, *and* Mary Magdalene. It was she who came early on the first day of the week to the sepulchre. It is written, *"But Mary stood without at the sepulchre weeping . . . Jesus saith unto her, Mary. She turned herself and said unto Him, Rabboni; which is to say, Master."* This is the beginning of the "line of splendor,"—the clear, unequivocal response when He utters our name,—Master!

A minority with a mission and a Master.

A fisherman with a sword, a coward shouting denial. And later, the risen Lord spoke to him, "Lovest thou me?" Shortly thereafter: here is the record, *"But Peter, standing up with the eleven, lifted up his voice, and said unto them, Ye men of Judea, and all ye that dwell at Jerusalem, be this known unto you, and hearken unto my words."*

A minority with a mission and a Master and a message.

[1] "Foreign Missions in Battle Array," by Vachel Lindsay. See his *Collected Poems* (Macmillan, 1925), p. 338. Used by permission of the publisher.

Peter was to stand until they nailed him like his Lord to a Cross. "Endless line of splendor!"

A tentmaker, Roman citizen, scholar, builder. *"Who art Thou, Lord? I am Jesus whom thou persecutest."* He had heard Stephen declare, *"I see the heavens opened, and the Son of man standing on the right hand of God."* He had heard Stephen's prayer, *"Lord, lay not this sin to their charge."*

A minority with a mission, and a Master, and a message, and the certainty of life everlasting.

When they sought to put the faith into the compass of a creed, they said, "I believe . . . in God" . . . yes, and "in Jesus Christ, his only Son Our Lord" . . . yes, and in "the life everlasting." "The line of splendor" moved up.

The Christians, in fearless splendor, faced the arena and its beasts, lash and faggots, hunger and thirst, and even the cross. They knew they would never die.

The Roman citizen wrote: *"I have been often at the point of death; five times have I had forty lashes (all but one) from the Jews, three times I have been beaten by the Romans, once pelted with stones, three times shipwrecked, adrift at sea for a whole night and day . . . through labor and hardship, through many a sleepless night, through hunger and thirst, starving many a time, cold and ill clad, and all the rest of it."* . . . *"I have been initiated into the secret for all sorts and conditions of life."* . . . *"I was not disobedient unto the heavenly vision."* . . . *"I have fought a good fight, I have kept the faith."* . . . *"Faith and hope and love last on, these three, but the greatest of all is love."*

And thus they marched, "an endless line of splendor, these troops with heaven for home."

Augustine, Francis and Bernard; Luther, Wesley, Livingstone.

The oratorio that was sung in Baltimore upon the occasion of the Sesquicentennial of the founding of Methodism in America was composed by Van Denman Thompson. The lyrics were written by Ethel Arnold Tilden. The Church and DePauw University, one of our church schools, can be rightfully proud of this distinguished composer whose genius has won international recognition; and of this wife of a great professor. Mrs. Tilden's poetry is her credential, opening the doors to the hall of fame. The oratorio was sung first in the lovely living room of the President's home at DePaūw. Dean Robert Guy McCutchan and the University choir, with Van Denman Thompson at the piano, sang it for us. In one section, Mrs. Tilden tells the story of a continent that beckoned the pioneers. She speaks first of the call of a "vast and golden land . . . such a land as strong men covet." It is the story of "frontier men . . . who have forgotten God." She tells of Methodist preachers meeting at Lovely Lane Chapel in Baltimore who were summoned to take "Christ's presence and Christ's power to mountain, valley, plain." And finally, the song of victory, "the whole land blessed."

From beyond the far blue mountains,
Calls a vast and golden land—
Grass green plains, and purple prairies,
Watered valleys, wooded hillsides—
Such a land as strong men covet—
Such a land as strong men take,
Stalwart men and dauntless women.
Brave adventure, riches challenge!
Heedless, unafraid of danger,
Or of death or savage torture,
Bent to conquer rush the settlers.
Westward, armed with ax and rifle,
With their ox teams and their wagons,
Sets a mighty tide of brave men,
Sets the tide of pioneers.
By the Boone Trail, by the Old Trail,
On and onward, laughing, cursing,
Ever fighting, brawling, singing,
Comes the flood of frontier men;
Masters forest, river, desert,
Raging beasts, and warring red tribes,
Gains dominion over all things—
All things only save themselves!
They are slaves of greed and passion;
Violence and mad lusts drive them—
Victors who have lost their birthright—
Men who have forgotten God.

. . . .

Go forth, thou little band,
Blest now at Lovely Lane;
Take thou Christ's presence and Christ's power
To mountain, valley, plain—
Go forth, as Christ would go,
To world-worn men take rest,
To sin-bound captives take release—
Bless thou, as thou art blest.
From this thy meeting House,
Go into all the land
And preach the gospel of thy Lord—
Go forth, thou little band!

. . . .

Now all the land with strong new life is filled—
Through farm and hamlet rings the rousing call
Of preaching men gone forth to plant Christ's Church,
Forsaking for His sake their worldly all—
And where the circuit riders' feet have trod,
Behold the footprints of the feet of God.

Now is the whole land blessed—
Now are men lifted up,
And into paths of light
Now are men's children brought.
Wide scattered frontier hearths
Are altars burning bright
Unto high learning, flares to guide
Youth's eager groping sight.
Stanch teaching-preachers give
Themselves, nor count the cost,
That faith, and learning may out-live
The frontier holocaust.
Love tends the sick and old;
And healing skill and prayer
Build temples where crushed, broken men
Find help and care.
In turn, her grateful sons
Go out through all the earth
Proclaiming, for their fathers' Church,
Christ's promise of rebirth.

Mary, the mother of Jesus, Mary Magdalene, Peter, Paul, Augustine, Francis, Bernard, Luther, Wesley, Livingstone, the men of the saddlebags, the classroom, the laboratory, the sick room and the surgery, the legislative hall, the market, the mine and the mill, the men and the women who were determined "that in the name of Jesus, the world shall be reborn."

The line moves up. One of the precious experiences of the ecumenical movement is that of fellowship with great souls: Berggrav of Norway, Niemöller, Lilje, Canterbury, the Metropolitan of Edessa. At Toronto, in a moment of recess between sessions of the Central Committee of The World Council of Churches, I stood by one of the tennis courts watching Frederick Nolde, a top-flight tennis player as well as the Director of the Commission of the Church on International Affairs. A quiet man, a German, stepped up with a word of greeting. It was Reinold von Thadden. His face and body still bore the marks of suffering. He was a lawyer who entered governmental service and later took over the administration of his family's estates in East Prussia. He became one of the great Christian lay leaders of Europe. He dared to oppose Hitler, was denied the privilege of attending Oxford; in fact, he was a prisoner of the Gestapo when the Conference was in session. He was called up for Army service, was occupation commander of the Belgian city of Louvain, and after the war was welcomed back to a grateful city which learned that even an enemy could be a Christian. Perhaps this was the reason why the Nazis dismissed him from the army as politically unreliable. He was captured by the Russians, and suffered the concentration camps near Archangel. After the war, he struggled back, reaching Berlin just before Christmas in 1945. His sister had been executed by the Nazis. Today he serves as a member of the Central Committee of the World Council of Churches. It was he who organized the great Church Day, or Kirchentag, held a few weeks ago in Essen. More than 150,000 laymen from all parts of Germany assembled. The President of Germany and Chancellor Adenauer were there. Once again, the broken bread and the wine of Holy Communion. Together they studied the great theme "Save the Man." It had four divisions: (1) Save his Freedom; (2) Save his Homeland; (3) Save his Family; and (4) Save his Faith. They sang "A Mighty Fortress Is Our God."

An endless line of splendor,
These troops with heaven for home,
. . . .
This is our faith tremendous,—
Our wild hope, who shall scorn—
That in the name of Jesus,
The world shall be reborn!

Yes, a minority with a mission! "An endless line of splendor!"

Evangelism was the dominant emphasis at Amsterdam. The Assembly said, "The evident demand of God in this situation is that the whole church should set itself to the task of winning the whole world for Christ."

Strangely enough, the "line of splendor" has at times lost its lustre as the minority approached majority. While never a majority, the Christians became respectable. The Roman emperors bowed before the Cross. Property, power and prestige came to the Church. There were those who held that they must rule in order to serve, forgetting that the order is reversed in Christian thought, and we must serve in order to rule. *"He who would be greatest among you must become a servant." "I came not to be ministered unto but to minister."* The Church took too much thought for the morrow. Losing its life for His sake became a counsel of perfection, and practical men sought incorporation papers for a new partnership: God, Mammon and Company. Church leaders became cautious. They identified the status quo with the gospel. The kings ruled by divine right, the division of society into nobles and serfs was ordained of God. Thus feudalism became sacrosanct and advocacy of change blasphemy.

But always within the fellowship there were those who caught anew the spirit of the Lord. The Nazareth sermon was repreached. The Savonarolas, the Joans, and the Husses were burned; the Franciscans and the Wesleyans were born. The scorn of tyranny leaped again from the preaching of the prophets, from the Magnificat of Mary, and the Acts of the Apostles.

Anne Morrow Lindbergh in a sensitive and significant story tells of her visit to the reconstructed twelfth-century Cistercian Abbey of Boquen in Brittany. She stood looking up at a stone madonna, "sheltered in her niche above the low arched door of the monastery." Beside her was Dom Alexis, the monk, the founder and head of the Abbey, himself one of the "endless line of splendor." The madonna had been carved by one of the brothers, he told her. "We call it Our Lady of Risk." Mrs. Lindbergh was startled by the unconventional title. "You mean," she said, "one must risk one's life—one must lose one's life to gain it?" "Yes," Dom Alexis answered, "Our Lady of Risk."

One of the chief characteristics of the "endless line of splendor" has been the readiness to risk all.

I received a card from one of our preachers a few days ago. He had been seriously ill this summer, and had endured painful surgery. He is home again, and ready for work. He sent the card to his people, and was thoughtful enough to send me one too. It announced his sermon topic for Home Coming Sunday. The message reads:

At Home
— September, 1950

Dear Trinitonians:

The experience of this summer was for me, your Pastor, a great adventure. It spelled renewed physical health, achieved by a combination of the many elements of God's healing process—skillful surgery, sympathetic nursing, freedom from financial worry because of your generous gift, willing coöperation, confident faith. Underlying all, was the mighty power which comes alone from love and prayer.

As I begin a new year of working with you, I realize that there is no limit to the adventure which the Church must undertake. Our day places on Christians the obligation to venture in unlimited spheres. Our eyes must be forward. Although grounded in "the distances of history," our vision must be lifted to new levels of moral and spiritual truth—to the forward reaches of eternity.

The "Home-Coming" events are listed on the back page of the folder. I look forward to greeting you, our Church family, deeply grateful that I may take up my work again.

Faithfully,
Your Pastor

The subject of his sermon was "Adventure Unlimited." Above the topic is a picture. A young man and a young woman sit upon a great rock. Before them the sea, the waves breaking high upon rocks just off shore, and beyond, the horizon and the sky. Beneath is this verse:

As boundless as the sea . . . and timeless . . .
So let our dreams be of eternal things . . . a venture,
Reckless . . . exploring . . .
Until they grasp the mind of God.

I wrote a brief note telling him I was glad he was home, and said, "I ought to know who wrote those moving lines, but I do not." He replied, "There is no reason why you should know them." They had been written by another in the "endless line of splendor," that minister's wife.

As boundless as the sea . . . and timeless . . .
So let our dreams be of eternal things . . . a venture,
Reckless . . . exploring . . .
Until they grasp the mind of God.

It is the adventure of saying "Yes" to Jesus. Amsterdam declared, "We have to learn afresh together to speak boldly in Christ's name both to those in power and to the people, to oppose terror, cruelty and race discrimination, to stand by the outcast, the prisoner and the refugee. We have to make of the Church in every place a voice for those who have no voice, and a home where every man will be at home. We have to learn afresh together what is the duty of the Christian man or woman in industry, in agriculture, in politics, in the professions and in the home. We have to ask God to teach us together to say No and to say Yes in truth. No, to all that flouts the love of Christ, to every system, every program and every person that treats any man as though he were an irresponsible thing or a means of profit, to the defenders of injustice in the name of order, to those who sow the seeds of war or urge war as inevitable. Yes, to all that conforms to the love of Christ, to all who seek for justice, to the peacemakers, to all who hope, fight and suffer for the cause of man, to all who—even without knowing it—look for new heavens and a new earth wherein dwelleth righteousness."

But how do we say Yes? Evangelism, yes! The new birth, yes! But after evangelism and the new birth, what? The world must be reborn. How?

We have "conquered earth, and charted sea, and planned the courses of all stars that be," but we have not yet learned how to live together. The "endless line of splendor" reaches from occupied territory to beachhead—but the bases at home, are they secure?

Today, the line confronts another line, dynamic, determined, unprincipled. Communism seeks to emancipate man by fundamental changes in the relations of property. But man's problems, involving property as they do, are far, far deeper than property. They root in the sinfulness of the human heart. The classless society cannot be established by men whose hearts are full of greed, whose minds justify the lie to achieve predetermined ends.

The Bishops in The Episcopal Address said:

> We reject communism, its materialism, its method of class war, its use of dictatorship, its fallacious economics and its false theory of social development; but we know that the only way to defeat it permanently is to use the freedom of our own democracy to establish economic justice and racial brotherhood. It is the man who is not exploited who is deaf to the slogan, "Abolish the exploitation of man by man." It is the man who knows he is treated justly who refuses the sinister suggestion of revolutionary activity to win justice. . . .
>
> Religious leaders and men charged with political responsibility must face a fundamental fact. Ideas cannot be destroyed by military force. It is possible to destroy the cities of an enemy, to bring his armed forces to surrender, in a word, to defeat him as far as the physical power to resist is concerned. But an ideology cannot be suffocated by poison gas nor demolished by atomic bombs. Ideas are conquered by better ideas whose truth has been revealed in practices that enrich personality. . . .

Let us not forget that it is less the materialism, dictatorship, and Marxian economics that have attracted masses of men than the promise of more abundant living, the abolition of the exploitation of man by man, and the setting up of a classless society. It is the hope that the masses behold in communism that hypnotizes them. These goals cannot be attained by the communist method of class war and the use of dictatorship. They can be won for man in the freedom of democracy. . . .

Communism moves forward with power because its leaders know what they want and believe they know how to get it. They have kindled fires of enthusiasm in the hearts of their youth, and these young people, united in a common cause, become a conflagration sweeping through the forests of exploitation. We reject communism. We know that within the freedom of democracy we can build a society at once just as brotherly, in which creative talent may be fully evoked and human beings may live in peace and security, enjoying fearless leisure and fruitful labor, and in which the impulse to creative action and service will be stronger than the acquisitive impulse. Let us recruit youth, acquaint them with the thousand-year struggle for political freedom, fire them with passion to preserve it, and, more, to use it to bring equality and fraternity to their fellows. But they must do more than sing "Let Freedom Ring." They must move out resolved to labor, to face the problems with cool and resolute mind, ready to live and to die that a society may emerge fit to be called the Kingdom of God.

Assuming the change of heart, the new birth, all that is done in bringing to the individual assurance of sins forgiven and the certainty of new life in Christ Jesus; after such experiences, whether they come quickly in conversion or more slowly through the processes of growth—after evangelism, what?

The communist proclaims an ideology that he affirms can be applied. He moves from its acceptance to its practice. But we of the Christian faith too often give pledge to the faith in our Affirmations on the Sunday but, in mental reservation, say, "Jesus preached perfectionist ethics, beautiful, but, like the evanescent images projected upon a silver screen, unreal." Too few of us really think the Sermon on the Mount can be applied in international and economic life. The dynamic ideology of the communist must be met not only by the dynamic ideology of Christianity but by "an endless line of splendor" that believes it can be practiced. The Church must come to grips with reality. It dare not content itself with the proclamation of the generalities accepted by all and in fact rejected by all.

The great issues of our day must be faced. Beneath the surface lie the issues of power and of justice. Power must be brought under democratic control, all kinds of power—political, economic, ecclesiastical. Justice must be established by the democratic process. The maintenance of freedom is prerequisite to both. Power cannot be brought under democratic control without freedom; justice cannot be established democratically save under liberty.

We must reconcile the necessities of technology and the necessities of brotherhood. Here the driving force that lies at the heart of the Christian gospel becomes essential.

The question of motivation must become central in our thinking. Some think that men can be made to work by putting a carrot in front of them, and that they, like the donkey, will move forward biting at the carrot that forever eludes them. Others think the whip is the better way. Neither the carrot nor the whip is the answer. There must be something alive in the heart of man that drives him to serve. It is the something in which Jesus grounded religion, love of God and love of man. Men who love God will seek to do His will. Such men see other men as brothers, beings of infinite worth, immortal. Men who love their brothers want for them what they want for themselves. It was out of deep meditation upon this theme that the central principle of Schweitzer's ethics emerged, namely, "reverence for life." The obligations will be better understood when we turn from such generalities as justice and brotherhood, equality and service to the concrete situations that must be faced as we think in terms of applying the general principles.

The "endless line of splendor" must advance. There are positions to be taken before we can answer, "Mission accomplished."

Dr. Ralph Bunche, distinguished American Negro, serving the United Nations, wins the Nobel prize for Peace. Splendid! The same Dr. Ralph Bunche is shown by the steward of an American dining car to a Jim Crow table, and because he will not eat under such conditions and lose his self-respect, he returns to his compartment, and is hungry. It simply will not do for Christians to talk about the "line of splendor" in India, China, Africa, and acquiesce in such a violation of faith. Our missions become mockery, and the communist who has never accepted Jesus rejoices in our rejection of the Lord.

Bishop Mondol of India writes me a letter full of heartbreak. He tells of his people starving, and begs that a delegation wait upon our President so that our food surpluses rotting away may be sent to save life. How do we face up to the real issues that lie beneath President Truman's Point Four, Senator McMahon's proposal relative to Atomic Energy, and Walter Reuther's plan for economic reconstruction?

If these proposals do not strike us as constructive, are we prepared to name others? Love and brotherhood become meaningful when we take the words of Jesus seriously, *"I was in prison, and ye visited me; naked, and ye clothed me; hungry, and ye fed me; thirsty, and ye gave me to drink."*

We are being mobilized in a war on communism. Some churches cry out for holy war. Hysterically, we pass laws that move us closer to totalitarian control, that endanger our liberties. The Communist party is a conspiracy, and communist aggression must be stopped and traitors must be brought to justice. But in the negatives, we may miss the great positives.

Let me ask a pertinent question. Suppose every Russian in the world were to die tonight. Suppose the Kremlin walls were to disintegrate and become dust. Suppose Russia as a nation were to cease to exist, her military might disappear. Would our problems be over? Would security and peace be our lot? No, the spectacle of a world in part overfed and in large part hungry; a world with its differing standards of living; a world with millions who have never known the meaning of liberty; a world hungry, ignorant and diseased, would still be a restless, seething world. The revolutionary surge would still confront us. The world must be reborn.

The "endless line of splendor" must move up. It must move into our own life. The minority with a mission must bow again before the Master, and moving out with a message incarnate, certain of life everlasting, take up the Cross anew, risk all, and thus save all.

COMMUNION

Communion Meditation

REVEREND HARRY V. RICHARDSON, PH.D., D.D.

A Minister of the Methodist Church;
President, Gammon Theological Seminary,
Atlanta, Georgia

Harry V. Richardson is nationally known as a preacher, teacher, and maker of ministers. In the years of his ministry, he has achieved a high place in American Christian circles.

A native of Jacksonville, Florida, he studied at Western Reserve University, took his S.T.B. at Harvard Divinity School, from which he graduated with honors, and his Ph.D. from Drew University.

In 1932 he went to Tuskegee Institute in Alabama to teach history for a year and to assist in religious work. At the end of the year, the trustees unanimously elected him chaplain of the Institute in which position he served for fifteen years. A modest, mild-mannered man, he has achieved a high place for himself and for the Negro in American life.

It was at Tuskegee that he became nationally known as a worker with young people and as a preacher. Under his ministry the Tuskegee Chapel became one of the outstanding pulpits of the South. He serves regularly as a preacher in both the University and the National Christian Missions conducted by the National Council of Churches.

For the past seven years he has directed an in-service training program among the rural churches of the South. He has been active in interracial work, organized a state-wide interracial ministerial alliance in Alabama, and now serves as a director of the Southern Regional Council.

Dr. Richardson has written widely. His articles have appeared in the "New York Times Magazine" and in learned journals. He is author of Dark Glory, *a picture of the church in the rural South, and of a section in* The Christian Way in Race Relations. *In 1948 he was elected to the presidency of Gammon Theological Seminary in Atlanta, Georgia.*

Sermon Fifteen

> TEXT: And as they did eat, Jesus took bread, and blessed, and brake it, and gave to them, and said, Take, eat: this is my body. And he took the cup, and when he had given thanks, he gave it to them: and they all drank of it. And he said unto them, This is my blood of the new testament, which is shed for many.
>
> MARK 14: 22–24

CHRISTIANS AROUND THE WORLD are sharply divided on points of faith and points of polity. In America alone these differences are expressed in more than two hundred and fifty separate denominations. Throughout the world there are many more.

But there is at least one point at which all Christians are united. There is at least one common element in our religion which runs through all the differing units, holding us in common unity, making us one faith, one body, one church. That point of union, that binding factor is the central sacrament of Holy Communion, the Lord's Supper. To be sure, there may be minor variations of administration or interpretation, but we are one the world over in our appreciation of the Sacrament's inner meaning and in our acceptance of its social implications.

As to the inner meaning of the Sacrament, we Christians are agreed, first, that the Sacrament begins with the open humility of admitting one's sins and being repentant for them. Unless we first have attained the purging humility of the honestly contrite heart, the Sacrament will never be able to bestow its blessing upon us.

Christians agree, secondly, on the spiritual meaning of the Sacrament, that mystic union of partaker and Master when by submitting ourselves to Him, His Spirit becomes our spirit, His aims our aims.

And thirdly, we agree on the rededication of will. This must always happen in the Sacrament so that we arise from the Table "intending to lead a new life, following the commandments of God." These are inner meanings of the Sacrament, and these we all accept.

Christians are equally united, however, in their acceptance of the outer or the social implications of the Sacrament. For instance, we do not dare to come to the Table except we first are "in love and charity with our neighbors." That is, as preparation for the Sacrament we must be willing to accept and must be committed to achieve a life of conscious Christian brotherhood in this world, with all that its achievement implies. We know that in this modern world, such achievement is difficult, to say the least. Yet living in love and charity with one's neighbors is the inescapable prerequisite.

Hear again the words of the General Invitation which set the requirements all over the world: "Ye that do truly and earnestly repent of your sins, and are in love and charity with your neighbors, and intend to lead a new life, following the commandments of God . . . draw near with faith and take this holy sacrament. . . ." Here is the heart of the Christian religion, here is the binding element which makes us all one.

As we gather at the Lord's table on this Worldwide Communion Sunday, the Sacrament seems to come to us with especial fitness at this time, almost as though it were intended for today. In the quiet sanctity of this service, we are likely to forget that the Supper was instituted in a time very much like our own. It was instituted in a setting of deeply disturbed feelings, of severe anxiety and much heaviness of heart.

We must not forget that it was on the night in which he was betrayed that he broke the Bread and passed the Cup. And even as they ate and drank with him, he announced that that night one of their own number would betray him. So deep was their anxiety, so fearful were they of possible error even within themselves that one by one they lifted their eyes to him anxiously asking, "Lord, is it I?"

What could be more bitter than this atmosphere of treason and mutual suspicion and the depression that comes from facing the end of one's world. For the disciples had left friends and fortune, everything, to follow him. Now in the face of his approaching failure, they faced the collapse of their hopes, their dreams, their world.

Are not these our own feelings as we look upon this troubled, modern world? Anyone who takes his life or his religion seriously is inevitably depressed at what he sees. The stark tragedy of wars present and wars to come, when we know that they are all to no purpose; the renaissance of nationalism, the persistence of group hatreds, the presence of crime and dishonesty even in highest places; the disregard for morals, the weakening of families and social foundations, the self-centered spirit of so much of modern life—these are destructive forces and they make us anxious, fearful, afraid that we are witnessing the impending collapse of our world. Here in our own social setting is the modern betrayal of our Lord.

To make it worse, we know that we cannot stand apart in self-assessed righteousness, pointing accusing fingers at others of our fellowmen. The same wicked passions that are wreaking so much havoc in the world run rampant in our own breasts, too. Like the disciples, in simple honesty, we lift our eyes to heaven, asking anxiously, "Lord, is it I?"

Now, it is just to these unexpressed, heavy forebodings that the Supper comes to speak its word of redemption and power. Remember that it was deliberately instituted to overcome such feelings of despair and defeat. Jesus knew the anxiety the disciples were suffering, and he knew the greater hurt that would come when he should be taken away. He, therefore, devised an instrument that would transcend disappointment, one that would bind them to him in an unbreakable triumphant fellowship that would overcome all their fears of the moment and make them equal to any trials yet to come.

This was the purpose of the Last Supper. It has worked wondrously well. It did bind the little group of depressed, simple people into a hopeful, radiant company who eventually found power to move out courageously to transform the world in his name. The fellowship has continued. Our presence at this Table this morning is witness to its existence in the hearts of believers all over the face of the earth. The fellowship is growing. Our number is larger than it has ever been, it will be larger still.

Therefore, brethren, as we commune this morning, we shall lift our heads, we shall strengthen our hearts. The paralyzing fears, the timid inactivity of so many men are not for us. The times are not hopeless, destruction is not inevitable. All our troubles have their roots in the hearts of men and, thank God, these can be changed.

We do not work alone, and we need not try to do so. The task is too great, the fight too hard. No one of us can save the world, but each of us can save his own world. No one of us can save all men, but each of us can save some men. This is the virtue of the fellowship, each working in his own little sphere yet all working together, joined in the holy communion of spirit and in the patient determination to attain righteousness within ourselves and to achieve brotherhood within the world. This, then, is the triumphant fellowship of two thousand years' existence which neither war nor sin nor social disturbance has yet been able to destroy. Yes, this is the worldwide fellowship moving steadily on to victory, led by the unconquerable spirit of Jesus Christ, our Lord.

The Spirit of the World

Reverend Robert I. Gannon, S.J., S.T.D., Litt.D., LL.D., L.H.D., D.Sc. in Education

Sometime President, Fordham University (Roman Catholic), New York
A Member of the Society of Jesus; Director of Retreats,
Mount Manrisa, Staten Island, New York

Father Gannon is one of the truly great preachers of our day, with the combined gift of sound scholarship, homiletic ability, and a balanced sense of the dramatic method of delivery. Wherever and whenever he speaks congregations listen; lawyers, business men, educators have him as their guest preacher whenever they can.

Born in New York in 1893, he was educated at Loyola School, New York, and Georgetown University, Washington, D.C. In 1913 he entered the Society of Jesus. From 1919 to 1923, he was instructor of English and philosophy at Fordham College.

After leaving Fordham he made his theological studies at Woodstock and was ordained in 1926. Following his ordination he was sent abroad for special studies, taking his S.T.D. from Gregorian University in Rome in 1927, and his M.A. from Cambridge (Christ's College), in 1930.

In 1930 Father Gannon reopened St. Peter's College, Jersey City, which had been closed during the war, and became its dean. He opened Hudson College of Commerce and Finance, of which he was the first dean from 1933 to 1935, and remained as dean until his appointment as president of Fordham University in June, 1936.

His accomplishments in education, religion, and the humanities have been recognized by great universities and colleges with a long shower of honorary doctoral degrees by Georgetown University, Litt. D., 1937; Columbia University, 1941; Manhattan College, LL.D., 1938; Holy Cross College, 1938; Bowdoin College, 1941; New York University, 1942; Rutgers University, 1945; Hobart College, 1947; Colgate University, Detroit University and Alfred University, 1949; L.H.D., Boston College, 1939; Hofstra College, 1948; D.Sc. in Education, Lafayette College, 1946.

In 1937 he went to Venezuela on the invitation of President Lopez Contreras for consultation on school problems, and in 1942, received the award of the New York Academy of Public Education for distinguished service in the field of Education. He is a trustee of Town Hall, an elective manager of the New York Botanical Garden, a trustee of the New York

Zoological Society, a director of the Netherland-American Foundation, and a member of the Committee for International Economic Reconstruction.

Father Gannon's most interesting pulpit assignment was in air-raided London, where he preached the Lent in 1943 at Westminster Cathedral as the guest of the late Cardinal Hinsley.

He was president of the Association of Colleges and Universities of the State of New York, 1946–49, was appointed director of retreats at Mount Manrisa, Staten Island, in 1949.

This Lenten sermon has the spirit of a great soul woven into it, for Father Gannon has the simplicity and power that make him a preacher among preachers. No one can read this sermon without feeling the need of a cleansing from the evil of the spirit of sin in our world. "The Spirit of the World" was delivered on the Church of the Air program of the Columbia network.

Sermon Sixteen

TEXT: I pray not for the world, but for them whom thou hast given me. JOHN 17:9

WHEN THE MAGI arrived in Jerusalem and went to the Court of Herod seeking "Him who is born King of the Jews," they found an interesting group in the palace. For the King, an Arab monster who had strangled his own sons, murdered his wife and most of her family and was at the time half dead with an unspeakable disease—this "Herod the Great," as he is called, was surrounded with faithful Scribes and Doctors of the Law, Pharisees and Sadducees looking for advancement. Among them was a man of middle age named Annas who had not long to wait before reaching the office of High Priest. Doubtless if we had looked among the group we could have found another familiar figure, a young Levite of twenty or twenty-five, who had one eye on the King and the other on the daughter of Annas. His name was Caiphas and he was there to learn his way about in the world—to make his noviceship in the spirit of the world.

The day the Magi came to this edifying court they created quite a stir, for they brought with them some fantastic story about a star and a new King of the Jews. Caiphas may have taken one look at the decayed mass of flesh on the throne, and thought to himself that the Jews could stand a new King, but he agreed with all the rest that no good could come out of any more Messianic nonsense. Come to think of it, there had been a Messianic story around the temple only a few days before. He would check up on that. Something about a couple of Galileans who had brought in a baby to be presented as usual, and old Simeon and Anna had made quite a scene. The priests who were on duty had told him how they had "prophesied" some rigamarole about the redemption of Israel. Prophecy, indeed! If he had been there he would have given them a lesson in prophecy.

Still, the old King was worried—he worried about everything except his soul—and insisted that they look up the ancient documents. It was then the turn of the priests to be worried, for lo and behold, the time had come, according to the most authentic calculations, and Bethlehem of Judah was the place. So their duty as priests was plain enough but not pleasant enough. For this new King was a disturbing figure. He might be going in for reform, and besides, Herod would certainly punish anyone who followed Him. So then and there Caiphas made his choice. He cast in his lot with Herod and a comfortable Jerusalem. He cast in his lot with the spirit of the world. So the Magi went on their way, the King's uneasiness was removed by a general massacre of children in Bethlehem, and Caiphas returned to the more important work of shaping his career. In time the daughter of Annas proved amenable and became his wife, and not long after Annas proved to be high Priest—which showed that Caiphas' judgment was worthy of higher place.

It was no surprise, then, a few years later, when his father-in-law had been deposed, to find Caiphas himself entering the Holy of Holies, going in before his God and bearing on his soul all the sins of his people. Of course, the price of admission had been rather high. He had bought the Sacred Office from the Roman governor—Josephus tells us that—and Roman governors were notorious for driving bargains. Still, Caiphas knew what he was doing, paid cheerfully and looked to various little ways of making ends meet that were often open to men in high office. The Temple, as a matter of fact, under his management proved to be a gold mine. There was, for instance, the one little item of turtle doves, the offering of the poor. It doesn't sound like much, but thousands of people visited the temple every day, and thousands of small orders make big business. Of course, the High Priest could not countenance ordinary turtle doves in that holy place. They had to be the purest kind of doves, which were raised—as it happened—only on the farm of Annas, his father-in-law. Then there was another item: lambs for sacrifice, and oxen—flocks of them, herds of them. Many priests were kept busy all day just slitting their throats. So for the greater convenience of the people, stalls were built all around the great court of the Gentiles. It *was* something of an innovation. David and Solomon, when they planned and built the temple, had not anticipated a county fair, but it showed the peculiar wisdom and talent of the High Priest. For first he got the rent, and then the little extras that are not unknown even in modern markets.

But the best source of revenue in the place, perhaps, was the money changing. According to the rule, all tithes and offerings on Mount Sion had to be paid in the coin of the temple, and as this coin was not used in the outside world, there was a little profitable transaction each time the tithes were paid. Caiphas, to make sure that all things would be done unto edification, placed the money changers' tables in charge of his own relatives and found them entirely worthy of his trust.

In fact, in a very few years, things were working nicely. The common people complained that religion was getting a little expensive and the devout Jews were scandalized at the absence of the old spirituality, but the men higher up were satisfied and nothing else mattered to Caiphas. For he had been well schooled in the spirit of the world by Annas. In fact, he was a model of worldliness. He was a symbol. He stood for the world that Christ refused to pray for; for the world that is only at home with the flesh and the devil; the world that concerns us most in the Holy Season of Lent. What world is that?

It is not the world of the Universe—the good, green earth, the sun and moon and stars—they are the footprints of God and give Him glory through the day and night. Neither is it the world of men. For most of the human race is lovable enough. As we go through life we come to know so many of God's little saints—patient, humble and charitable. In fact, it is a humiliating thing sometimes for a priest to hear confessions. He meets so many people who are holier than himself, and even most of the sinners are not so bad. Weak men usually wish that they were different and long to be more worthwhile. Many a silly and frivolous woman can be brave and generous when a crisis comes, and they say there is honor even among thieves—at least among thieves of the poorer class.

This is the world that Christ came to save. This is the world that gets away from God through its passion and pitiful weakness. Then what is the world that he refused to pray for? What is the world that is symbolized in Caiphas? As old Father Blount used to say, it is human society in so far as it ignores God's claims and lives to please itself; in so far as it treats God's creatures as its playthings; in so far as it refuses to recognize the existence of sin and sets up concupiscence as the rule of life. What was it that shattered Christianity in the sixteenth century and would have wiped out all traces of Christ's Church if that Church had not been Divine? The spirit of the world. It wasn't primarily Luther or Calvin or Henry VIII or Cranmer. It wasn't any of the forces on the Protestant side. It was primarily certain Popes and certain Cardinals and certain Bishops of the Renaissance who buffeted the face of Christ and tore His seamless garment to shreds. Too many of them could have doubled for Caiphas.

As in the Church, so in international and domestic affairs. What guided the Treaty of Versailles at the end of the First World War? What strangled poor Austria and Hungary, plundered the land and sold millions of the people to neighboring states? The spirit of the world. Who presided at Versailles anyway? There was talk of having the Vicar of Christ, so that justice might be done and peace restored. But of course that was foolish. Caiphas was the obvious choice. He was elected unanimously—and when he took his place at the head of the council table, he found himself surrounded by his cronies. It must really have seemed like old times to Caiphas, for there before them, manacled like a thief, was the spirit of Christ. Once more they had come together, not to try Him but to condemn Him. For they had all agreed ahead of time that the spirit of Christ must be killed in international affairs. So they had their way and got what they wanted—plunder and revenge. What did humanity get? A troubled armistice that lasted until Caiphas could arrange a pact between Hitler and Stalin. The spirit of Versailles marched on. At Teheran, Yalta and Potsdam no principle was ever allowed to complicate the discussions. Caiphas saw to that. And today we face fresh crises that are stale, every one of them, with the odor of worldliness.

But why go abroad for examples of worldliness? We have them all around us here at home. Caiphas is the busiest man in the United States of America. The elder Morgan, in his time, was a director in more than three hundred corporations. Who can count the corporations of Caiphas—the corporations, I mean, that are built solely on the spirit of the world? No thought of right or wrong, only of dividends. Who can count the law firms of which Caiphas is the senior partner—unmentioned, perhaps, but very influential. And in politics—who is the strong silent man in every community who never has to be elected? Caiphas. Who rules the underworld—High Priest of the Racketeers? The same High Priest who sent Christ to the cross. For all these classes of society are living for themselves alone. Moral ideals are a joke to them. All that remains of religion is a pose of respectability, a veneer, a pretense. Statesmen pretend to be patriots. Big business throws out a cloud of philanthropy. And through it all walks—Caiphas. A symbol of the world, he goes about in every generation, marked like Cain with a mark of blood, but the blood is the blood of the second Abel, the blood of Christ.

Caiphas never realized, of course, that he was a symbol. To him the trial of Christ was a very personal matter. This Galilean had done two things, both of them unpardonable. He had unmasked the Temple gang and he had ruined big business on the Holy Hill. He had called the friends of the High Priest "hypocrites and whited sepulchers full of dead men's bones." And not content with words, He had made a scourge of little cords and driven the traders and money-changers from the Court of the Gentiles, the customers and relatives of the High Priest! This in itself condemned Him to death. The fact that He was foolish enough to talk about being the Messiah would be a convenient excuse. They could slay Him now out of patriotism and one man should die for the people.

So at last they stood face to face. Two irreconcilable forces—Christ and the spirit of the world. And victory seemed to go once more to the spirit of the world. For in the end the High Priest could rend his garments, a picture of injured piety, and smiling in his beard, cry to the Sanhedrin, "He hath blasphemed. What think ye?" Guilty—of course.

Guilty of death! They had their own reason, those seventy-one holy murderers. But what reason could we have had for joining them? Yet sometimes, in a kind of nightmare, we have heard the sound of our own voices shouting "He is guilty of death." He hath stood between me and my unworthy desires. If he is the real ruler, I must conform to His will and sacrifice some of the things I love. My pleasures, my health, my revenue, my place in the party. I would rather have Him die to me, crucified again in my heart, than take the spirit of the world at its true value and despise it.

Yet we know, as we feel it actually at work all around us, that this world which Christ refused to pray for is a horrible thing. It has its attractive side. It can be polished and sophisticated and amusing when it wants to be, but under the surface it is ulcerous. It stinks with the odor of death. Pray then, especially now in the holy season of Lent for the Grace to be unworldly; Christ said, "I pray not for the world but for them whom Thou hast given me."

Palm Sunday

REVEREND RICHARD M. STEINER, D.D.
Minister, First Unitarian Church
Portland, Oregon

Dr. Richard Steiner is one of the favorite preachers of Portland, Oregon, where his preaching has drawn to his church a number of the leading people of the city.

Born in Sandusky, Ohio, where his father was at that time minister of the Congregational Church, he was reared in a college town, Grinnell, Iowa, and attended Grinnell College. Before completing college he had a fling at newspaper work, serving as night police reporter for the Chicago News Bureau in 1919. "It was for me," he writes, "an invaluable contribution to my career as a minister. It taught me to work under pressure and to accept interruption of one's work with equanimity. It made me conscious of 'deadlines.'"

He took graduate work at the University of Michigan, taught English and Journalism at Washington State College, 1925–26, and Bradley College, Peoria, Illinois, 1926–28. He soon became the unofficial pastor of his students, decided to transfer to the department of religion and ethics, went to Chicago Theological Seminary to get his B.D. and graduated there in 1933. He had no intention of going into the ministry then, but Dr. Ozora Davis, later Dr. Albert W. Palmer, and Dean Charles Gilkey of the University of Chicago Chapel, were insistent that he had the makings of a preacher. At the time of his graduation his father underwent a serious operation, requiring a year's convalescence and he took over his father's work at Grinnell College in the Chair of applied Christianity.

When First Church, Portland, the largest Unitarian Church on the Pacific Coast, and one of the largest in the country, fell vacant through the retirement of Dr. William G. Eliot, Jr., Dr. Steiner was recommended for the church and in 1934 accepted the call. He has been there ever since as a happy minister with a happy congregation. He was ordained in his home church at Grinnell, Iowa (The First Congregational Church) a few days before assuming his duties in Portland. He is married and has two sons. Mrs. Steiner is a member of the Society of Friends (Hicksite) and he joined Purchase Meeting prior to their marriage, so that he holds fellowship with Congregationalists, the Friends, and the Unitarians.

To him a sermon is an art form. It can appeal directly to the emotions, or to the emotions through the intellect; he prefers the latter. His sermons are written to be heard rather than read. He belongs to the conservative order of Unitarians, being essentially a Christian Theist. Grinnell College conferred the honorary D.D. upon him in 1942.

This Palm Sunday sermon was preached March 18, 1951. It is an excellent example of Unitarian preaching, yet his Congregational background can be seen in some spots.

Sermon Seventeen

BEHOLD, A SOWER went forth to sow!

Jesus of Nazareth began his ministry by the Sea of Galilee, a small inland body of water set in the midst of the fertile crescent between Africa the unknown, Asia the inscrutable, and Europe, the womb of western civilization!

Jesus of Nazareth gathered to himself disciples, untutored fishermen, laborers and tax-gatherers.

Jesus of Nazareth resisted the temptation to be diverted from his self-imposed task of purifying the religious and moral life of his people.

Jesus of Nazareth preached and practised a gospel of a friendly universe, a companion God, a universal helpfulness that counted all men as sons of God whose obligation it became to observe the spirit of God's moral law, that in the fullness of time that spirit should become the ruling spirit on earth as it is in Heaven.

Behold now his triumphant entry into Jerusalem, where the populace, misunderstanding his mission to be the leader of a revolt against the Roman occupation, hailed him with shouts of "Hosanna!" and the strewing of palm branches in his path, a symbol of victory and devotion to the victor.

Behold Jesus of Nazareth crucified upon a cross between two thieves, a symbol of degrading rejection and failure.

Behold a sower! . . . and the fruits of his sowing.

In Christian tradition there are few more cherished memories than that of the triumphal entry into Jerusalem. Many Christian churches celebrate Palm Sunday with a pomp and circumstance second only to their celebration of Easter. For, to many Christians, Christ the King entering into the city while the multitude shouted their hosannas, waving palm branches and strewing flowers in his path, is a much more congenial picture than Jesus the meek, walking the dusty roads of Galilee from humble village to humble village, preaching his Gospel.

Yet one wonders; for it must have been an incongruous procession, Jesus and his disciples, thirteen men at most, their leader seated upon a young colt, their travel-stained garments fluttering in the spring breeze, no men of arms, no drum roll, no martial music, no banners flying, just a few men with dusty sandals or bare-footed and bare-headed, walking into a crowded city, expectantly waiting the Feast of the Passover in the shadow of the temple wherein God dwelt in His Holy of Holies under the deeper shadow of the occupying forces of Rome, the legions of Caesar. Who knows but that the hosannas may have had a touch of mockery, who knows but that the palm branches, the symbol of victory, may have been a kind of ironic joke, a cynical prank? But Christian tradition has insisted that it was no prank, that the people of Jerusalem, desperate under the demands of the temple tax-gatherers, desperate at the indignities which they had suffered under the Roman occupation, were ready and ripe to proclaim even this dusty and ragged band, this somber figure, riding on a colt, as their deliverers.

One cannot know, of course, what thoughts coursed through the mind of Jesus as he came through the gates of the city of Jerusalem, but we know enough of his intelligent awareness, of his sensitive realism to know that if he heard the shouts, he knew that they were not for him but for the aspirations of his people. Their hosannas could not have been music to his ears, for had they struck a responsive chord in his heart of hearts, he would not have ended the week upon the Cross. He knew that he was at last face to face with the ultimate act of his career. The stage was set for a temporal triumph and a spiritual débâcle, or a spiritual triumph and a temporal tragedy. He would prove within the space of a week whether he was a man of iron or a man of God. The temper of the people at that moment in Jewish history had been stretched so taut that within a week it was by no means inconceivable that Jesus could have become the dictator of Jerusalem.

Someone has said that to be a man on horseback does not make a man a dictator. What makes him a dictator is the shouting of the populace at seeing him on horseback. The hosannas were not for Jesus the meek, but for Jesus, the hoped-for Messiah. The flowers cast in his path *were* for the Saviour, not the Saviour of souls, but the hoped-for Saviour of a racial culture. The palms that were waved were the symbols of victory, not the victory of Jehovah, not the victory of the spirit, but a victory of the Jewish people over the garrison which was Caesar's.

We have no record of any inward struggle on the part of Jesus to succumb to this temptation. He had been tempted before, been tempted of the Devil, so the record says. His mind was made up. A great pity must have held him in thrall, a pity for a people who had ears to hear, but had not heard. He had spoken of the Kingdom of God and they had understood him to mean a Kingdom of the Jews for the Jews. He had spoken of conquering evil, and conquering evil meant conquering Rome. His thoughts may well have turned back to that day when he had taught them, saying, "Behold a sower went forth to sow," a parable which he ended with that cryptic phrase, "He that hath ears to hear, let him hear."

Had Jesus been a philosopher, more concerned with analysis than with synthesis, which is the mission of the religious leader, he might well have reflected that it is not the essential wickedness of humanity which causes humanity to do evil things, it is stupidity. Was it a naïveté or was it a real and deep wisdom which led Jesus to the conclusion that there is an essential goodness in humanity? Was it a naïveté or was it a deep wisdom which gave him the confidence that because man was essentially good the Kingdom of God must eventually come? Christianity has proclaimed for centuries that it was a deep wisdom. Yet the verdict of history has said that it was naïveté, for the Kingdom of God has not come, though better than nineteen hundred years have passed since Jesus sacrificed himself for its coming.

Which is right—Christianity or history? Who was wrong—Jesus or the generations which have come after him and of which we are a part? If history is right, if Jesus was hopelessly naïve, then the dominant ideal which has sustained and strengthened Christian men and women during all the centuries, during all the crises which have faced generation after generation, then that ideal is false and we who are gathered here this morning are devoting ourselves to that which is untrue. Far better that we join the ranks of the irreligious and unreligious, for to concern ourselves with that which is false is to utterly waste our energies. But what is it that history has proved? Has it really proved that Jesus was wrong? Or has history only proved that the coming of the Kingdom of God, that the establishment of his will is not an automatic process, that men and women must work at the establishment of a Kingdom, that it will not come just because men are essentially good, but that it will come only when that which is essentially good in men achieves its purpose due to the efforts of good men? Jesus knew that what men sowed they would reap. If they sowed tares, they would reap tares. If they sowed good seed, they would reap good harvest. For centuries men have believed that good can be accomplished by force of reason, but reason is no guarantee against stupidity.

I have been re-reading recently the "Little Flowers" of St. Francis and have been charmed again by the account of Friar Juniper, so lacking in intellectual capacity that it is no exaggeration to call him the "Dopey" of St. Francis's followers. Like Dopey of the Seven Dwarfs, so sensitive and lovable, so generous and good that one cannot but echo St. Francis's words, "Would that we had a whole forest of such Junipers." What Juniper had was not intellectual capacity, but the capacity to understand—an understanding that has no relationship to intelligence in the sense that it is a product of the intelligence.

Understanding is a product of devotion and of devoutness, of a simple faith in the goodness of humanity and in the goodness of God, and the necessity of translating that faith into simple deeds of goodness. Stupidity is no respecter of persons. The intelligent individual may rationalize, and it is because of his intelligence that he is able to rationalize his stupid deeds so that he believes that he has been acting wisely. There is a phrase in our Liturgy which reads, "From unrighteous anger and an impatient temper, from an uncharitable judgment and readiness to believe evil, from inordinate cares and needless anxieties, from complaints against Thee, and from rebellion against Thy Holy Will, O Lord, deliver us." This is a prayer to protect us from our stupidities. For it is stupid to be a prey to an unrighteous anger and an impatient temper. It is stupid to succumb to an uncharitable judgment and readiness to believe evil, to be the victim of inordinate cares and needless anxieties. And these are the things from which the world needs to be saved. For it is these things which keep the Kingdom of God from being achieved. Most of the peril which faces us today, the undermining of our free institutions which have been so painfully erected by the sacrifices of so many, can be traced to unrighteous anger, to an impatient temper, to uncharitable judgment and readiness to believe evil.

Yet the very fact that there is *something* which is in peril, the very fact that we recognize and are able to understand the causes of our peril is evidence that if we believe history to be right in judging Jesus to be naïve, we are wrong. All that history has proved is that where men have worked, and men *have* worked, to establish the Kingdom we have been able to catch glimpses of its reality. The seed that was sown nineteen hundred years ago whenever and wherever it has fallen into good earth, whenever and wherever it has been cultivated by men who have refused to spare the sweat of their souls, the triumph of Jesus' confidence, of Jesus' concern has been given reality.

It is not the world as a whole which continues to place its faith in those things which are evil. Ever present in the world has been the Church to deny the value of force and violence, to deny the worth of greed and selfishness, to deny the effectiveness of imprisonment and death to kill an idea. He who counts Jesus a failure, who says that he was naïve, reveals his own spiritual failure, his own historical naïveté. He either fails to see or refuses to see the oceans of self-sacrifice that have been poured for nineteen hundred years upon the sands of time to make a sea, a reservoir of life-giving waters to nurture the growth of the human conscience in its dealings with the oppressed, the underprivileged, the sorrowing and the suffering. He refuses to see or fails to see this fruit borne of the seed, scattered by the wayside among the rocks, the thorns and the good earth nineteen hundred years ago in Palestine.

The acres of earth, purged of worldly ambition, that have been cultivated for the glory of God and the enrichment of the human spirit by those whose prayers and pleas to the living God have wrought a solid substance, a visible manifestation of mankind's yearning for peace, this, too, has been a fruit of the seed of yearning planted nineteen hundred years ago by man inflamed with a passion for peace. Yet, having said this, the fact remains that no success is permanent, no victory perpetual. Jesus may yet be forgotten. His victory may yet prove hollow. The seed which he planted may be destroyed by the devouring fowls of the air, whose flight swift as rockets, may destroy, with atom bombs, the culture that bears the name of Christian. The gentle roots may yet meet the upthrust impenetrable rock of a hard and sterile secularism which will wither into extinction man's faith in himself and in his God. Or the thorns of totalitarian materialism may choke and strangle all liberty by which and through which men's spirits grow. We cannot ever be sure that as the earth turns and the seasons come and go, utter and complete spiritual famine will not spell the end, will not write "finis" to the Christian hope, the Christian dream planted nineteen hundred years ago by a sower who went forth to sow.

If this Holy Week is anything more than an anniversary of events that took place in a time remote, a civilization decayed, if this week is to have meaning in the midst of this century of violence and of crisis, it will only be because you and I will give it meaning. If the victory of Palm Sunday, which was only the prelude, the overture, to the greater victory of Easter, is to find revival in our heart of hearts and in the heart of the world, it will only be because it is revived in us.

The prophets of doom, the scientists, the politicians and the preachers who have painted for us the agony of a world devastated by radioactivity and biological warfare, have, in no small measure, overreached themselves. They have inured us to fear.

We who live in the cities are the prisoners of our cities. We either cannot or will not move away. We have adopted the philosophy which says "kismet" to our fears: What will be, will be. This is not courage. This is not cowardice. This is not victory. This is not defeat. This is not human. It is an animal acceptance of life as it is. It is not a divine acceptance of a challenge to make life what it ought to be. It was that acceptance by Jesus of Nazareth which has brought us to this hour of worship. If future generations are to continue to worship and to work for the establishment of the moral law of God as the ruling law of earth as it is in heaven, then we shall have to sow as other men have sown since the dawn of conscience.

The task of the sower is never done. Each year as the snow recedes and the warm earth stirs with life, the sower must put his hand to the plow and, having plowed, scatter the seed. When the burden of life is finally laid down, his son takes over the land to plow, to sow, to reap, and then *his* son, and his son's son. So long as this occurs, the harvest is brought in. Should it cease the sower would starve. Humanity would die.

Jesus, the sower, spoke many parables, but he spoke no greater parable than his journey from the avenue of the Palms to Golgotha, the place of skulls.

"He that hath ears to hear, let him hear."

HOLY WEEK: Wednesday of Holy Week

The Word of Life: The Word, Power for Life

REVEREND MICHEL RIQUET, S.J.
Member of the Society of Jesus and Conferencier of Notre Dame Cathedral, Paris, France

Père Riquet's Lenten sermons have become the high light of the annual services in Notre Dame Cathedral in Paris. He speaks there as a witness to the things he knows out of the experience he has had with faith and life. This sermon for Holy Wednesday brings the power of the true love of God and Christ to the mind and heart of his hearers. Love, he shows as the way to eternal life with God.

Michel Riquet was born in Paris in 1898. A brilliant student, he served in the infantry in 1917–18, was cited for bravery, was decorated with the Legion of Honor in 1945, and was later awarded the Medal of Freedom with Silver Palm.

Father Riquet worked with the French Resistance until the Gestapo arrested him on January 18, 1944. He was first imprisoned at Mauthausen, then transferred to Dachau. He was kept a prisoner until the allied armies released him in May, 1945. After his return to France, Cardinal Suhard asked him to preach a series of sermons on "The Christian Confronts the Ruins" ("Le Chrétien Face Aux Ruines").

Thus Père Riquet began his series of Lenten Conferences at the Cathedral of Notre Dame in Paris in 1946. A second series was preached in 1947, a third in 1948, a fourth in 1949, on the theme, "The Christian Confronts Authority." A further series was preached in 1950 and 1951, with the great cathedral packed at each service when he speaks. The translation of "The Word, Power for Life," was made by Dr. Clayton Williams of Paris and appreciation is gratefully expressed to Père Riquet and Dr. Williams for their coöperation in making it possible to bring this message from France to American readers.

The conferences of Notre Dame were established in France in 1820 and the greatest preachers of France were brought to Notre Dame for a course of sermons. The Jesuit de Ravignan, the Dominican Lacordaire and the great Montsabre made the conferences famous. The conferences have continued until now in spite of wars and difficulties. Father Riquet has added new glory to these famous conferences.

When Père Riquet speaks of the Word of the Cross and the Word of Life he has something to say to every heart in our world today. A sincere effort has been made in the translation to retain the full French flavor of the sermon.

Sermon Eighteen

MY BROTHERS. Undoubtedly we all aspire to live well, to live intensely and to live eternally. But there are laws of life, and exigencies of life. To him who seeks baptism and, through it, life eternal, the Church answers: "If you wish to enter into Life, follow the Commandments," for fidelity to the words, the Commandments of Jesus, remains the essential and indispensable condition for fulfilling our desire to live forever, eternally happy. We know quite well and have repeated it previously in this pulpit, that all the Commandments which are found in the Law and the Prophets may be summarized in a single precept: "Thou shalt love."

But to love with a true love is not only to have sympathy, to manifest a platonic goodwill but rather, to give oneself, sacrifice oneself and to serve and help others, as Monsieur Vincent said, "with all the strength of our arms, and the sweat of our brows."

It is then that we hesitate, falter, and try to bargain with this gift of ourselves. Whether we are actuated by love or by fear, the Law requires renunciation and sacrifice. We do not always have the inclination for this, or even the courage.

St. Paul experiences in the presence of the Law sentiments and a spiritual state which we all know well. "The inner man within me," he writes, "takes satisfaction in God's Law, but I witness in my bodily members another law, which struggles against the law of my spirit and which holds me captive under the law of sin which is in my members."

"To will Good is within my reach, but not to accomplish it. For the Good which I will, I do not carry out; and the Evil which I do not will, I commit." [1]

Like St. Paul, every man recognizes and admits his inability to follow, by himself, and above all, with constancy, the prescriptions of the Law. Moreover, our Master has pointed out to us that, if the path to perdition is wide and comfortable, straight and strict is the way that leads to Life.

"Whosoever wishes to be my disciple must take up his cross daily, and follow in my footsteps." This is what He offers us.

To the young man, rich and well-born, who asks him, "Good Master, what must I do to share Thy Life Eternal?" Jesus answers first, "You know the Commandments: Thou shalt not kill, thou shalt not commit adultery, thou shalt not steal, thou shalt not bear false witness, thou shalt not harm thy neighbor, honor thy father and thy mother." All of this the young man had carefully observed since his childhood. Then Jesus turned his regard toward the youth and conceived a love for him, and then he said: "You lack one thing! Go, sell your belongings, give them to the poor, and you will have a treasure in heaven; then come and follow me." But the youth became sorrowful upon hearing these words, and he went away very sad, for he had much wealth.

Then looking about him, Jesus said to his disciples: "How difficult it is for those who have riches to enter into the Kingdom of God!"

His disciples were stupefied. But Jesus persisted and began again: "My children, how difficult it is to enter into the Kingdom of God! It is easier for a camel to pass through the eye of a needle than for a rich man to enter into the Kingdom of God!" Thereupon, his frightened listeners thought to themselves, "Who then can be saved?" Jesus answers them, "For men it is impossible, but not for God; for with God everything is possible." [2]

[1] Romans 7:18–23
[2] Mark 10:17–27

Here we are at the heart of the paradox of Christianity. To love, one must die, extinguish oneself, extinguish bodily egoism, extinguish the temptation of the world, in order to live for God, in God, with God, eternally.

As Paul explains it, it is a question of "forcing the death of the works of the flesh, through the spirit," of ridding oneself of sin, with Christ, in order to relive, with him, a new life. "His death was the killing of sin, once and forever; his life is a life for God. May sin no longer reign in our mortal flesh, nor make you obey its lusts. Put not your members to the service of sin as instruments of injustice; give yourself up to God, on the contrary, as do the dead who are reborn to life, and give your bodies to the service of God, as instruments of justice." [3]

St. Paul is not a Manichean who attributes to the principle of evil, to the devil, the creation of the body and the natural functions which assure its growth, its subsistence, and the reproduction of human life. He opposes those who condemn as impure wine, meat and married life. For him, these are creations of a good God and must be used in rendering thanks to God.

The flesh, the dead body, and the terrestrial members are, in his statements, symbolic expressions of moral realities: the flesh, the body, and the members, in so far as they are instruments of sin, are the repositories of evil concupiscence.

Accordingly Paul says to us: "Mortify, therefore, your earthly members: fornication, impurity, passions, evil desires, as well as cupidity, and idolatry, for all this draws the wrath of God." [4]

But he says, also, that our members are "the temples of the Holy Spirit" and that after having "served injustice, they must serve justice." [5] That our bodies are the members of Christ and that we must glorify God with our bodies [6] but, precisely by not satisfying the lusts which agitate the body and attempt to make it serve impurity and sin.

In any case, we must struggle, renounce, sacrifice, mortify and constrain. And that will always be painful, difficult, at times superhuman for us. "Those who belong to Jesus Christ have crucified their flesh, with its passions and its lusts." [7]

This simple phrase of St. Paul's finally determined Augustine to turn completely to Christ by renouncing all those passions which had dragged him along with his carnal instincts, and it will always be painful for us to hear him remind us that there is no Christian life without effort, without struggle, without sacrifice, without some heroism.

But Jesus reassures us. To his disciples who are amazed, who are afraid when they see that it is so difficult to enter the Kingdom of God, more difficult than for a camel to pass through the eye of a needle, Jesus answers:

[3] Romans 6:10–14; Ephesians 2:1–19; Colossians 2:9–15
[4] Galatians 5:19–22; Colossians 3:5–6
[5] I Corinthians 6:19; Romans 6:13–19
[6] I Corinthians 6:15–20
[7] Galatians 5:24

"For men, it is impossible, but not for God; for everything is possible for God." [8]

"Everything is possible for God." This is what our Lord had answered Abraham when Sara refused to believe that she, old and until then sterile, could give birth to Isaac. This is what Job humbly recognized after having begun to argue about the wisdom of God. This is what God said to Zacharias when He told him of His intentions to raise Jerusalem from its ruins: "If it is a difficult thing in the eyes of the rest of these people, in these days, should it also be difficult to my eyes? Yea, I shall save my people and I shall lead them, and they shall live in the midst of Jerusalem, and they shall be my people and I shall be their God, in truth and in justice." [9]

God wills that all men might be saved. But it is necessary that they coöperate in this, that they consent to this, that they do not refuse his Grace, which calls them.

The value of faith consists not in exerting superhuman efforts, by oneself alone, but in trusting confidence in God. We must believe that with his Grace, which He never refuses, we shall be capable of accomplishing what He demands of us and face what causes us to be afraid.

"In all things, God coöperates toward the good of those who love him." [10] But we must love Him, and believe and place our hope in Him. This is the primary condition, indispensable, but sufficient. The rest follows. When we love, everything becomes easy, both because love lightens every burden and because God never fails those who love Him. It is not that He spares them the trials through which they grow to conform more and more closely to the image of the beloved Son, but because He gives them the courage to surmount these trials, if not always in triumphal glory, at least in the serenity of a filial heart.

Remember the cry of Saint Paul: "What will force us away from the love of Christ? Tribulation? Distress? Persecution? Hunger? Famine or the sword? For it is written: "Because of you they put us to death throughout the whole day; they look upon us as scraps of the butcher. But withal, we shall triumph through Him who has loved us." [11]

The word of God which will give us, more than any other, the courage to live to the full our Christianity, to remain constant in it, with the assurance of our salvation and our resurrection in Christ Jesus; the word which will preserve us both from all illusions and all deceptions, the word which fulfills all the conditions and the exigencies of that Life Eternal to which we aspire, this is the Crucified Word, Verbum Crucis.

The Cross is the last word, the last statement of Jesus, our Savior. Any conception or interpretation of Christianity, of either its morality or its theology which in any way eliminates the Cross is but a lie!

[8] Mark 10:27
[9] Zacharias 8:7–8
[10] Romans 8:28
[11] Romans 8:35–37

On the contrary, the authentic sign of true piety, of true faith, of true charity, of all genuine saintliness, is the Cross accepted, carried, and embraced.

The Cross accepted in a complete and manly resignation to the will of God. Not slavish submission, but acquiescence to the Will of God, whom we believe, whom we know, who loves us and whom we love!

The Cross carried courageously, without any heroic ostentation, rather in the resolution to accomplish one's sacrifice as God requires, without turning back, without cowardice, without running away toward worldly and deceiving consolations!

The Cross embraced with the same enthusiasm which inspired St. Andrew to say as he lovingly saluted the instrument of his torture: "O good Cross, which received thy beauty from the members of our Lord, desired so long, loved so anxiously, sought so unceasingly and at last prepared for the soul which desires it: receive me from the hands of men and restore me to my Master, so that, through Thee, He may receive me, He who through Thee will free me."

In the calm and sweet serenity of the contented Theophane Venard, writing tranquilly to his father, from his prison at Hanoi, we find these words:

"It is nearly midnight. Around my wooden cell are lancers and long sabers. In one corner of the room, a group of soldiers is playing cards, another group is playing at dice. From time to time, sentinels beat on the tom-tom and drum the night watches. Two meters from me, the electric bulb projects its wavering light toward my sheet of Chinese paper and allows me to scratch these lines. From day to day, I await my sentence. Tomorrow, perhaps, I shall be led to my death. A happy death, is it not? Death which is desired, for it leads to life!

"An easy saber blow will remove my head, like a Spring flower which the gardener picks for his own pleasure. We are all flowers, planted on this earth, flowers which God picks as he wills, some a little sooner, some a little later. One is the purple-stained rose, another the Virgin Lily, another the humble violet. Let us all, each according to the perfume or the brilliance which is given us, seek to find favor with our sovereign Lord and Master.

"I wish for you, dear Father, a long, peaceful, and virtuous old age. Carry gently the cross of this life, following Jesus, up to the Calvary of a happy end. Father and son will see each other again in Paradise. As for me, little ephemereal one, I am going away the first. Adieu!"

These are not the words of a desperate person or one disillusioned by life. As another martyr of Christ, Apollonius some eighteen centuries earlier said to the Prefect who was questioning him:

"I love life, Perennis, but have no fear of death through my love of life. Nothing is more precious than life, except the Eternal Life, which assures the immortality of that soul which has lived rightly in the present life."

This clarifies and explains all. Christ died only to live again. The Christian lives, acts, loves, and suffers in the hope and expectation of a death which will accomplish all his vows, because it will unite him, forever, to the Eternal Life of his Master and Savior, Jesus Christ.

With this prospect, should he not have the courage to bear all, to sacrifice all, to undertake all, and to dare all, from the moment that, in the midst of his suffering, his duties, his renunciations and his deprivations, he recognizes the path which brings him to his Savior and permits him to experience and enjoy the love which gives to life an eternal and divine meaning?

Do not conclude from this that the Christian should avoid the human tasks whose goals do not surpass the horizons of this world, or that the summit of perfection or the normal ideal for Christians is to be attained through a mystic evasion of this world. It is not, for the Christian, a question of detesting, much less of destroying mundane activities, even temporal ones, for they are precisely the indispensable instruments through which we make effective our love of men, for, as we have often said, it is in this love of men that the sincerity and the truth of our love of God are experienced and realized. Once more, love and charity furnish us with the ultimate standard, the divine formula which is capable of harmonizing those apparently contradictory exigencies of Eternal Life and of life on this earth.

On the one hand, everything must be subordinated to Eternal Life; and yet Eternal Life does not exist without relationship to and continuity with the present life. Transfigured by the Cross and by charity, this life with all its multiple activities, all the tasks which make up the life of a normal man, is the necessary prelude to the Eternal Life. One cannot love without acting, without serving, without helping, therefore we must have the power to do this, we must acquire the means and utilize all we are and all we have at our disposal not for ourselves but to serve others.

So, precisely, the Word of the Cross is a Word of Life, for in the final reckoning, the men who serve humanity most effectively, are not those who think only of satisfying their own egoism, but rather those who by renouncing themselves, spend themselves in service to others. Any concern for comfort, advancement, or temporal gain, thrusts man into mediocrity, leads to complacency and makes him seek the easiest solutions and choose what is the least painful and the least costly! On the other hand, the spirit of sacrifice which the Christian finds in his contemplation of his crucified Master renders him audacious, courageous and capable of any heroism. Thanks to this humanity is saved from sinking into universal corruption, from losing itself in venal beastiality.

So be it!

Good Friday Meditation

Reverend Pierre Maury, P.C.N., D.Th.
Pastor of the Reformed Church of the "Annonciation,"
Passy-Paris, France

One of the leading French theologians and preachers of contemporary France is Pierre Maury, pastor of the famous Reformed Church of the "Annonciation" in Passy-Paris. His preaching is sought by thoughtful people and his congregations are made up of French people who seek the best in the French Protestant tradition.

Pierre Maury was born in Nîmes (Gard) in 1890, studied at the secondary school at Montauban, the Sorbonne, 1907–09, with a stay in Berlin, the University of Paris 1909–10, took Theology at Montauban 1910–13 (a stay in Harvard 1913), and received the Doctor of Theology, honoris causa, *from St. Andrews, in June, 1946.*

Dr. Maury has had a brilliant career in the Church. In 1919–25 he was secretary of the French Students Federation of Christian Associations, Paris. Director of Semeur; Pastor at Ferney-Voltaire (Ain) from 1925 to 1930; he was consecrated on May 30, 1926, at Ferney. Then he went to Geneva to be secretary of the World Students Federation of Christian Associations, 1930–34. He was Director of "Foi et Vie," 1930–45. He has been Pastor of the Reformed Church of the "Annonciation" (Passy-Paris) since October 1, 1934, and is in charge of Teaching the Reformed Dogma in the Faculty of Theology in Paris. In June, 1950, he became president of the National Council of the Reformed Church of France at the National Synod at Nîmes. He is one of France's leading Barthians.

His principal publications are Three Spiritual Stories; St. Augustin, Luther, Pascal; The Great Work of God (*Careme 1937*), Je Sers, *translated into German;* The Holy Lord's Supper in the Sacraments (*Je Sers*); Jesus-Christ, *the* Unknown (*Careme 1948*), *Oberlin, translated into German and Dutch;* The Protestant Position (*Plon*); The Sermons d'Advent, 1940 on Essay XI, *published in German. He is the translator of Karl Barth's* The Word of God and the Word of Man (*Je Sers, 1933*) *and translator of Thurneysen* Dostoievski or Man's Limitations (*Je Sers, 1934*).

He earned several military titles in World War I, started as a motorcyclist, finished as Second Lieutenant in Aviation, and won the Croix de Guerre with bronze star and a palm. In the second World War, he held the rank of captain. He was made a Chevalier of the Légion d'Honneur, 1933, and an Officer of the Légion d'Honneur in 1951.

Sermon Nineteen

TEXT: . . . you were ransomed from the futile ways inherited from your fathers . . . with the precious blood of Christ.

1 PETER 1:19 (Rev. Stand.)

IN THE MIDST of the joys or the sorrows of life, in the web of our days filled with things that are useful, exacting, or bothersome, we are at times seized by a strange anxiety. It is as though a hand were suddenly placed upon our shoulder, stopping us. It may happen when death comes to a very close friend, or when we are informed that we are suffering from a very serious ailment, or it may be when some friend causes great sorrow to another. It may also come, simply, when a thought arises out of the very depths of self. We are confronted, irritatingly, exasperatingly, with the sudden question: "What's the use? What use am I?" Our very existence is questioned, our presence among those who surround us, our relationship to those who have gone before and to those who come after.

Quite instinctively we answer: "I'm not useless, rather I'm needed by those who are dependent upon me. I am certainly useful to those causes to which I have given myself." Then what happens? When trials suddenly come those who seemed to depend upon us forget our usefulness. We are deeply grieved and upset about it. Or it may happen that these causes for which we were working and were willing to give our enthusiasm collapse, and we are again seized by this sense of uselessness. If, on the other hand, that particular endeavor meets with success there is really little difference for we discover that success in its brevity is just as disheartening as failure. And together with Ecclesiastes we say: "Vanity, all is vanity."

Why so? Because death is inescapable. All men are, as their name implies, mortals. And death makes all things useless. At least that is what men think and say. Some put it bluntly in everyday speech: "I've caught on! I've caught on to one thing, and that is that life is useless."

Are we the only ones who claim that life is not worth living? Have not all pastors, standing by open graves, repeated Paul's statement: "If Christ be not raised, your faith is vain, if in this life only we have hope . . . we are of all men most miserable! Let us eat and drink, for tomorrow we die!"[1] All that dies is vain.

All is vanity. God says so too. Yet, at the same time He says more. He proclaims: "God has ransomed your life from the futile ways inherited (it's

[1] 1 Corinthians 15:17, 19, 32

not your fault that you are that way) from your fathers. He has ransomed your life from its futility.

He has *ransomed* it. The word has become trite in Christian speech; it is used in season and out of season. If to some the word is not trite it is either revolting, or mysterious, and there are those who ask: "To whom could God pay the ransom about which Jesus spoke in that memorable phrase: The Son of man came to give His life a ransom."[2] When Paul uses the word, ransom, he is thinking of the ransoming of a slave.

God is like somebody walking past a slave camp, today we would think of a concentration camp. As God looks He sees men behind barbed wires, living futile, worthless lives; to this He will not consent. It may well be that these slaves are guilty, and not merely born in slavery. They may have been reduced to slavery because of theft, like the slave Paul mentions in his letter to Philemon. Or perhaps they are innocent slaves, prisoners of war. . . .

One thing is certain, God does not consent to men living futile lives in the slavery of a concentration camp. So he pays. He must pay. As God He wants men to be free.

So God pays. But what price?

The Apostle says that God pays a price which is above human calculation. God could, of course, have paid with silver or gold. He is master of all silver and all gold, of all the treasures of the world. These treasures of the world, however, cannot pay the ransom for a man living in the concentration camp that is his own life and death. All the treasures of the world eventually end up by being devaluated by death. God paid with Himself, with that which is priceless, both to God and men. That is the ransom paid for our life and the lives of all those imprisoned in this horrible camp.

Is all this pious rhetoric? No, all this has happened to you. On a certain Friday in history the entire ransom price was paid with the life and blood of one man, the Son of God Himself. This has happened. It has happened to *us*. Before we knew about it, without our knowing it, perhaps even without our realizing that we were in a slave camp. It has happened. That is the Good News, the Gospel.

All else that is not related to this event is vanity, chasing after the wind. Henceforth your whole existence is wrapped up in this event. The Lord Jesus Christ himself the precious victim; more precious than all our church offerings and donations, more precious than anything in heaven or on earth. As a matter of fact it was heaven itself, the very heart of God that ransomed our lost life.

If that has happened—and it has—what are we to do about it? What does God expect us to do with our ransomed lives? The answer is simple, through this redemption. He wants to make us useful. Useful, to others,

[2] Matthew 20:28

so that we will no longer be men and women living next to each other, but for each other.

When we come to understand what has happened we will no longer, like the Ecclesiastes, see the lives of our neighbors in an unending circle of worlds moving toward an all-embracing nothingness. We will see them as beings bearing names who need our presence and whose presence we need; they will have really become neighbors, *near* [3] to us. That is the new usefulness for which our lives have been ransomed by the death of Jesus Christ.

God, however, does not only tell us that we are to be a neighbor to someone. That is secondary in a life ransomed by Jesus Christ. God speaks about another usefulness. We are to be useful to God. He has paid such a high price because He wants our life for Himself.

True, He did not need our life in order to be God. He was not unhappy when we were not in existence. And yet, mysteriously, somehow He desired the need for our love. For He loves us. And now he expects of us that we give Him this ransomed life, ransomed from all slavery and futility. He wants our life, our person. He is longingly waiting for us, giving us a *rendez-vous*. He wants to meet us in our offerings, in our prayers, in the things we undertake. And so we become useful to God! "Pleasing God." What a wonderful expression! To be able to give pleasure to God, to give Him the joy of our repentance. The kind of joy about which Jesus spoke when He said the angels of heaven rejoice when one poor little earthling, guilty and rebellious, repents and comes to God with a cry for pardon.[4]

Not only is there the joy which comes to God from our repentance, there is also the joy that comes from the simple existence of our love for Him, and our prayers to Him.

There is also the joy of our daily worship, and the joy which may be caused tomorrow when, called upon to choose we choose that which will be pleasing to God. There are the joys about which our Confession of Sins speaks when it refers to "fruits of justice and saintliness." [5] There is the joy of being present at His Holy Table where He invites and welcomes us.

To know that our ransomed life can be put to such great usefulness thrills us and gives us hope, even when we may be moving toward the grave. *Our hope is in the living God.* Yes, but our hope is not in the living God as long as we do not agree to be ransomed by Him, and made useful.

[3] *Prochain* in French, meaning neighbor, comes from *proche,* near. Translator's note.
[4] Luke 15:7
[5] *A Confession of Sins,* by Théodore de Bèze. Used every Sunday in the Reformed Churches in France. Translator's note.

The Inescapable Cross

REVEREND WALTER N. ROBERTS, PH.D., D.D.
A Minister of the Evangelical-United Brethren Church, and President of Bonebrake Theological Seminary, Dayton, Ohio

Dr. Roberts has divided his life and professional career between Ohio and the Philippine Islands. Born in Lewisburg, Ohio, in 1898, he attended Otterbein College, studied for his divinity degree at Bonebrake Theological Seminary, did research at Yale Divinity School, and took his Ph.D. at Hartford Seminary Foundation (Kennedy School of Missions) in 1932. In 1939 Lebanon Valley College conferred the honorary Doctor of Divinity degree upon him.

He was ordained a minister of the Church of the United Brethren in Christ in 1924, and went as a missionary to the Philippine Islands from 1925 until 1930; he was superintendent of the mission the last three years he was there, and was also professor of philosophy and religions in Union Seminary, Manila, from 1928 to 1930. In 1939 he was acting executive secretary of the National Christian Council of the Philippine Islands.

He returned to the United States to become pastor of Fort McKinley Church in Dayton (1932–38), and was part-time professor of practical theology at Bonebrake Seminary 1935–38 and has been president of the Seminary since 1938. In keeping with the modern trend to keep one's studies up to date, Dr. Roberts took a year at Union Theological Seminary, New York, 1946–47. He is a member of the general board of Christian education of the Evangelical-United Brethren Church and represented his church at the first general assembly of the World Council of Churches at Amsterdam in 1948. In 1936 he wrote The Filipino Church.

In this sermon for Good Friday, Dr. Roberts discusses the age-old problem of the necessity of Christ to die on the cross, a problem which has vexed mankind for nearly twenty centuries. The sermon was also published in "The Pulpit."

Sermon Twenty

Text: If thou be the Son of God, come down from the cross. . . . He saved others; himself he cannot save. Matthew 27:40, 42

A PERSON CANNOT preach on the cross without feeling his utter unworthiness and tragic inability to fathom its deep significance. Yet, each year, as we enter into the Lenten season, we feel constrained to try again.

The religious leaders of the time expressed in taunting and scornful words what many of us would like to ask: Why must the Son of God die on a cross? Could this God-man have saved himself from the cross? Was the cross necessary? Is there something in the moral structure of the universe that makes it inescapable?

These questions are not only contemporary; they have been asked from the days of the early Church. St. Paul expressed the attitude of Jew and Greek toward the cross when he said, "We preach Christ crucified, unto the Jews a stumbling block, and unto the Greeks foolishness."

One of the major problems in Christian theology, from the days of Paul to our own time, has been this question of the meaning of the cross. It seems to be contrary to all the customary ways of thought—utterly unreasonable and foolish!

On the other hand, when once the Christian gets a glimpse into the meaning of the cross he sees how utterly inescapable it is. Like Paul, he sees in it a demonstration of the wisdom and power of God.

The cross was necessary to break the power of sin. The twenty-seventh chapter of Matthew tells the story of the growing mobilization of the forces of sin: A motley crowd that included some of the best people of the times, innocent bystanders like Simon of Cyrene, the Roman centurion, who said, "Truly this was the Son of God." On the other hand, there were those of the populace and the ecclesiastical authorities who indulged in the bitterest taunts and cruelest words conceivable. They had heard the accusations against Jesus before the Sanhedrin. While he was hanging there on the cross, they sank to their lowest and hurled insults into his helpless, bleeding face: "Aha! You who would destroy the temple and build it in three days, save yourself, and come down from the cross! . . . He saved others; himself he cannot save."

In that thoughtful little book entitled *The Meaning of the Cross,* Henry Sloane Coffin calls attention to the people who had a share in crucifying Jesus. There were, first and foremost, the religious leaders, the Pharisees. Another group were the Sadducees, who inherited a lucrative commercial privilege by controlling the Temple area. Pilate, the representative of the imperialistic government, had his share. There was also Herod Antipas, a man of the gay world. Judas, a disillusioned idealist, played his criminal part. The crowd was there. A guard of soldiers was at hand.

After this survey of the people who had a part in bringing Christ to the cross, Dr. Coffin makes the comment: "This was the world which executed the Life subsequent generations until this hour revere as the best earth has seen. And plainly it is the world in which we still live. All these forces are present and active in our society—religious intolerance, commercial privilege, political expediency, pleasure-loving irresponsibility, unfaithfulness, the mob spirit, militarism, public apathy." [1]

These are the forces, in every age, that have sought to kill God and nail his Son on a tree. Wickedness is a powerful, dynamic force. Sin has a death-grapple hold upon man. These people were not the most wicked people who ever lived. They were some of the most respected people. Yet, in people of this type, sin may even dare to challenge God. Whoever raises a determined voice against such forces will hear the echo, "Away with such a fellow from the earth! Release unto us Barabbas!" It is a sobering fact that the sin of good people may do such deadly things.

In all history, whenever men have thrown their whole lives against such forces of sin and wickedness, there has been a cross and a crucifixion. Yes, even yet, whenever men dare to oppose sin and wickedness, sooner or later they find their Golgotha. The forces of sin are still determined.

It took a Christ and a cross to break the power of sin. Somehow the situation has been changed. Sin is still powerful, but there was One whom it did not conquer. In his victory the power and dominion of sin were broken. He alone could do it, but it led him inevitably to the cross.

There was no other good enough
To pay the price of sin;
He only could unlock the gate
Of heaven and let us in.

We do not like to think of evil as having such power, but there is no other way to account for the world's supreme tragedy. It took the Christ on a cross to break the power of sin, for the cross was inescapable even for the Son of God. It was unbelievable to the mob that the Son of God, the Messiah, should suffer and die. They did not believe he was the Son of God. That is why they taunted him as they did: "If thou be the Son of God, come down from the cross." Those Jewish leaders, chief priests, scribes and elders, who had heard the accusations made against Jesus in the Sanhedrin took up one of the supposed claims. While they did not believe that he had saved people, they knew that others thought he had. It was so utterly inconceivable to them that one who had the power to save others

[1] Henry Sloane Coffin, *The Meaning of the Cross* (Charles Scribner's Sons). Used by permission of the publisher.

would not first of all save himself. Hence their jest: "He saved others; himself he cannot save." You can almost hear them chuckle among themselves at this remark. How ridiculous in their eyes he seemed to be!

These jesting ecclesiastical leaders, all unbeknownst to themselves, had spoken a profound truth. "He saved others; himself he cannot save"—because he came to fulfill the plan of God the Father. If he had saved himself he would have defeated the eternal purpose he was determined to carry out. He wanted to live, but in the choice between life with compromise and death with loyalty, he chose to be loyal to his Father, even though it meant the shameful cross.

Our easy-going religious fancy prompts us to imagine how much better it would have been if Jesus had lived to a ripe old age; if he could have trained thousands of apostles instead of twelve and sent them out to all parts of the world; if he had continued his ministry of healing, of teaching and preaching, and had thereby blessed untold numbers. We would like to picture what he could have done if he had followed *our plan.*

Some would even venture still further in religious imagination. Why did he go to Jerusalem at all? Why didn't he withdraw to a quiet place and continue his life work where he would not have to encounter the opposition of the religious leaders at the nation's capital? Surprisingly, we are not the first to conceive such a plan. Peter suggested it. No doubt Judas had some such plan in mind, and likewise the other disciples. But Jesus was determined to go to Jerusalem. Any suggestion of a contrary course was, as far as he was concerned, the tempting of the devil. Notice how sharp was his reply to Peter's suggestion that he should not taste death: "Get thee behind me, Satan: thou art an offence unto me: for thou savourest not the things that be of God, but those that be of man."

Jesus went to Jerusalem perfectly aware of the issues and dangers ahead. He was confident that it was his Father's will that he go. For him to pursue any other course would be utter disloyalty to God. He followed his convictions. He went to Jerusalem. It was a showdown between Jesus with all that he was and taught as over against the Jewish leaders and their ecclesiastical system. If the struggle in Jerusalem ended in death for him, as he foretold it would, he was confident that his death would not defeat his cause. He was sure that a coward's attitude was unthinkable and intolerable. If the cross was inevitable and inescapable he was ready for it.

When he arrived in Jerusalem he was more persuaded than ever that he was doing the Father's will. With this conviction was another just as sure—namely, sure death lay ahead. He did not go to Jerusalem to die. Jerusalem might repent. He went to deliver God's message, regardless of the outcome for himself. But before long there was little doubt what the results would be.

As one looks back over the stirring days of that first Lenten period and views the life and work of Jesus in its long perspective, he is led to the conclusion so well stated by the late W. E. Orchard: "The principles he adopted for his life led inevitably to his death, and his death revealed and summed up the purpose of his life."

We have seen that the cross was necessary to break the power of sin and that it was inescapable even for the Son of God. But we must not fail to apprehend its personal implications for ourselves.

The cross is essential to our salvation. We have heard this so much that it has become trite.

H. R. MacIntosh said: "The great reason why we fail to understand Calvary is not merely that we are not profound enough, it is that we are not good enough. It is because we are such strangers to sacrifice that God's sacrifice leaves us bewildered." It is a sober fact that we cannot understand the meaning of the cross until we experience it in our own living. Is it possible that we cannot begin to grasp what Christ has done for us until we in a Christlike spirit "take up his cross," and follow him? That seems to be precisely what the gospel implies.

What Christ has done "once and for all" for all men, must be reënacted in each human life, if one is to have life and have it abundantly. To believe on Christ, to accept him as one's personal Savior and Lord, is to die to self and live for God regardless of the consequences. To live the Christian life means that one must "seek first the Kingdom of God." There is no possibility of becoming a Christian without this crucifixion of the self. The cross is essential to our personal salvation. The Christian gospel has stressed this fact in many different ways. When Peter preached on the day of Pentecost, he said: "Repent, and be baptized every one of you in the name of Jesus Christ for the remission of sins, and ye shall receive the gift of the Holy Spirit." They repented, were baptized and they received the gift of the Spirit. When the Philippian jailer asked, "What must I do to be saved?" Paul answered: "Believe on the Lord Jesus Christ, and thou shalt be saved."

Central in apostolic preaching was the doctrine that Christ died for our sins. Along with that is the condition that we must believe and accept to be saved.

The Church has held the cross in a central position in its theology from apostolic times to the present day. We are not saved until we have lost our lives in Christ, through an act of complete self-giving.

Not only is the cross essential to our salvation, but the vicarious way of life is of the very essence of Christian living. Every life has a cross and must be lived vicariously.

It is easier to talk about the cross and sing about it and think of it in semi-magical terms than it is to live it. A leader of the labor movement, who knew what it was to bear a cross for that cause, declared: "Years ago I recognized my kinship with all human beings, and I made up my mind that I was not one whit better than the meanest of the earth. I said then, and I say now, that while there is a lower class I am of it, while there is a criminal class I am of it, while there is a soul in prison I am not free." We may not agree with his political philosophy but one can be sure that this vicarious spirit has a remarkable similarity to the highest and best in the Christian life.

In one of his books E. Stanley Jones illustrates the vicariousness of our own physical bodies in these words: "Let an infection come upon the arm—does Nature say, 'There is nothing can be done—once infected always infected'? On the contrary, the healing, forgiving, atoning processes set in at once. I say 'atoning,' and I mean it, for the white corpuscles rush with reckless sacrifice and die by the millions, that the infection may be repelled and the rest of the body saved. The pus that comes off is the corpses of the white corpuscles which have sacrificed their lives for us. So the Atonement was not merely on a Hill two thousand years ago—it is in your very blood—therefore inescapable."

It is only as we enter into the vicarious way of life and take upon ourselves Christ's sufferings that we begin to apprehend the meaning of the cross. Someone has said: "Our great task in reference to Jesus is to understand him. We understand him insofar as we are able and willing to share his experience so that we see from within his own desires and purposes." Perhaps that is what it means "to know Jesus."

The cross is essential to our salvation. It was inescapable even for Christ. It was necessary to break the power of sin. In the deepest sense one must conclude that those who taunted Christ on the cross were utterly wrong. He did save himself by giving himself completely to God. It is the only way in the world that Christ, or you or I, can be saved; "For whosoever would save his life shall lose it: and whosoever shall lose his life for my sake shall find it."

An American church leader was in Japan several years ago at Easter time. He preached a sermon on the sufferings and love of Christ. After the service a Japanese woman came to him and said: "To live is to suffer. To live supremely is to suffer supremely. There must be someone in the world who lives supremely because he suffers supremely." That Someone is a loving God who revealed his love on a cross.

There is a green hill far away,
  Without a city wall,
Where the dear Lord was crucified,
  Who died to save us all.

May our hearts respond to his love!

Love so amazing, so divine,
  Demands my soul, my life, my all.

Gamblers at the Cross

Bishop Gerald Kennedy, Ph.D.
Resident Bishop of the Portland Area of the Methodist Church, Portland, Oregon

Bishop Kennedy is one of the dynamic young preaching bishops of the United States. He loves to preach and considers it the great function of the pastor-preacher although he has always taken care of the other duties in his own pastorates. He urges the ministers in his conferences to be both good pastors and excellent preachers. He made a world tour recently to study the needs and work of the Methodist Church in foreign fields. His preaching has a rare combination of spiritual insight, scholarship, courage, vision, and faith in Christ the Savior. Beginning his preaching as minister of First Congregational Church, Collinsville, Connecticut, 1932–36, Dr. Kennedy was called to become minister of Calvary Methodist Church in San Jose, California, in 1936, and went to First Methodist Church, Palo Alto, in 1940, at the same time becoming director of the Wesley Foundation at Stanford University. During 1938 to 1942 he was acting professor of homiletics at the Pacific School of Religion, and was called to be minister of St. Paul Methodist Church, Lincoln, Nebraska, in 1942. The same year he was made lecturer in religion at Nebraska Wesleyan University.

In 1948 he was elected Bishop of the Portland Area of the Methodist Church, which includes Oregon, Washington, Idaho and Alaska. His election was the first to reach across jurisdictional boundaries since the unification of the three main branches of the Methodist Church, and he is the youngest bishop in this denomination. His great success as pastor, preacher and administrator drew such favorable attention to his work that it was inevitable that he should be selected for this high office in his church.

While he was pastor of St. Paul Methodist Church, he became known as the radio preacher on The Methodist Hour in that section and as the speaker every Sunday on the voice of St. Paul's. On Wednesday evenings he gave the popular radio program, "Adventures Along the Book-Shelf." He was educated at the College of the Pacific, the Pacific School of Religion, and took his Ph.D. at Hartford Theological Seminary in 1934.

He is a contributor to national religious journals and is the author of The Pause for Reflection (*sermons*), His Word Through Preaching (*a fine handbook about preaching*), The Lion and the Lamb (*a study of the parodoxes of the Gospel given on the Peyton Lectureship at Southern Methodist University in 1949*). *He is the editor of* The Best of John Henry Jowett *(sermons). He has written the first of a series of books for the church to be called the "Know Your Faith" series; his volume is entitled* I Believe.

This sermon is an excellent example of his ability to preach on a subject that will attract attention and to give the mind and soul of his hearers food to take home. It was published in "The Pulpit" in August, 1951, and is used by their permission.

Sermon Twenty-one

TEXT: And they crucify him, and part his garments among them, casting lots upon them, what each man should take. MARK 15:24

I WANT TO speak in favor of gambling, and it is my hope that some of those listening to me may be encouraged to become gamblers. My thought springs from the scene of the crucifixion, so let us turn back to that tragic event in which Christianity finds the source of its power and the clue to its truth. You remember the scene. At the foot of the cross the soldiers wrangle over the clothes left by the man being killed. They are not very valuable and there are not many of them, for he was a poor man, but someone might as well have them. There is no point in cutting them up and dividing them. Why not gamble for them and thus provide some amusement and let somebody profit from the deal? So they cast lots for the garments of Jesus.

But in contrast to this cheap, heartless, tawdry scene, the strange man on the cross is making the heroic gamble. He is betting his life that love is stronger than hate, that life will conquer death, that God seeks and forgives. It will be too bad if we ever lose sight of the risk he took at Calvary, for the stakes were never higher and the daring courage shown never greater. When you think about men running risks for higher stakes, think of Jesus.

Let us begin by saying that gambling is a part of human nature—indeed, an inevitable part which can never be eliminated. This sounds to many people like a shocking thing for a minister to say, because gambling is associated with vice and racketeers. It has come to mean an ignoble attempt to get something for nothing and a preying on the weakness of human nature. It means broken homes and ruined lives. That, of course, is precisely what it does mean when this human tendency is directed toward cheap goals. America is indulging in this vice more than it ever has in all its history, and Americans are spending about thirty billion dollars a year in this activity. It is estimated that forty-five percent of the adult population participates in this racket. There will be increased pressure upon us to legalize gambling as taxes get higher, for many people will see an easy way to pay national bills without recognizing that in this direction lie national corruption and moral débâcle. There are tremendous profits in degrading people and in corrupting governments, and in these rotten endeavors the gambler is always a key figure. Those soldiers at the foot of the cross symbolize an ancient human sickness and the perversion of a human impulse.

But this is a perversion, for the greatness of humanity lies in its irresistible impulse to risk something for unprovable goals. Every great achievement was made possible by men who dared to commit their lives and fortunes to an enterprise they thought worthy to have a legitimate claim on their final devotion. Think of all the lonely men who have stood against the mob, risking their lives for the sake of righteousness and the future. If you want a real picture of gambling, look at the man on the cross and not at the poor fellows beneath it.

Studdert-Kennedy understood this as well as any man, and he wrote:

> How do I know that God is good? I don't.
> I gamble like a man. I bet my life
> Upon one side in life's great war. I must,
> I can't stand out. I must take sides. The man
> Who is neutral in this fight is not
> A man. He's bulk and body without breath,
> Cold leg of lamb without mint sauce. A fool.
> He makes me sick. Good Lord! Weak tea!
> Cold slops! [1]

If it is shocking for Christians to hear that man is a born gambler, how much more shocking to hear a minister say that God is a gambler too! You can hardly escape this conclusion, however, if you choose the God of the Bible over against the God of philosophy. The philosopher prefers to believe that God is a cold principle, objective, mechanical and without personality or emotion. T. H. Huxley portrays the human relationship with such a God as a chess game, in which our invisible opponent is fair and just, but never overlooks a mistake. The one who plays well is rewarded but "one who plays ill is checkmated without haste, but without remorse."

There are some Christians who seem to prefer this concept of God over the biblical interpretation. They picture the Atonement as if it were a cold, legal deal, worked out by an unmerciful judge. They seem to believe that God is caught in his own laws and so he must insist on the best man who ever lived dying in agony to satisfy his demands. Let us make what may seem a foolish suggestion. We ought to begin all our thinking about God by assuming that He is at least as good as we are. Such a simple assumption would help us escape much unchristian theology. I would not keep any man in endless

[1] *The Best of Studdert-Kennedy* (Harper, New York, 1948), p. 147. Used by permission of the publisher.

torment, no matter what he had done. I think he ought to have a taste of the suffering he meted out to others, but to roast eternally, as Jonathan Edwards intimated would be the fate of sinners, is certainly overdoing it. I would never send unbaptized babies to hell, but there are theologians who insist, even today, that God does it. This is to make God more vindictive than men and utterly unworthy of our worship. The Father of our Lord Jesus Christ does not fit such a pattern.

There are Christians who think history is a closed system, foreordained, predestined, with no room left in it for freedom. So they draw their charts showing that the end of the world is coming this year or next, and proving that some contemporary man or nation or institution is the anti-Christ. Once again, the God of the Bible, who is Lord of history, is not like that. He breaks into the human situation in marvelous new ways and he leaves the door open always for the generation and the man who will repent. God has made history an open door, and freedom is an essential part of its structure.

Once you take a mechanical viewpoint, then you are ready to insist, as a popular evangelist did, that we will eat on transparent gold plates in heaven, and there will be no labor unions there. It could be that this man has a source of information not available to the rest of us, but probably such a naïve picture of heaven is an inevitable result of an unintelligent, inadequate and in many ways unchristian theology. Such people have taken God's freedom from Him and have failed to comprehend that He too, being free, makes high ventures.

You do not find the Bible speaking of God as Superman, but you hear it speaking of a Father strong enough to become weak, good enough to share our sorrow, and loving enough to descend to our suffering. You see God daring to grant men liberty that they may become creatures in His image, capable of hearing His voice and speaking with Him. He knew that such a venture would mean that men could become more evil than the beasts, but He risks even that for the sake of creating free persons and great souls. His universe is not a machine, it is a dream, and whenever we call Him Father we are confessing our faith in our God's willingness to run grave risks.

Martin Luther talked so much about the clergy marrying that he was finally forced to practice what he preached. He married a nun, Katherine von Bora, and although it was no love match they were happy together. When a daughter arrived Luther wrote to a prospective godmother: "Dear lady, God has produced from me and my wife Katie a little heathen. We hope you will be willing to become her spiritual mother and help make her a Christian." [2] And God is like that. He takes us as we are—hesitating between the mud and the stars—and, like a father, he gambles on his children by giving them freedom.

[2] Roland Bainton, *Here I Stand* (Abingdon-Cokesbury, 1950), p. 293.

The cry of our day is for security, and we are overdoing it. Let it be made clear at once that this does not mean we should go back to a time when a poor man was at the mercy of the rich and the powerful. That is what too many people mean when they criticize security. But when a nation grown wealthy makes its foreign policy a desperate attempt to hold on to what it has, it is growing senile. When a society barters justice for stability, it is already on the decline. Someone said that an insurance salesman is a fine servant of society, but it is tragic if he becomes a symbol of society. The most insecure people in the world are the ones depending on some legality to hold their jobs. The only real security is of the spirit.

Now this brings us to the necessity of faith if we are to live with any dignity or meaning, and faith implies risk. No man can live ten minutes on the basis of what he can prove. The many dangers which face any man going home from church can be emphasized until we conclude that it is hazardous to start toward the door. We might slip on the stairs, a car might hit us, an earthquake might topple a wall on us, or a bomb might explode as we go by. But most of us assume we will get home and usually we do. Business lives on credit, which is to say, business lives on faith; and if American business had to operate on a cash basis it would be crippled. Every businessman has to risk continually, and no man can play a part in this most practical, hardheaded activity without being a man of faith.

Science lives on faith. We have talked much nonsense in our day about science's dealing only with what can be proved. Every scientific theory is a leap of faith, a gamble, and the so-called laws of science are generalities based on a certain number of experiments and hardly ever final. When it comes to the ultimate nature of the universe, science, like everything else, has to move out beyond the facts.

Consider the men who are the great heroes of the Bible. They are not always morally perfect, for even the greatest of them were guilty of moral breaks and disgusting sins. They were not the safe, respectable men who valued security above all else. They were the men of faith, the gamblers for high stakes—justice, mercy, righteousness. They would hazard the loss of their lives for God's sake, and the greatest of them all was the Galilean who dared to believe that his death would be redemptive.

Thoreau's essay "On the Duty of Civil Disobedience," a document looked upon with some suspicion by the professional patriots of our day, says:

> I saw that if there was a wall of stone between me and my townsmen, there was a still more difficult one to climb or break through before they could get to be as free as I was. I did not feel for a moment confined, and the walls seemed a great waste of stone and mortar. . . . As they could not reach me, they had resolved to punish my body. . . . I saw that the state was half-witted, that it was timid as a lone woman with her silver spoons, and that it did not know its friends from its foes, and I lost all my remaining respect for it and pitied it.

So much of the craven fear of our time is slavery. How terrible is the need of our civilization for the great men of faith who will take the risk of obeying God rather than men. And I believe with all my heart that our civilization cannot be saved without men who will risk their lives for the preservation of personal and spiritual freedom. In this shabby period, God send us men of faith, which is to say, gamblers for the human spirit.

The world was never in greater need of men who will gamble for Christ and His Kingdom. In a day when we suffer a pathological fear not only of our enemies but of our own brethren if they are slightly nonconformist, we are in desperate need of confidence. In this crucial time when we seem to distrust the democratic way and foolishly put our trust in military might, we need encouragement to risk something for God.

Christians ought to be willing to gamble for brotherhood, and theoretically we are. We have no difficulty in loving humanity, but it is often beyond us to love individuals. The idea that prejudice and hatred arise only out of ignorance is nonsense. They arise out of our irritations with people who live too close to us. Dostoievsky confessed that within twenty-four hours he began to hate the best of men. Brotherhood is a risky thing to believe in if you really expect to begin practicing it on your neighbor. We can believe in it in the Constitution and the Bill of Rights, but when a civil rights bill is put before the voters to make discrimination unlawful in a particular city, we are not so sure about it.

Across the world and in our own nation men aré rising up and demanding equality of opportunity. The safe and secure way to deal with this movement, according to the champions of the status quo, is to crush it if we can. In the Orient our policy sometimes seems to be to support any reactionary regime which screams loudly that it is against communism. What would happen if America should say, "We were born in a revolution, too, and we are on the side of all who hunger and thirst for a more adequate life. Let us give you a hand. We are gambling on the kind of world where all men are regarded as brothers"? Would this not be the real sentiment of a Christian people, and isn't it worth risking all it will cost?

Paul Geren, in his *Burma Diary,* tells about the American ambulance driver who was coming out of Burma heavily loaded with men. When the procession was halted, some wounded Chinese soldiers climbed all over the vehicle because it was their one and only chance to get away before the Japanese arrived. But the order had been given that no others could be taken and the driver got out and pushed the wounded men off the ambulance. Later he decided to go back to China, because, as he said after relating this experience, "I owe the Chinese a debt." Perhaps we owe all men a debt, and in taking the risks involved in paying that debt we might find our souls again.

We ought to be gamblers for truth. Dr. Samuel Johnson would never allow his servant to say he was out just because he did not want to be disturbed. He told Boswell that, while a philosopher might be able to distinguish this from lying, he did not want to damage his servant's sense of the truth. Not many of us go that far. Dr. Wallace Hamilton said that a recent book full of lies was excused on the ground that it was lying in a good cause. Too much of our propaganda is spoken with the same excuse. We cannot get back to any firm foundation for our social life until we are willing to bet that truth is the only way. What a difference it would make in our legal affairs, for example, if every lawyer should assume, as Gandhi did, that his duty is not to prove the guilty innocent but to help the court arrive at the truth.

How truthful and honest was Jesus! He never once appealed to the lower part of man's nature and he never presented his case on the basis of an unworthy motive. When he called his disciples, he told them the truth about the hardships and the dangers. Yet he believed in men and bet his life on their hunger for goodness which would be decisive soon or late. Tempted to lose faith in men, I turn back to Jesus and find that faith restored. But even more wonderful than that, when I lose faith in myself He looks at me with understanding eyes and, in spite of knowing all about my betrayals, I know He still believes in me. Perhaps this is his greatest miracle. Are we willing to take our place by His side and look at men through His eyes? Perhaps if we would gamble on truth, as He did, we too could be redeemed by His faith in us and in all men.

Wanted: Christian Gamblers who will not sit by the campfire but will push out into the social underbrush where it is dangerous to go and where a man has to risk his life. Wanted: Christian Gamblers who will scorn taking cheap chances for the flashy impermanent rewards of the world and will bet their lives on God and His purposes. Wanted: Christians who will gamble that Jesus is right and has a claim to their complete allegiance.

Easter

THE MOST REVEREND FULTON J. SHEEN, PH.D., D.D.
Titular Bishop of Cæsariana and Auxiliary Bishop of New York
National Director of the Society for the Propagation of the Faith
New York City

For years Bishop Sheen has been one of the outstanding preachers of the United States; his is one of the important living voices of our day. His forceful and convincing preaching makes every listener eager to catch each word. To hear him is to realize that here is a man who believes what he says and who has thought out his message calmly and prayerfully. Year by year his following increases and the sphere of his influence becomes larger.

After graduate work at the Catholic University of America, the University of Louvain, Belgium, and Angelico University, Rome, he was ordained in Peoria, Illinois, in 1919. Step by step he has risen from very modest places in the Church to be one of the most honored Catholic preachers. He taught at St. Edmund's College, Ware, and the Westminster (London) Diocesan Seminary in 1925–26; in 1926 the University of Louvain, recognizing his genius, awarded him the Cardinal Mercier prize for International Philosophy, the first time this honor was ever given to an American.

Before World II, he was called to preach in Europe nearly every summer from 1925 to 1939, speaking in London at Westminster Cathedral and St. Patrick's Church, Soho Square; at the University of Cambridge Summer School, at Glasgow, in Rome, and elsewhere on the continent. In 1934 he was named a Papal Chamberlain of the late Pontiff, Pope Pius XI, with the title of Very Reverend Monsignor, and the following year Pius XI made him a Domestic Prelate with the title of Right Reverend Monsignor. At a special service in Rome in June, 1951, he was consecrated Titular Bishop of Cæsariana, and appointed Auxiliary Bishop of New York.

For a number of years, he taught philosophy at The Catholic University of America in Washington, D.C., and was in such demand as a speaker that he gave more than one hundred sermons and addresses each year, speaking in almost every major city in the United States to secular and religious groups who thronged to hear him. For years he has also been the regular Lenten preacher on the Catholic Hour of the National Broadcasting Company and at St. Patrick's Cathedral, New York.

He has written some thirty books on philosophy, religion, morals and socio-economic questions, including Freedom Under God, Whence Come Wars,

Philosophies at War, Peace of Soul, Lift Up Your Heart, *and* Three to Get Married.

This Easter Sermon was given March 25, 1951, on the Catholic Hour over the National Broadcasting Company's network, under the sponsorship of the National Council of Catholic Men, Washington, D.C. The Bishop sees the trial and crucifixion of Jesus and its message for our day. What the world needs more than anything else, the Bishop points out, is faith in the Risen Lord of Life.

Sermon Twenty-two

THERE IS NOT the joy this Easter that there ought to be. The enemies of God are too optimistic, and the friends of God are too pessimistic.

The enemies of God are *too optimistic* because they believe that they have conquered. At the close of the last century, Nietzsche boasted: "God is dead." Since that time, the enemies of God have made a formidable progress. Thirty-seven out of every one hundred people in the world today are either pounded by the hammer, or cut by the sickle of atheistic communism. The Charter of the United Nations makes no mention of God and His Moral Law. The last international convention of one of the great world organizations given to humanitarian service cut out from its Preamble the mention of God. The enemies of God can boast that there are nine countries in which the Gospel of Christ may not be preached, so that Christ is recrucified, not in the three languages of Hebrew, Latin, and Greek, but in most of the languages of the world.

On the other hand, the friends of God are too pessimistic. Seeing 13,000 missionaries banished from China and their work of centuries undone; seeing Russia, the land that once was holy and is now made unholy by dictators who bomb their way to proletarian thrones; seeing Poland, which once was the Ireland of the East, now made a puppet whose strings are pulled by atheistic hands; seeing the Red octopus reach out its tentacles to minds, poisoning truth into a lie and calling darkness, light—the lovers of the Christ crucified, in their despair, are tempted to cry out: *Domine usquequo.*

What is all of this false optimism of the enemies of God and this *unwarranted despair* of His friends, but a repetition of what transpired in the last few days of the earthly life of Our Lord, when His enemies were too optimistic, and His friends too pessimistic?

The enemies of Our Blessed Lord were too optimistic! Thanks to mass propaganda and organized demonstrations before a Governor's palace, they convinced a time-saving politician that "we shall not have this man rule over us." When finally, they reduced Him to a common criminal, four taunts they hurled at Him on the Cross, and all of them were boasts about their victory and His absolute defeat. First, He said that He would destroy the Temple, and then rebuild it; but the Temple was still standing as a reproach to His boastfulness. Second, He saved others, but now could not save Himself. Third, He said He was a King, but He was proven to be a mock King with a crown of thorns for a diadem, a nail for a scepter, blood for royal purple, a crucifixion for a coronation. Finally, His claim that He was the Son of God was now a stupid lie, for if He was, why did not God deliver him?

When He is taken down from the Cross, Joseph of Arimathea boldly goes to Pilate to ask for the body of Our Lord. The Greek word which the Gospel says Joseph used was *Soma,* which is the word of respect for a body. Pilate was too optimistic that the power of Caesar would no longer be challenged, and he told Joseph that he would give him not the *soma,* but the *ptoma,* which means cadaver or rubbish. The final optimism of the enemies was the setting of the guards, not to prevent the Resurrection, but to prevent the apostles from stealing the body and saying He had risen from the dead. Finally, they rolled a great stone in front of His tomb. This was the final victory! He who had called Himself "the Rock," is now rock-bound in a tomb—never to rise again. Even before Nietzsche wrote his blasphemous lines, the enemies had scored their apparent victory; for God is dead.

On the other hand, the friends of Our Lord were too pessimistic and despairing. Though they had heard Our Lord say that He would rise from the dead after being in the belly of the earth for three days, they still did not believe. The women go to the grave on Easter morning with spices which they had prepared, not to greet a Risen Lord, but to anoint a dead body. Not in the least expecting the Resurrection, they ask: "Who will roll us back the stone from the door of the sepulchre?" Mary Magdalene herself, who had risen from the deadness of sin into the newness of Divine life, and who had heard Our Lord say that He was the Resurrection and the Life, comes also with spices and weeping, not with joy in anticipation of a Resurrection, but with sorrow, for the Beloved is dead.

When Magdalene finds the tomb empty, instead of believing that He has risen, she says to the Angel who asks her why she weeps: "Because they have taken away my Lord, and I know not where they have laid Him." When Our Lord finally appears before her in the garden, she does not even look up. Seeing a figure whom she mistakes for a gardener, she calls Our Lord "sir," and asks: "If thou hast taken Him hence, tell me where thou hast laid Him, and I will take Him away." She is not prepared to face one who conquers death, but rather to find the corpse and re-bury it. Finally Our Lord speaks to her, and she recognizes Him, saying now not "sir," but *Rabboni*—Master! She runs to tell Peter and John, saying: "I have seen the Lord." But they, hearing it, do not believe it, saying it is a woman's tale.

Easter afternoon, when Our Blessed Lord becomes the fellow traveler of

His disciples on the way to Emmaus, He finds them also downcast with despair, because they had hoped that it was He who would have redeemed Israel, but now it is three days since He is dead.

Seven days later, Thomas the Apostle, still refusing to believe the good news, says that he will not believe until he can put his fingers into His hands and his hand into His side. In that moment Our Lord appears: "Put in thy finger hither, and see My hands and bring hither thy hand and put it into My side and be not faithless, but believe." Apparently the one thing that the apostles and lovers of Our Blessed Lord were not expecting was His Resurrection from the dead, and when He appeared in their midst, He said to rebuke their fears: "Why are you troubled and why do doubts arise in your hearts?"

Well, indeed, may Our Lord say on this Easter Day to His friends: Why are you troubled in heart, despairing and cast down? Grant that there is persecution and darkness over the earth, did not the Master say that as they persecuted him, so would they persecute us? Have we lost the Christian virtue of hope? Why should our attitude be any different now than the attitude in the First Century? They too faced the world with misgivings; they looked upon it as something perishable, expecting momentarily the coming of Christ, Judgment, and the dissolution of the world. But they faced it bravely because, knowing that Christ had risen, they sought the things that were above.

But in our day, too many are seeking security, rather than the happiness of the Resurrection. We are like those who, taking an ocean journey, are more concerned with the life-belts than with the cabin; or traveling by air, are more interested in the parachute than the beauty of God's sky; or traveling the highway, are less happy about the ride than looking for first-aid stations. Rather with St. Paul, should we say: "If Christ is not risen, then we are of all men most miserable." Shall we believe that God reserves all the mourning for His sons and all the joys for His enemies? Are we condemned to hang our harps upon the willows, and sing nothing but doleful dirges, while the children of Satan are to laugh with gladness of heart?

No, rather we have received not the spirit of bondage to fear, but the spirit of adoption whereby we cry out: "Father." Fear not! Realize that He Who went into that grave is Truth Itself, and Truth crushed to earth will rise again. Dostoievsky tells us the story of two men looking at the painting of Holbein, "The Taking of Christ from the Cross." One said: "I like looking at that picture." The other said: "Some people's faith has been ruined by that picture." And right he was; that picture would ruin the faith of a materialist, an atheist, a communist, and for all who believe that there is nothing after this life. If there is no Resurrection, but Christ is dead, one cannot believe either in the Goodness of God or in the goodness of man. But if He Who took the worst the world had to offer and conquered it, then evil shall never be victorious again. Rejoice, for He that is dead now lives, and though the bells are tolling for the execution of the Church, the execution will be eternally postponed.

You, therefore, who believe in the Resurrection, be not disheartened. Remember that the Church, like Christ, is not so much a continuously living thing, as something that has survived a thousand crucifixions through a thousand Resurrections. Though the Iron Curtain be rolled down against the Gospel of Christ in Russia and the Bamboo Curtain against the Church in China, be assured that He Who split the Rock, and gave the earth the only serious wound it ever received, that of an empty tomb, will one day roll up the curtains like darkness before the light, and He Who you thought was dead will be walking with the wings of the morning.

Be not cast down at all the talk of the atom bomb, asking yourselves in despair: "Must we die?" But rather, in the light of the Resurrection, ask: "Must we be reborn?" Though the scientists steal an atom from the sun, split and fission it, as does the sun, remember that when the Lady of Fatima visited the earth, she brought the sun with her, clothing herself with it as a garment, to remind us that the sun and its fission belong to her and to life, not to the godless and to death.

When you hear of the diabolical wickedness of the men whose flag is red, because drenched in the blood of its victims; when you hear of those who martyr not only the body, but soul, and who make what we might call the dry martyrs, like the Mindszenty's and the Stepinac's and the Beran's, be assured that these broken bodies and minds will cry to Heaven for greater vengeance than the blood of Abel cried, and that a new day of hope will dawn, when these men will chant a requiem over the graves of those who won the battle, but lost the day.

God never permits an evil without good coming from it. Communism is an evil, but in the Providence of God, it may be the fertilizer of a new civilization, the death that is spread over the world in the winter of its discontent to prepare the dead earth to tell its secrets in flowers in the new springtime of the spirit.

It may very well be in this second millennium of Christian history, that the world is in the throes of a new birth, when the Christian message shall pass from the Western to the Eastern world. Within a short time the Mystical Christ, Who has been crucified, will take His bleeding Hands to the Japanese, who will lay their lotus flower upon them to change the wounds of hate into scars of love. To the Chinese, He will bring His bruised and torn body, and the halt, the lame, and blind, the hungry and the famished, will bring their healing hands and cover up from sight those imprints of a night forever past. To the people of India, He will show His open wounded side, and they who sought their peace in a Nirvana of unconsciousness, will be drowned at last in the Sacred Heart, in a love that is a healing of the soul. Finally, He Who went into darkness through a crown of thorns, will turn to Africa and the people of the midnight Madonnas, and they shall pluck out His thorns, and crown Him with blossoms and with flowers as white as their souls, as perfumed as their faith.

Despair not then, for your King was one who stumbled to His throne, who allows evil its hour, but wins the day. Say then with Paul:

"Who then shall separate us from the love of Christ? Shall tribulation? or distress? or famine or nakedness? or danger? or persecution? or the sword?

"For I am sure that neither death, nor life, nor angels, nor principalities, nor powers, nor things present, nor things to come, nor might,

"Nor height, nor depth, nor any other creature, shall be able to separate us from the love of God, which is in Christ Jesus Our Lord."

God Love You. Pray for our missionaries.

EASTER

The Birthday of Our Eternity

REVEREND LLOYD R. GILLMETT

Rector of the Church of Saint John the Evangelist, Protestant Episcopal, St. Paul, Minnesota

The Reverend Lloyd R. Gillmett has been rector of the Church of St. John the Evangelist ever since he returned from service in the United States Navy as a chaplain during World War II. In addition to serving as rector, he is secretary of the Standing Committee of the Diocese of Minnesota, a member of the Diocesan Council, examining chaplain, a member of the boards of Child Welfare, Council of St. Paul, Goodwill Industries, St. Luke's Hospital, and a member of the Board of Trustees of St. Mary's School, Faribault, Minnesota, and Seabury-Western Theological Seminary in Evanston, Illinois.

While he was a student at Western Reserve University he played football and participated in other athletics, was elected to Phi Beta Kappa and was graduated with his A.B. Degree, magna cum laude. *He received his B.D. from the Episcopal Theological School, Cambridge, Massachusetts,* cum laude, *and served as assistant at Emmanuel Church in Boston. After leaving Emmanuel Church he became Rector of St. Paul's Church, Duluth, Minnesota, and left there to become a chaplain in the United States Naval Reserve, serving in the States, in Hawaii, and on board attack transport which took part in the Okinawa campaign, and ended at Tokyo. After being released to inactive duty, he became Rector of St. John the Evangelist in January, 1946.*

This Easter sermon has a fresh approach to a great subject, is informal, conversational.

Sermon Twenty-three

TEXT: Dying, and, behold, we live. II CORINTHIANS 6:9

"THE BIRTHDAY of Our Eternity" is what some people have called our final going away from this earthly realm. St. Paul, for example, knew that death's door opens not to a dusty dungeon but into an eternal life for he said we die and "behold we live."

"Some day," said Dwight L. Moody, "some day, you will read in the newspapers that Dwight L. Moody of Northfield is dead. Don't you believe a word of it. At that moment I shall be more alive than I am now." There, you see, was a man who lived with the Easter Faith in his heart. When he got up in the morning it set him on his feet and in the evening, worn out by the hard work of the day, it put him to sleep in peace. That is what the Easter Faith does for a person, it gives him strength and peace and courage because, working and playing and living, he knows that come what may, as we die "behold we live."

That is our hope and that is our faith and that is our surety on this Easter morning and always. But I can see someone almost standing up in the pew and wanting to speak out, "What are the reasons for the faith that is in you?" Well, this morning I am going to tell you some of them.

For one thing, there is that longing in my heart and in your heart and in everybody's heart to go on living forever. Everybody has that longing. Napoleon once voiced it as he was looking at a valuable painting, and he asked the artist this question, "How long will that painting last?" "Perhaps it will last about eight hundred years," replied the artist. "Such a poor immortality!" exclaimed Napoleon. "I want to live not for eighty times eight hundred years, but forever." Well, you see, he longed for that continuance of life as you do, and I do, and all people do.

And I say that that longing is significant, most significant. It is a sign that death is merely a birthday into life. And why is that true? Because, so far as I am able to see and understand anything at all in this world, for every real longing which is universal there is a reality which meets and satisfies that longing. You long for and need food, and it is provided. You long for and need truth for your mind to grapple with and grow by, and truth is here in the world waiting for you to discover and understand it. You long for and need fellowship, and fellowship is available. For every real longing which is found universally in human hearts there is a reality to meet it and satisfy it, and that is at least one reason why death must be a birthday into life. God has provided a reality to meet and satisfy that longing for as we die "behold we live."

Another reason for the faith that is in me is that this life always remains incomplete. Work remains incomplete. If you have ever been in Spain you perhaps saw some of the paintings by the artist known by his nationality, El Greco. In one of the cathedrals there you can see his paintings of Apostles and Saints. Some of these works are merely bare sketches in outline. He expected to complete them but his life here on earth came to an end before he was able to do so. Bacon started an essay on the subject "Fame." He wrote only a paragraph or two and then at the end of the page he left a note saying, "The rest is unfinished." And only the other Saturday afternoon I happened to listen to the broadcast of an opera and during the intermission in which a quiz was held I learned something about that particular opera which I had not known before. Toscanini on one occasion when he was conducting a performance of the opera stopped the orchestra and dismissed the audience when the opera was about two-thirds of the way through. You see, the original composer of the opera had not been able to finish it before his life's journey had ended and it had been finished by someone else. So, Toscanini stopped the performance, dismissed the audience by saying, "Here is where the master laid down his pen." Work remains unfinished! It does not make any difference who you are or how many talents you have. You will come to the end of your life here on earth with only a portion of your work done and only some of your talents developed. Always unfinished, you see. And it is impossible to believe that God would begin such an amazing work which He has begun in you by bringing you into life and giving you a chance to begin to grow, without also making provision for your continuing development. Because of that I say we die and "behold we live."

Then also a continuing life is just the kind of wonder that we have to expect in this wonderful universe. There is something great and marvelous going on in this world, and I tell you that the ultimate goal toward which the whole creation is moving must be something wonderful too, greater than your mind and my mind and all people's minds can really understand and fathom. I remember well some words written by H. G. Wells, who certainly was not any sentimentalist nor conventionally religious person. He said, "All this world is heavy with the promise of greater things." Now that sounds like something. That is a statement of fact which ought to inspirit and compose your soul. "All this world is heavy with the promise of greater things." And when you think of this fact you see what he meant, namely, always the wonders of the world as they have become understood have turned out to be more wonderful than people have ever dreamed of. It wasn't long ago, you know, that people thought that they were living on a flat earth in a small universe, but now we know that this earth on which we are standing is part of the Milky Way, which is so big that it takes light seventy-eight thousand years to travel across it, and the Milky Way is just a speck of cosmic dust in all of creation. The universe is more marvelous than people ever dreamed of. And the small things are marvelous too. When the molecular theory of matter was first discovered people thought that they had hit upon something great, but little did they suspect that the atom was such a marvel in itself. You see, as soon as one door has been opened which reveals a wonder, immediately more doors stand before you waiting to be opened, doors behind which are additional and greater wonders. Always the wonder turns out to be greater than people expect or even dream of. "All this world is heavy with the promise of greater things." And a continuing life for you and for me and for every child of God is just the kind of a wonder that you ought to expect in this wonderful universe. I suspect that when your birthday comes you will be surprised and say, "Why, of course, of course, this is it." We die and "behold we live."

A birthday into eternity is promised also because of the fact that in science and in every realm of life nothing gets lost in this world. Not really! Our shortsightedness and the smallness of our minds sometime make us think otherwise, but it is not so. Nothing gets lost! You can take a sledge hammer to the statue Venus de Milo and you can smash it into a thousand pieces and then take the pieces and grind them into a powder but not a single particle of the statue at which people have gazed for years has been lost to the world. It is in another form but nevertheless it is not lost. That truth holds on wherever you go, and do you think that God would preserve a speck of dust and not preserve a human soul? I suppose it was the realization that nothing gets lost that led Alfred Whitehead, that giant of a philosopher at Harvard University, to the conclusion which he reached. "How can a person best describe God?" he asked himself. And he kept asking that question so long that ultimately he reached this conclusion—the best concept under which you can think of God is that of "a tender care that nothing be lost." Well, do you think that you are going to get lost? Of course not! Do you think that that little child you laid away in that never to be forgotten plot of land is going to be lost? Of course not! Do you think that even the downmost child of God will ever get lost? Of course not! God provides a birthday for all. For we die and "behold we live."

I expect that great birthday also because of a truth which was once expressed by William James. He said, "The best argument I know for an immortal life is the existence of a man who deserves one." Now that is saying something significant. "The best argument I know for an immortal life is the existence of a man who deserves one." It is quite obvious to anybody with any discernment of mind that if a soul is not worth preserving then it is hard to believe that that soul is immortal, but when you think of some people, at least, you know that they are worth preserving. When you think, for example, of that man whose statue is now brooding over our nation's capital, Abraham Lincoln, you know that he is worth preserving. When you think of Shakespeare, of his mighty mind and genius, you know that he is worth preserving. Or when you think of a simple, good, and transparent person such as St. Francis, you know that in him there is much which is worth preserving. But not only the great and well-known people are worth preserving, but also the large numbers of unknown and unrecognized souls who have left no memorials behind them, people who by their kindnesses and sacrifices and faithfulnesses have sustained and leavened all of life—when you think of such you know that they too are worth preserving. Yes, when you think of all such people, great and small alike, you cannot help but believe that they are still going on. What kind of a business do you think that God is in anyway? Do you think that He is something like that artist that I once saw over in London? On the banks of the Thames River there, I saw an artist drawing on the sidewalk in colored chalk the portraits of people. For a few shillings he made an excellent likeness of you and then during the night the tramping feet of pedestrians and the rain washed the pictures away. Well, do you think that that is the kind of business that God is in? Do you think that He brings human souls into being, gives them minds to think with, and hearts to love with, and hands to work with, and then lets the rains and the feet of time wash them away as if they had no value at all? That is incredible. Nothing gets lost. A great birthday awaits you. We die and "behold we live."

That is what the Disciples learned on that first Easter Day—that nothing gets lost, that death's a birthday into eternity. The realization of that great truth came to them after Good Friday. Who are these men walking along the street in Jerusalem with heads bowed? Oh yes, Peter and James and John. Why don't they speak? Why are they so downcast?

"Peter, what's the matter?"

"Jesus is dead."

"James, why so dejected?"

"Jesus is dead."

"John, why are you so hopeless?"

"Jesus is dead."

Right after Good Friday that is the answer that they gave and that is why they were so without hope and without purpose.

"Peter, where are you going?"

"I go afishing."

"James and John, where are you going?"

"We are going back to mending nets."

You see, that is where they were going, back to their old jobs. That is what the Disciples had decided to do, leave Jerusalem and take up their old work. Their dream had burst. Their Master and friend had been put to death and they were all through.

But you meet Peter and James and John three days after the crucifixion and what a change you see in them!

"Peter, why are you so glad?"

"Jesus lives."

"James, why are you so jubilant?"

"Jesus lives."

"John, why have you not gone back to your old job?"

"Jesus is alive."

You see, there was a complete change in the hearts and the minds of the Disciples. You may not be able to understand the stories of the empty tomb, but you simply cannot argue away or account for the change of mood in the Disciples without saying that in their experience they learned that Jesus was not dead but was alive!

And there is the promise for you and me and for every child of God in this world in which nothing gets lost. "Because I live, ye shall live also."

EASTER

Goings Out and Comings In

REVEREND JAMES GORDON GILKEY, D.D., LL.D.

Minister, South Congregational Church, Springfield, Massachusetts

Men and women find comfort and help in the sermons of Dr. James Gordon Gilkey. He combines religion, psychology and personal living into messages which assist those who are emotionally upset.

In 1916 he was ordained as a Presbyterian minister and served as assistant pastor of Bryn Mawr (Pennsylvania) Church in 1916 and 1917. Since that time he has been minister of South Congregational Church, Springfield, Massachusetts. This church has a membership of nearly eighteen hundred and he has one of the largest Sunday morning congregations in New England. Dr. Gilkey was one of the first ministers to preach on the radio, broadcasting his morning service regularly as far back as 1923.

Since 1927, he has published fourteen books dealing with problems in religious thought and everyday living. Among them are Solving Life's Everyday Problems, *which has been transcribed into Braille for the use of blind readers,* How to Be Your Best, God Will Help You, When Life Gets Hard, *and* Here Is Help for You.

During the summers preceding World War II he traveled widely, visiting places as remote as Australia and Russia.

In this Easter sermon Dr. Gilkey draws on his wide experience in the Christian faith and his knowledge of men, to present a fresh approach to the Easter story.

Sermon Twenty-four

TEXT: The Lord shall preserve thy going out and thy coming in
From this time forth, and even for evermore. PSALM 121:8

ONE MORNING more than two thousand years ago some villagers in Palestine started on a pilgrimage. They were of course travelling on foot, and their destination was the distant city of Jerusalem. For decades and centuries pious Jews like them had been making that journey—making it once, twice or even three times a year. As those men and women trudged along they sang the pilgrim-songs which had become traditional among their people. One of those songs was reserved for use at a particular point on the long journey. As the pilgrims finally neared the Holy City and caught their first glimpse of the hills on which it was built, the men in the group would sing:

I will lift up mine eyes unto the hills
From whence cometh my help!
My help cometh from the Lord
Who made heaven and earth.

Then the women and children in the group would make this reply:

He will not suffer thy foot to be moved,
He that keepeth thee will not slumber.
Behold, He that keepeth Israel
Shall neither slumber nor sleep.

Then the structure of the song changed, and the affirmations and responses became much shorter. The men would sing:

The Lord is thy keeper,
The Lord is thy shade upon thy right hand . . .

To this the women and children would reply:

The sun shall not smite thee by day
Nor the moon by night . . .

Then the men would take up the song:

The Lord shall preserve thee from all evil,
He shall preserve thy soul . . .

Finally everyone would join in the song's climactic affirmation:

The Lord shall preserve thy going out and thy coming in
From this time forth, and even for evermore!

The origin of the 121st Psalm? Now you see what it was. That Psalm was a pilgrim-song, sung year after year as the Jews of long ago journeyed to Jerusalem. From them, and in an English translation, it has come down to us. From us it will go on to our children, and from them to their children. As everyone now realizes, it is one of the noblest religious songs our race possesses.

Think now about the final couplet in that ancient song. You have heard it, maybe you have repeated it, many times:

> The Lord shall preserve thy going out and thy coming in
> From this time forth, and even for evermore.

You see of course the original reference there. Most of the pilgrims in ancient Palestine "went out" from tiny villages in which they had lived for many years, and in which everything was thoroughly familiar. Then they "came in" to a city in which everything was strange. The walls and gates and towers of Jerusalem, the people and the customs and the surroundings there—all were unfamiliar. Would God protect the pilgrims as they made their trip from village to city? Would He guard them as they moved from one world to another? This old song gave its brave answer:

> The Lord shall preserve thy going out and thy coming in
> From this time forth, and even for evermore.

So, heartened by that assurance, the pilgrims pushed on—pushed on toward the strange and the unknown.

Is it possible that this familiar couplet from the 121st Psalm has a wider reference? Is it conceivable that these old words hold the answer to the riddle of life and death and survival? Before you make any reply consider the experience of two imaginary individuals.

The first is a little girl who yesterday celebrated her first birthday. Her mother had made a cake for her, and on it had placed a solitary candle. Then, as the baby's two older sisters watched intently, the mother put the cake before the baby and lighted the candle. The baby gazed at the cake and the glowing candle with fascinated eyes. What were they? What did they mean? Where did the bright light come from? Now that the baby has begun her second year she will presently "go out" from the little world of infancy, and "come in" to the larger world of childhood. She will learn to walk and talk, then read and write. Finally she will understand what birthdays are, why there are birthday-cakes and birthday-candles, and why the number of candles on the cakes changes from year to year. Silently but inevitably she will thus leave the first world she knew, and enter a second one—the larger world which her older sisters already know. Is she sure to make that journey safely? Is she certain to enter the world of childhood after she has left the world of infancy? The experience of our race gives a clear answer. God has so arranged the pattern of life that, as Time carries children onward from their first year, they advance safely from the tiny village of infancy to the larger community of youth. The Lord, working through the mysterious life-process which He conceived and then made an integral part of the vast scheme-of-things, preserves their going out and their coming in. They grow up—grow up silently, safely, inevitably. They make, without realizing it, the transition from one world to another.

The second imaginary individual I mention is a boy of nineteen. Behind him is the somewhat restricted world of adolescence: before him waits the larger and freer world of maturity. How many fine things the boy will discover amid those new surroundings! Within himself the ampler strength and endurance and resourcefulness of the twenties and thirties. Around him the larger opportunities and more significant responsibilities of manhood. Strangest of all, that world ahead probably holds for him a mate and love and finally a home. Even now, at nineteen, he glimpses those splendors from afar.

A heavy sky and a drizzling rain
And the lamps in rigid rows,
Long smears of light far down the street
Where a lean cat stalking goes.
Blank—save for a glimmer here and there—
The gaunt, dark houses stand . . .
But a boy and a girl against a gate
Are whispering hand in hand.

There is a tiny dripping-sound
Of rain from off the roof,
And gleaming like black armor goes
The policeman's waterproof.
He crosses the street to give them room
As he trudges his evening beat,
For he too knows that heaven may look
Like a rainy London Street.[1]

A glorious world ahead . . . and is the boy sure to enter it safely? The experience of our race gives its clear and heartening answer. God has so arranged the life-process that, as Time carries young people onward, they move safely and inevitably from the realm of youth to that of maturity. Generation after generation millions of boys and girls make this mysterious transition. They grow up—the Lord preserving their going out from one stage in their development, and their coming in at the next.

Surely you can now foresee the next point in my sermon! It follows logically from all I have said so far. Suppose God does enable us to advance safely from infancy to childhood, and then from youth to maturity. Suppose He enables us to move on, in the course of time, from maturity to old age. What then? What can we infer about a further transition after old age?

[1] "A London Idyll," by John Presland. From *Poems of London and Other Verses* (Macmillan, 1918). Used by permission of the publisher.

I think we can infer that, when the years finally bring us to the far edge of old age, God will enable us to reach in safety the next stage in our continuing existence. Once again the Lord will preserve our going out, and then our coming in. Have we any idea what that transition from old age to whatever-is-waiting-beyond-old-age will be like? Here, stated clearly and candidly, are the convictions which many of us now hold. As you see, they rest on the assumption that the transition from old age through death to whatever-lies-beyond will be similar in nature to the transitions through which all of us pass on our way to old age.

Many of us now lay aside, reverently but resolutely, the sensational pictures of death and the future which originated many centuries ago. The picture of human beings suffering as they die, the picture of the dead rising mysteriously from their graves, the picture of a terrifying Last Judgment, the picture of the "saved" ascending into a jewelled heaven while the "lost" sink down into a flaming hell—all these pictures we quietly but firmly lay aside. We think they were drawn by men of long ago whose knowledge of God and life and the pattern of human existence was pitifully limited. The new pictures we substitute for the old? They are the pictures which this sermon has tried to draw. The picture of a person's life, each person's life, as a series of stages-of-development. The picture of a quiet, easy, safe advance from one stage to the next. And above all human beings—whatever their race, creed, color, social position, moral record—the unfailing wisdom and power and love of our Father in heaven. He has so planned the pattern of our existence that all of us are preserved in each of our successive goings out and comings in.

In October, 1800, a boy named John Todd was born in Rutland, Vermont. Shortly afterward his family moved to Connecticut and settled in the little community of Killingworth. There, when John was only six years old, both his parents died. The children in the family had to be parcelled out among the kind-hearted relatives, and John was assigned to an aunt who lived ten miles away in the village of North Killingworth. With her he lived until he went away to study at Yale College, and later prepare for the ministry. When he was in the early forties, and minister of the Congregational Church in Pittsfield, Massachusetts, his aunt—now a woman of advanced years—fell seriously ill. In great distress of mind she wrote her beloved nephew. Suppose she died: what would death be like? Would it mean agony, terror, maybe annihilation? Here is the letter John Todd sent in reply.

It is now nearly thirty-five years since I, a little boy of six, was left quite alone in the world. You sent me word you would give me a home, be a kind mother to me. I have never forgotten the day when I made the long journey of ten miles to your house in North Killingworth. I can still recall my disappointment when, instead of coming for me yourself, you sent your colored man Caesar to fetch me. I can still remember my tears and my anxiety as, perched on your horse and clinging tight to Caesar, I started for my new home. Night fell before we finished our journey, and as the darkness deepened I became more and more afraid. Finally I said anxiously to Caesar, "Do you think she'll go to bed before we get there?" "O no!" he answered reassuringly. "She'll sho' stay up fo' you. When we gets out of these here woods you'll see her candle, shinin' in her window." Presently we did ride out into a clearing, and there—sure enough—was your candle. I remember you were waiting at the door of your house, that you put your arms around me, that you lifted me—a tired and frightened little boy—down from the horse. There was a fire on your hearth, a warm supper on your stove. Then after supper you took me up to my room, heard me say my prayers, and then sat beside me till I dropped to sleep.

You undoubtedly realize why I am now recalling all these things to your mind. Some day soon God may send for you, to take you to a new home. Don't fear the summons, the strange journey, the dark messenger of Death. At the end of the road you will find love and a welcome, you will be safe—there as here—in God's love and care. Surely He can be trusted to be as kind to you as you were years ago to me!

FAITH

The Tamest of Passions

REVEREND PAUL E. SCHERER, D.D., LL.D.[1]

A Minister of the Evangelical Lutheran Church and Professor of Practical Theology, Union Theological Seminary, New York

With his mastery of words and language Paul Scherer could have been a great novelist. Instead, he uses his genius for great preaching. He himself insists that it is mostly a matter of hard work in the preparation of each sermon. His sermons of the "Great Preacher Series" in Reading, Harrisburg, and other cities; his Lenten and Easter sermons in Detroit, and his sermon on the 400th anniversary of Martin Luther have the elements of greatness.

Born in Mt. Holly Springs, Pennsylvania, in 1892, Dr. Scherer studied for his B.D. at the Lutheran Theological Seminary, Mt. Airy, Philadelphia, and was ordained a minister of the Lutheran Church in 1916. During 1918–19 he was assistant pastor of Holy Trinity Church, Buffalo, New York. He taught at the Mt. Airy Seminary for ten years—from 1919 to 1929—and was pastor of Holy Trinity Church, New York, from 1920 to 1945. In addition he has preached frequently at colleges and universities along the eastern seaboard, in England during the summers of 1930 and 1931, and on N.B.C.'s Sunday Vespers program. At the August Conference in Northfield he has served as vice-chairman since 1937 and as dean since 1942.

For years he has been in constant and wide demand as a speaker and preacher for universities, summer conferences, and Lenten services in leading

1 Sermons by members of the Advisory Committee were contributed at the request of the editor and are included on his responsibility.

cities. His work has been recognized with the honorary D.D. by Roanoke College, the LL.D. by the College of Charleston (from which he took his bachelor's degree), and the Litt.D. by Wittenberg College.

He is the author of several excellent books, including When God Hides *(sermons),* Facts That Undergird Life, The Place Where Thou Standest, Event in Eternity, For We Have This Treasure, *and* The Plight of Freedom.

His sermons show his fine ability to analyze a passage of Scripture and develop its essential meaning for his hearers. A preacher of real ability, he now gives much of his time to training the ministers and young men who attend Union. His emphasis is upon preaching methods, sermon construction, textual preaching, written sermon manuscripts, delivery of the sermons with criticisms for constructive development. This work is bound to show in improved preaching during the coming years as his students go to important churches over the country.

This is distinctly a sermon for today. Dr. Scherer looks at love and faith, current doubts, sins and unbelief, and by his own magnificent technique, leads the hearer and reader to see the way out of sin and unbelief. He creates a mood in men's hearts and minds to lead them to make their own decisions, knowing that all lasting conversions or conclusions must come from within the man himself. What a faith he preaches for our time! God has not gone away and left us alone.

Sermon Twenty-five

TEXT: O God, of Thy wisdom teach us, and of Thy grace replenish our souls: that Thy will may be in our hearts a song, and our ways be held steadfast before Thy face forever; through Jesus Christ, our Lord. *Amen.*

I WANT to speak to you this morning of those moments—if indeed no more—which I dare say come to all of us, when the zest seems to slip out of the Christian faith; and whatever is left must call itself *faith* by courtesy and *Christian* from habit. It can't make heads or tails of the noun, and wonders what there is about the adjective to put a glint in anybody's eye.

The text is from the prayer Paul used to offer for these people at Colossae. It repays study. He always knew what to pray for. He knew that for some of them the salt had begun to lose its savor. Little groups, meeting in houses, talking in the marketplace, chatting two by two, arm in arm, on the streets, had come to have their doubts about Christianity. In their exciting world, after all, wasn't it pretty dull? You couldn't exactly care for it any more—

not as you once did. It didn't have quite the appeal of the mystery religions from the East. There now was something to whip up any man's jaded taste! A dash of Greek philosophy, perhaps, to make everything intellectually respectable; a touch of Oriental speculation—very satisfying emotionally; sprinkled generously with Jewish asceticism—the kind of rigorous self-discipline that everybody knows is morally most stimulating, fairly packed with vitamins: What a dish!

And from his prison at Rome, maybe at Ephesus, the apostle tells them how it is with him. He tells them of the glory forever burning in his mind. He wants to kindle it again in theirs. He tells them of the joy endlessly at home in his heart, swelling like organ music. You can't miss the sound of it: "Filled with the knowledge of his will . . . that ye might walk worthy of the Lord unto all pleasing, being fruitful in every good work . . . ; strengthened with all might, according to his glorious power, unto all patience and long-suffering with joyfulness; giving thanks unto the Father . . . who hath delivered us from the power of darkness, and hath translated us into the kingdom of his dear Son."

It's almost a doxology, and I wonder how many of us feel like joining in? Or is this religion of ours, on the whole, just a stupid business, day after day? Well, if not stupid, at least humdrum, with not much lilt in it anywhere; nothing to keep your heels from coming down flat on the pavement when you start out after breakfast, and when you come home for dinner at night, no spring in your step? I don't need to describe the mood. You know what I'm talking about.

And it isn't just a mood. It comes too near being a kind of all-pervading atmosphere! Some years ago Will Durant wrote an open letter to a few of his "famous contemporaries." He had seen what was happening to religion, what we had allowed Science to do for us; until as a result civilization seemed to be toppling over into chaos, threatening with barbarism all our happy dreams of a brave new world. And he didn't know the answer. "What is the meaning of human life?" he wrote. "Thought seems to have destroyed itself. Knowledge has brought disillusion. Truth has not made us glad. Why did we hurry so to find it? It has taken from us every reason for existence except the moment's pleasure and tomorrow's trivial hope." And Eugene O'Neill wrote *The Ice Man Cometh,* and Walter Lippmann took up the chorus: telling us of brilliant atheists who had become very nervous; of women who had emancipated themselves from the tyranny of their homes and with the intermittent but expensive help of the psychoanalyst, were enduring liberty at last as interior decorators; of young men who were world weary at twenty-two, and crowds enfranchised by the blood of heroes who could not be persuaded to take an interest in their destiny. "These," said he, "are the prisoners who have been released. The prison door is wide open. And they stagger out into trackless space under a blinding sun." In that world Nazism sprang up like a weed. It sprang out of those noisome depths. Hitler knew that men and women who were fed to the gills with the whole sorry mess were ready to take a chance on anything. Stalin knows it too. That's what we've got to "contain" if we want to "contain" communism! And the Christianity with which we are familiar, by and large, is simply no match for it.

Most of us are a decent sort. America is a Christian nation—half and half anyhow. "Blessed be the God and Father of our Lord Jesus Christ," we are not like some we could mention! How many Thanksgiving sermons will handle that theme? We're free, and don't have to lie about everything, and don't have to grab everything in sight. You can trust us a little farther than around the corner. Of course—there are slums in Harlem. Half the people of the globe and more are victims of tyranny and starvation. Just on the edge of our lives, and farther away than thought can reach, there is incredible tragedy. We know that. And the prospect is not too good. Nobody can be sure of anything much. The future, even day after tomorrow, is your guess, and my guess. About all we can say is, "I believe in God the Father Almighty" —and so on through the rest of it: Jesus Christ, his only Son, our Lord, Good Friday, Easter, the Holy Spirit, the Church—things like that, from childhood. And there's comfort in them. But this violent paroxysm of the psalmist, what does it mean? "My soul longeth, yea, even fainteth for the courts of the Lord: my heart and my flesh cry out for the living God." Not many of us can honestly claim that there's any such turmoil inside of us. It's all right as poetry. The prose of it is that everything is comparatively quiet even now on the Western front. Except when we stand up and sing about the faith of our fathers, or feel the rough stones of *Ein' Feste Burg,* few of us ever really miss a pulse beat. Say it's human, and we can't be at fever pitch all the time. The other fact we've got to face, though, is that the world is being swept off its feet just now by a passion; and ours seems to have been tamed for quite a while. When you open the cage, it's altogether too likely to lie there, with its head between its paws, and look up at you. Nice lion! If you're ever inclined to laugh at it, there has to be a sob in your laughter!

And all the while this perfectly ridiculous little Jew, cooped up in jail, writing his letters, flinging them around as if they might amount to something! He said one day that God has turned the years into a constant pageant for him; then went on to pass the whole parade in review: "Of the Jews five times received I forty stripes save one. Thrice was I beaten with rods, once was I stoned, thrice I suffered shipwreck, a night and a day I have been in the deep." Listen to the bands playing: "In journeyings often, in perils of waters . . . of robbers, in the city . . . in the wilderness." After that, the tramping of feet: "In weariness and painfulness, in watchings often, in hunger and thirst, in fastings often, in cold and nakedness." And he was trying all along to get rid of some thorn in his flesh, with no answer from God but this: "My grace is enough for you." Just a level cupful it was, no more. "Enough"—that was all. Except that when the promise had been made good, all back and forth around the rim of that Mediterranean world, it seemed to Paul like a rushing torrent: "The grace of our Lord flooded my life."

Do you suppose any of us could hoist the flag to the top again, instead of flying it at half-mast? What the apostle really says in this prayer of his for the people of Colossae is that they have come to feel as they do about the Christian faith, lost the thrill of it, started looking about for something more exciting, because for one thing they have run off from what knowledge they had of the will of God, and so quit "being fruitful in every good work." "We . . . do not cease to pray for you, and to desire that ye might be filled with the knowledge of his will . . . ; that ye might walk worthy of the Lord unto all pleasing." It's clearly the place for everybody to begin.

What flattens out Christianity for most people, and their own lives with it, is that they either refuse resentfully, or thoughtlessly neglect, to associate the merry-go-round of human existence with the will of God. No wonder it isn't merry any more: it only goes round now. They plod through the same old routine all week long, with the same old strain pulling on the muscles, and the same old worry plucking at the sleeve. The chief end of man may be as the little boy said, "To glorify God and enjoy himself forever." But it's one thing to glorify God by throwing a bridge across the Hudson, for instance; that's conceivable. It's another thing to glorify Him in the back hall with a broom. That isn't imagination: that's sentimentality. Besides, whoever in a world like ours can fool himself into believing that even when he *is* fruitful of a good work or two it's going to matter very much? Good or bad, nothing notable seems to come of it; while

> Tomorrow, and tomorrow, and tomorrow
> Creeps on in this petty pace from day to day.

Certainly the way to get one foot at least out of that bog is to quit taking uncritically such pale and tallow-faced nonsense. Granted the smothering nightmare of little jobs, and the ruin they can work in human life. The Bible yet does its best for a man, page after page after page. May be it is emancipation he needs. The charter for it is there. But maybe he needs to understand, too, that before and after there is a will which can turn every bit of what he does into a sacrament, and provide him with things he never thought of doing, to lift the stature of his soul. That's the nub of it.

Karl Barth has much to say of the strange new world we seem to enter every time we open that Book: new and strange simply because that world is inhabited, and not empty; instinct with God, nothing like the place we live in, where little is little, and big is big, and ninety percent of almost everything is nothing! Where Abraham lived you didn't pack up and move on the first of May. You went out, not knowing whither, looking for a city which hath foundations. And Isaac lived there, with God moving around among his plans to get married; and Jacob lived there, with God turning up one night by the side of a river. Read it all. You call it the history of salvation. You think it had something to do with an infinite mind and the yearning of an eternal heart. All right. But strip it, the next time you read it. Strip it as we strip our life, with that mind away somewhere, and that heart a grand Perhaps. Instead of keeping it all up there, the way things used to be, bring it all down here, the way things are now: where a sower sowing his seed is a sower sowing his seed, "this it is and nothing more"; where a dead sparrow is a dead sparrow, and the man on the street wants to know who in hell cares, or in heaven either; and towers fall, and people build houses on the rock, and nothing is a parable. What has God to do with any of it? Read it that way, this afternoon, and you'll be back in the world to which you've grown accustomed. And it's flat, and it's monotonous, and it's intolerable, and it makes people sick. Why on earth don't we gather up our things and get out of it?

It isn't a desert, unless we make it so, a dead stillness all around, its lone and level sands asking nothing, holding nothing, promising nothing, like some insane devil watching you with its staring eyes. It's the place where all of them lived, there in the Book. And when they looked they saw something. And when they listened they heard something. That's the only difference. Something God wanted of them: and they got their fingers on it and went through with it, and laid it down with a tired sigh at night, and went to sleep. We run out of faith because we run out of love: "being fruitful in every good work." And we run out of love because we run out of God, and "the knowledge of his will."

But let's move on, get another foot out of the bog. The phrase now is "strengthened with all might . . . unto all patience and long suffering." In a world that for them had become meaningful again, even the rough going, Paul implies, would carry with it a kind of relish; and it sounds like a silly thing to say. Or does it? Surely there isn't a great deal of flavor about these words in English: "patience" and "longsuffering." Whoever wants them can have them, as far as we are concerned! The very idea! To insist that God Almighty, with all His resources, is at their disposal; not just enough this time of what He has to meet their needs, but everything He has, there in His hands, and the upshot of it is that you can put up with anything that happens! you can even keep from flying off the handle and getting nasty! sit down in a rocker on the porch. Now that Heaven is ready to empty itself wrong side out, you can look around in a kindly way, with a genuine glow of relief, and get along all right! It isn't exactly romantic is it? And it isn't likely to be what Paul has in mind.

He's talking about the courage to hold on, even when everybody else is going to pieces, and not let go, and not give way, staying right where you are. Look at that same bridge across the Hudson in a storm. I love to watch it with the wind blowing, and the waters all churned up. And I love to think of a life like that, with its traffic moving! Paul's may be. Or somebody next door in a wheel chair—or across the street, who can't be whipped. He's talking about courage, not cowardice—red courage, with the steady strain on it. He's talking about the great heart inside of a man that doesn't know what it is to sneer at life and snap back at the things that go wrong. And here is no gloominess in it, and there's no sourness in it: that's why Paul says, "longsuffering with joyfulness." He knew the other kind! Nothing but "a buoyant sense of mastery," and nobody can look down his nose at it! That's romance!

And the difficulties Paul faced were part of it. They helped to whet his appetite. Possibly it isn't so silly after all! He never spoke lightly of the rough places, or wasted any time pointing out the "bright details in other parts of the landscape." Carlyle used to charge Emerson with doing too much of that: called it a cheap and easy and dishonest optimism; said he was like some fellow away up on the beach "chattily throwing a cheery word or two to poor souls wrestling for their very lives in great dark deeps, with thundering billows knocking the breath out of them." Paul knew what that wrestling is, and how tired people get. A friend of mine called me up this week, his job gone, and his wife gone, and sixty-five years of his life gone. It was just too much. What was there that could keep him going? Paul knew how utterly beyond all reason it seems at times, so that you simply stand there at the sight of it, fooling around with your hands, but he knew too that there is something inside of us—there was inside of him, and inside of my friend—which is forever wanting to match itself against the odds, keeps thrusting at all there is in the world that tries to stop us; and when there isn't anything left to hold off at arm's length, or climb over, we either invent something, or turn on the radio because we are bored to death!

That's why the New Testament doesn't address itself to any of your puny virtues, asking of you what you can do by turning your hand. It addresses itself to all your chivalry and valor and dreams of greatness. A first-rate artist once remarked that if you wanted to paint a picture of Christianity you'd have to show a meek and aged woman leading a child into a pew by candlelight. And a first-rate poet chimed in: "Thou hast conquered, O pale Galilean; the world has grown grey from thy breath." But somehow, there in the Gospels, that doesn't seem to be what's done it! With Jesus of Nazareth looking at us, and always speaking to the hidden grandeur in our souls; to what we are not, and want to be, and may yet one day become! Taking what little loyalty he could find, and shaking it free of everything else, until in a desperate world it was holding on grimly to God with its finger-tips. Taking the love men offered him, and squeezing out of it every last dripping adjective, to leave it lean and taut, ready for the fire or the sword or the cross! Getting faith to tense its muscles. Maybe that's all it needs to keep from going stale! Getting hope to set its teeth! "Strengthened with all might, unto all patience and longsuffering with joyfulness."

And this to top it off—or to undergird all which is better still—"Giving thanks unto the Father." Apparently next Thursday has more to do with the Christian life than one might normally suppose. But see why. This is not likely to appear in the proclamation. "Who hath delivered us from the power of darkness, and hath translated us into the kingdom of his dear Son." In a world where everything mattered, because God hadn't gone away and left anything; in a world where all He had, up to the level, not of their need, but of what they were able to take and use, was held out to them; in a world where above all else He had willed and done this thing on a cross that was the turning point of human history.—You couldn't fly a flag at half-mast any more, not even theologically: as if nothing would happen, so evil is it all—nothing, whatever you might do!

Every once in a while we seem to come out of our theology worse off than Hamlet. "The time is out of joint," and so say we all of us. The trouble with the so-called liberals is that they don't know how far out! But Hamlet had something to add: "O cursed spite, That ever I was born to set it right!" He at least had an idea that he could! We—don't know. That's a pretty liberal view of things too! He didn't exactly usher in the kingdom of God. With Ophelia dead, and Laertes dead, and the king dead, and the queen dead, and Hamlet himself dead, what was it precisely that he set right? But if you ask that question you don't understand life as Shakespeare understood it. And you don't understand tragedy. The very essence of tragedy is that you can bear to see the curtain go down on that scene. It isn't failure! Something is right now that was wrong before. No tit for tat nonsense. Something profound in the universe, that's worth all the waste!

Far, far more deeply and truly, it was that way on Calvary. "Delivered . . . from the power of darkness"—and all of us still here in the middle of it? "Translated into the kingdom of his dear Son"—and so little sweetness and light? What's the answer? The answer is that this isn't a place where you can whistle a tune, as if every story were bound to have your kind of happy ending, and everything at the last could be wrapped up in a neat bundle and put away on a tidy shelf. Too much whistling is nervousness! It is a place where we can sing, and have it make sense: *Before Jehovah's awful throne.* Do you notice how altogether inconsistent the second line seems? *Ye nations, bow with sacred joy.* The awful throne and the sacred joy hardly belong together, do they? Not until you have met in Christ all the goodness and severity there is on earth out of heaven, the beauty and the terror of life, its mystery and its majesty.

Then you, too, like Paul, will be stumbling on some doxology down every familiar road. He'll be walking along, wrestling with the weakness in himself, dismayed by the war that's going on inside of him, "O wretched man that I am!" When all at once he turns a corner and finds Another, stronger than he, ahead of him, waiting there, "I thank God"—and the disjointed bits fall into a pattern: the sufferings of this present time, every creature, and all things, as God works together for good with them that love Him!

It may be we don't care about it—this vast glory of Christ, these strange verses about an empty tomb, so strangely true about every life that wills it to be, because it was true of that life, all of it set here in the world's face. If we don't care about it, there's no use paying any attention to Paul, or even listening to him. But if we do—well, the Japanese, they say, have a very lovely way with their trees: they train and shape and force the trunks and branches into delicate patterns, coaxing the most unimagined harmonies out of Nature, as if she were a huge instrument for them to play on. "Delivered—from the power of darkness." If we care enough, it can twist back into some beauty every ugly thing there is!

Let us pray. *Grant us, O God, of the wonder of Thy love a deep and constant joy: that out of our acknowledged weakness we may stand fast in a strength far greater than ours, and through the busy tumult of our days share with Thee the peace of Thy steady purpose through Jesus Christ, our Lord.* Amen.

FAITH

The Church and the Western World: Its Belief

REVEREND EUGENE CARSON BLAKE, D.D.

A Minister of the Presbyterian Church, U.S.A., Stated Clerk of the General Assembly of the Presbyterian Church, U.S.A., Philadelphia, Pennsylvania

Dr. Blake is one of the younger ministers of force and power. He was born in St. Louis, Missouri, in 1906, attended Princeton University, then studied for two years at New College, Edinburgh, Scotland. He entered Princeton Theological Seminary in 1932, and received the D.D. from Occidental College in 1941.

He was a teacher at Forman Christian College, Lahore, India, in 1928 and 1929, and became assistant pastor of the Collegiate Church of St. Nicholas, New York, in 1932, where he remained until 1935. Then he went to Albany as pastor of First Presbyterian Church. In 1940 he was called to the Pasadena Presbyterian Church and in 1951 was elected Stated Clerk of the Presbyterian Church, U.S.A., and moved his offices to the national headquarters in Philadelphia. This is the most important executive position in the Presbyterian Church, giving him national and world oversight of the work of the church.

From 1938 to 1940 Dr. Blake was visiting lecturer on religion at Williams

College. He is a trustee of Occidental College, a member of the Board of Monte Vista Grove Homes, a member of the Board of Christian Education of the Presbyterian Church in the United States of America, and pastor of radio station KPPC. His religious views have been influenced by Professor Theodore Meyer Greene, the late Hugh R. MacIntosh, and by Reinhold Niebuhr. He is in favor of Church union to strengthen the Protestant churches.

In discussing the belief of the church in the western world Dr. Blake considers the early Christian Church, how Jesus worked with his first followers, the importance of Peter, and insists that the Church is important to man, and man to the Church. He says without hesitation that if Jesus had not been more than a prophet there would have been no Christian Church. It is a significant thing that a great preacher has been elected to head the entire program of the Presbyterian Church and would seem to indicate the importance Presbyterians attach to preaching.

Sermon Twenty-six

TEXT: And I say also unto thee that thou art Peter and on this rock I will build my church and the gates of hell shall not prevail against it. MATTHEW 16:18

THE chief difference between man and an ape is that the man is capable of giving his activities sustained direction by an idea which he holds in his mind. The ape, more than most of the other animals, does from time to time catch glimmers of ideas: a chimpanzee can be taught to open a complicated lock in order to get at a banana; he can even be mentally alive enough to solve new and unfamiliar locks in order to get at the bananas. But even the lowest order of intelligence in man is capable of being governed by mental images, ideas in his mind, far above the level of the chimpanzee. The dullest workman hears his alarm in the morning, looks at his clock, sees that the time has come to get up, performs a number of operations (very complicated for a chimpanzee) getting his breakfast, getting dressed, standing on a street corner to wait for a bus, and arrives at his factory in time to go to work. All this because an idea, a mental image furnished by his memory, is giving his various actions a sustained direction. Ideas (good and bad) govern all the activities of human beings above the level of instinct and habit. The ideas that men have determine the kind of life they live, the culture and civilization they develop, and the ends to which they give themselves. When an idea is held by a man over a long period of time, it becomes what we call a belief or a conviction. Thus the character and quality of the life of a man or a group of men, a nation or a church, depend more on belief than on any other factor. There was a day when belief seemed unimportant. Not now!

The subject of my sermon this morning is the belief of the Church and how that belief through all the vicissitudes of nearly two thousand years has created the Christian Church and gone far towards determining the whole civilization and culture of our western world. Although one can fill whole volumes with Christian theology and although in twenty centuries, scores of Christian confessions and creeds have been written, there is one basic and simple belief that made the Church, has held it together, and has at the same time been the most creative idea in determining the character and culture of our western world. Put in a sentence, that idea, that conviction, that belief is this: That Jesus of Nazareth was and is uniquely God's representative, revealing God's love, His saving purpose, His just will, and even Himself to man. Let us go back this morning and look at the moment when that idea became a conviction and so began its creative influence among us.

After some months of preaching, teaching and healing in his native Galilee, Jesus had turned his back on the cheering crowds and gone out of the country to the northeast and entered the Gentile region of Caesarea Philippi where he was quite unknown. With him he had taken twelve intimates whom he had selected out of the crowds. The purpose of the retreat was obviously to get away from the popular clamor and to set direction for the future. We may picture them walking along the Roman road together; young men they were at the threshold of their lives, glad in each other's company, enjoying the sights and scenery of a countryside until then unfamiliar. Do you suppose they sang some walking songs as young men do today? Did they have their private jokes, jesting at one another in good fellowship? We don't know the answer to these questions but we do know that from time to time they discussed as young men do quite openly the deepest questions of existence—all the more so in this company since they had been gathered together by the eloquence and personal attractiveness of one who had already shown himself to be a teacher of unusual power. Already they had been drawn to his allegiance sufficiently to leave their jobs to follow him. Some had been fishermen. One a tax collector. Doubtless part of their enthusiasm on this expedition was the freedom they had found from old routine. And the other part of their excitement was just as surely the expectancy they felt that great things were in the offing in the company of Jesus. I picture the season as spring, the winter rains having greened the hills and meadows; multicolored poppies brightening the margins of the road. I see them sitting near a running brook for midday rest after a frugal meal of bread and cheese or fish. I hear Jesus himself start the crucial conversation by asking them a question. "Whom do men say that I, the son of man, am?" This was not a hard question. The twelve quickly replied, reporting a variety of popular opinion which they had heard from the crowds down in Galilee. "Some say you are John the Baptist (come to life again). Some say you are Elijah returned to earth. Others Jeremiah. Nearly all agree you are in the great succession of the prophets of God come to our nation." This was much. Already with the crowd Jesus had won a most distinguished reputation. To be identified in the popular mind with John the Baptist, recently beheaded, already a religious and popular hero despite his harsh message of repentance—that in itself showed how far Jesus had come in the few months since he had laid down his carpenter's tools in Nazareth and set out on the open road to preach and teach and heal.

But even more was it to be connected in the popular mind with Elijah, popularly remembered as the rescuer of the pure faith of Israel when the foreign Jezebel (becoming queen) had tried and almost succeeded in substituting throughout the land the nature worship of the Baals of Tyre. And some (with even more insight) were paying him a higher compliment yet to connect him with Jeremiah. They had noted similarities in Jesus' warnings and in his very personal preaching of individual responsibility with that of Jeremiah who had spoken for God in a much earlier time when Israel was also under a hated foreign yoke. Yes, all agreed that Jesus was one of the prophets, a spokesman for God, in the great succession of their nation. Was not this enough? Why did Jesus turn his back upon crowds who thought so well of him?

From our vantage point of history we can say this: There would have been no Christian Church if Jesus was no more than one, even perhaps the greatest of the prophets of Israel. More than that, as we shall see, Jesus was already looking beyond his present popularity to the hatred and enmity which he saw would come. Not yet had the opposition gathered itself together, the pride of race, the concern with religious revenue, political fears which would soon combine to crucify him on a Roman cross. But Jesus knew that the cross was before him, that a hill outside Jerusalem was standing already stark against the sky only awaiting the erection of that cross to change its silhouette. A popular teacher men delight to listen to—but one hated, maligned, discredited—would not the crowds and cheers completely melt away?

As the conversation died down and there were no more suggestions of what people were saying or thinking of Jesus, I picture him pausing, looking at them squarely and with a deeper significance, asking them directly: "And whom do *you* say that I am?" They weren't so quick to answer this. This could not be answered by a recital of others' opinions. To answer was to state one's faith. In the back of all their minds had been hovering an idea, a possibility that none so far had quite dared to utter. For some generations the faith of Israel had had more and more of a future turn. The belief that one day God would send His very own anointed representative to redeem the nation—this belief had become more and more general and as human hopes of peace and prosperity and independence of hated foreigners had more and more betrayed them, the hope and expectation of a Messiah, a Christ from God was more and more their only hope.

Simon, who because of his over-common name, Simon, son of John, was nicknamed Rocky or Cephas, or Peter perhaps because of his unhewn size and strength—Simon was their leader. Naturally it was he who spoke—never one to guard his tongue, always one with quick insight and quick enthusiasm—he spoke out what doubtless others had been thinking but had not quite dared to put in words: "Thou art the Christ, the son of the living God." Until that moment this had been only an idea, an intriguing possibility, but once expressed, once spoken, it became a conviction—the conviction, the belief, the faith upon which the future turned, the future of Jesus and the twelve, of the Christian Church and yes, of the whole history of our western world.

At once, Jesus recognized that turning point. "Blessed art thou, Simon, son of John, for flesh and blood hath not revealed it unto thee but my Father which is in heaven." It was this faith, now crystalized by Simon, that he had been awaiting. This was the creative truth that would form the Church and change the world. Jesus went on to make the statement then of his greatest prophecy: "And I say also unto thee (Simon) Thou art Peter and on this rock I will build my Church and the gates of hell shall not prevail against it." Of course it is clear that the rock he spoke of was not Peter himself (much less successors) but this foundation faith which Peter had just now for the first time spoken: "Thou are the Christ, the son of the living God." The Christian Church has been built upon this foundation belief, this faith, this living conviction:

1. That there is a God, one God, Creator of heaven and earth, transcendent, not to be confused with what He Himself has created; a single God with His own purposes Who rules the universe He has made. That is the first part of this belief that has made the Christian Church and profoundly affected the culture and civilization of our western world. Through all the different centuries and in all parts of the world this belief has had varying rivals. There was and is the ancient idolatry which worships natural force and bows down before that. In olden times men made idols of wood and stone, of silver and of gold and bowed down before them. Today men make gadgets of steel and aluminum and plastic and give their lives to serve them. But all through the history of our western world against this natural materialism of man and his preoccupation with the things that he and nature make, the Christian Church has called him to worship one God, the creator of heaven and earth, spiritual in being, sovereign in power and almighty in His Purposes.

This faith I veritably believe was and is the necessary groundwork for all the best of our whole culture and civilization. Without this idea of God there would have been no science. Without this unifying principle, men would yet be living in the fears of Africa's jungles or the abject hopelessness of India's pantheistic confusion of good and evil, God and matter, and of being able to see little difference between the end and the beginning. This belief inherited from Israel and transmitted to our western world by the Christian Church in one transcendent and sovereign God, Almighty, Maker of heaven and earth—this idea that has ruled the mind of western man is part of the basic belief expressed that day first by Peter on a roadside in Caesarea Philippi nineteen hundred years and more ago.

2. But this belief has one other basic aspect: namely this: that this transcendent God, Alpha and Omega, sovereign Creator and Judge, that this God chose Jesus of Nazareth to reveal Himself to man. "Thou art the Christ," said Peter, "the son of the living God." It is this belief shared by all living branches of the Christian Church from the beginning until now that has determined the character of men and nations and set up the values towards which our life has moved. For the God Whom Jesus revealed is a God of Love, to be called our Father, concerned with every one of us, His children. For when Jesus of Nazareth was accepted by the twelve as their Christ, the anointed of God Himself, a whole new conception of the character of God became regnant in their minds. They did not get it all at once. Only then did Jesus *begin* to transform their whole idea of God's Messiah from that of a conquering king to that of a suffering servant. He began then to tell them why as God's anointed he must go to Jerusalem and suffer there and die. He began to teach them then the lesson so hard to learn and over which we all still stumble: "If any man will come after me, let him deny himself and take up his cross and follow me."

He began then to show them that the love of God would last beyond all tragedy and that the goal of God was to gather an increasing company of men and women who in their love for Him and each other would in their earthly pilgrimage transform the world from a snarling arena into a heavenly city, from wars and hatreds and enmities into a brotherhood of justice and of peace.

It is because we believe that Jesus is the Christ, that despite our sin and selfishness, despite all our injustices and all the chaos of hatreds and competitions and fears and greeds—despite all this and because of this one ruling conviction of the Christian Church that Jesus is the revealer of God, His anointed one, the Christ, there are today all through our world scattered brotherhoods that meet in His name; brotherhoods in Christ crossing all boundaries of race or class or nation. It is this belief, brought to us by the Church, that is the light in all our darkness and the hope against all our fears.

Christ, the Corner Stone

REVEREND TOYOHIKO KAGAWA, D.D.
The United Church of Japan
Tokyo, Japan

Considered a "saint" by thousands of his followers and invariably compared with Gandhi and Schweitzer, Toyohiko Kagawa is believed by many to be one of the world's foremost exponents of applied Christianity. His influence on the social, economic, and political development of his homeland has far exceeded that of any other native Japanese.

Kagawa accepted Christianity in his youth, knowing that conversion would cause him to be disowned and disinherited by his family. In self-dedication to the Master's example he worked and lived for fourteen and a half years among the poor of the dread Shinkawa slums of Kobe, where he suffered from tuberculosis, was nearly blinded by trachoma, and was attacked and disfigured by thugs.

He studied at Princeton Seminary, returned to Japan and soon emerged as the most successful native evangelist in Japan. In 1929 he organized and led the united church movement to win a million souls for Christ, and his own meetings resulted in 64,000 conversions. In the first three years after the last war his meetings brought forth declarations for Christ from 200,000 of his countrymen. He first entered the arena of national reform in 1919, when he helped found the Japan Federation of Labor. He started the first trade union in Japan, then launched the Japan Peasants' Union. Considered the father of his country's Christian cooperative movement, he is the founder of the Rural Consumers' Cooperative Association, the Central Japan Federation of Consumers' Cooperatives, the Kwansai Cooperative Association, and the Students' Cooperative Association.

He prepared the way for rebuilding the slums of Japan's six largest cities after the earthquake of 1923 and won enactment of the first unemployment insurance in the Orient during the depression years. In 1925 he helped launch his country's first peoples' party and in 1945 organized the coalition Labor Party which won control in the elections of 1947. He started the first night school for laborers in Japan and has organized approximately one hundred Gospel schools for farmers. In 1928 he founded the National Anti-War League and in 1930 was elected president of the Federated Peace Associations of Japan. He is currently chairman of the International Peace Association and vice-president of the Union for World Federal Government. Through the years he has personally maintained church settlement centers in Kobe,

Osaka, and Tokyo. He has written about one hundred and twenty books, one of which has sold more than three million copies.

Some 300,000 Americans heard his message when he returned to this country for a five-month tour in 1950. This sermon represents Kagawa's present thinking and faith. The translation was made by John Shields.

Sermon Twenty-seven

TEXT: He that loveth not knoweth not God. I JOHN 4:8

IT IS NOT ALWAYS easy to go among the people who have never known God and teach them the truth that God exists and to explain to them what God expects of us, His children. Even in Christian societies many people today question the existence of God. They become very skeptical because they cannot see God.

We are like a baby in the abdomen of its mother. As long as the baby exists inside the abdomen and cannot see its mother's face, it might exclaim: "I have never seen my mother's face, so I think mother does not exist." Or when the baby opens its eyes and sees its mother in the wall of the abdomen, it might decide: "Mother is the wall only."

Today we live, move, and have our being in the abdomen of God. We live and move and have our being in God even though we cannot see the face of God. Now the baby, having made these observations, may think, "Oh, I am indeed a great philosopher." But we know that baby is very presumptuous. So with us. As long as we remain materialistically minded, we are like that baby—very presumptuous.

This universe looks like a material world; but truly speaking this universe is a world of God. We are but finite realities in the bosom of God. Man has a consciousness, but our consciousness is only a part of the greater, all-inclusive consciousness of God.

We may content ourselves with being mere agnostic, self-centered men, or we may seek to achieve a higher consciousness that will make us greater than men, that will let us behold more than the eyes of man—so close to the earth—can see. Religion is nothing more than our awakening into this higher consciousness of God. The highest aspiration of our religious awakening is to reach God's level of cosmic consciousness. Then we can know God and know the Will of God. Then we can understand that God is all of life, that God is power, and that God is Love.

All of the studies that man pursues—science, economics, politics—are a part of this cosmic consciousness. Religion never contradicts science. Science is simply another window for consciousness. To me the theory of evolution never was contradictory to the idea of creation. The scientist may study the phenomena of life, but he cannot create or alter the laws and conditions governing this phenomena. God provided the laws and conditions. The scientist can only invent and create new phenomena by following the laws God originated.

To know God is to love God, and to love God we must love everything that is a part of the consciousness of God. Nothing would be created by God except out of goodness and love. Therefore, to share God's consciousness, we must share His love; we must be conscious about what God is conscious of and we must understand the agony of God when human beings fail. This is why it is written that "he that loveth not knoweth not God." Unless we are able to love others, we cannot really know and love God.

Some people have said to me, "But, Mr. Kagawa, it is very difficult to love other people so much. We are instinctively inclined to selfishness and egotism. Our instincts have a limit." So I say to them, please understand this: the aborigines of the Solomon Islands do not eat human flesh from instinct. Instinctively, we do not like to eat human flesh at all. But these cannibals had the superstition that by doing so they would become stronger, and after a while, because they all ate human flesh, they thought it was natural and an instinct to do so. As long as we think our superstitions and selfish ambitions are instinct, it is very difficult for us to find God. We must raise ourselves above the levels of such false "instincts." They are but the excuse for self-indulgence and self-interest. We must lift ourselves to God's level of consciousness and love. Then we can know God and can see through the eyes of God and let God move in our bodies.

God has provided the most wonderful laws of love. First, there is instinctive or physical love: the love of a mother for her children and of a man for his wife. Second, there is moral love: a love for the goodness in men, a love which makes us want to help sick people and old people. But these two are not enough. Having moral love only, we scorn sinners, we despise prostitutes, gamblers, and publicans. Therefore, we need a third and supermoral love, for we must love the corrupt sinner, even as God does.

Jesus was born on earth to teach us how to love even sinners. Jesus had a great consciousness of God, and in this great cosmic consciousness he could understand the failures of God's children and could feel the need of redemption. The suffering of sinners and the suffering of the poor was his own suffering. We know that God can preserve the bodies that mankind must lose, but it was Jesus who discovered God's principle of restoration of values that mankind has lost. Abraham taught God's method of Creation. Moses explained God's system of Preservation. Jesus illustrated God's law of Redemption.

This was the meaning of Christ's relationship to God and to man. Christ lived on earth as the perfect example of complete fulfillment of love for God. He taught us that to love God we must love all of God's creations, even the most despised of sinners. He showed us that redemption through the grace of God could be achieved only through self-sacrifice made in fulfillment of the will of God.

When we acknowledge the love of God we create new values in our own life and thus preserve the values that God has created. The more you love the more you will find that wonderful energy of God flowing into your consciousness and that God is trying to act through your consciousness.

It is then that we can become one with God and can make our mortal bodies the true instrument of the will of God. Then we have a great, overpowering urge to help other men also find this wonderful love of God and we draw closer to Him. We want to help men when they are tired and share their burden of poverty and despair. We want to cut them away from their chains of bondage and give them the true dignity God intended. We want to fulfill their faith in righteousness. We want to do this because we have compassion for all the suffering of mankind. We want to do this because we have such love for all the elements of God in man.

No one can say how this love should be fulfilled. Nobody can weigh the effort or the value of the sacrifice we make to help our fellow man. Only in ourselves will we know when God is satisfied with what we have done. Laws do not make a good person or a good society. It is spirit that makes people and society good. There is no use for outer disciplines, or laws made only by man. There is only one enduring constitutional law of society: the law of sacrifice through love.

Jesus died that our spirits might know this and live. We must experience the blood of Jesus. We must not shirk acceptance of the cross that Jesus bore. We do not take up the cross to punish our bodies, but to fulfill the will of God, to justify our being in the eyes of God, and to enable ourselves to be one with God.

The cross of Jesus means permanent progress toward God's true kingdom. It means a process of permanent sacrifice by the individual to help lift society to God's level of consciousness and love.

The cross of Jesus, and each individual's acceptance of that cross, is the foundation upon which humanity can develop. On this redeeming love of Jesus we can build a new and wonderful standard of morality. Here we shall have the spirit of service, the spirit of brotherly love, the spirit of sharing together, and the spirit of social solidarity. With this new standard of morality we can unite men and women; we can unite all classes of society; we can unite the different nations; and we can unite mankind with God.

Do You Believe in Love?

Bishop Herbert Welch, D.D., LL.D., Litt.D.
A Bishop of the Methodist Church (Retired), New York City

Bishop Welch has had a long and brilliant career in the Methodist Church in the United States and in China and Japan. His preaching has a quality that is a fine combination of spiritual insight, knowledge of men, and words that sing. Born in New York in 1862, he was educated at Wesleyan University, Drew Theological Seminary, and Oxford University. His work has been honored by Northwestern University, Western Reserve University, and the University of Vermont, Ohio Wesleyan University and Allegheny College, with the LL.D.; by Wesleyan with both the D.D. and LL.D., and by Boston University with the D.D.; the Litt.D. was conferred by West Virginia Wesleyan in 1928.

He entered the Methodist ministry in 1890 and held pastorates at Bedford, New York; St. Luke's Church, New York; Summerfield Church, Brooklyn; Middletown, Connecticut, and Chester Hill Methodist Church, Mt. Vernon, New York. He was elected President of Ohio Wesleyan University in 1905 and won wide recognition there until his election as a Bishop of the Methodist Church in 1916. He was resident Bishop of Japan and Korea from 1916 to 1928 and was decorated with the Third Class Order of Sacred Treasure by Japan. He was recalled to the United States to be Bishop of Pittsburgh (1928–32), then returned to the Far East to be Bishop of Shanghai (1932–36). He retired in 1936, but was recalled to be Bishop of Boston in 1938–39, became Chairman of the Methodist Committee for Overseas Relief for eight busy years, 1940–48. He has served on so many boards of the Methodist Church that the list of his positions would read like the activities of three men. In 1914 and 1915 he was President of the Methodist Educational Association, Vice-President, National Christian Council, China, 1935–36, member Board of Governors, West China Union University, 1936–48; Nanking Theological Seminary, 1937–43; Member Board of Managers, American Bible Society, 1936–48. In 1935 he delivered a series of College Lectures *in the Korean language.*

He edited Selections from the Writings of John Wesley (1901); *author (with others)* The Christian College (1916); That One Face; Men of the Outposts (1935).

This sermon discusses the spiritual significance of human and divine love, opens new channels of thought and meaning in an age-old subject.

Sermon Twenty-eight

TEXT: The greatest of these is Love. I CORINTHIANS, 13:13.

WITHOUT apology I return to a theme no doubt discussed in this place again and again through the passing years, because it is not some trivial or incidental part of religion and life, but the very heart of the whole thing. Christianity is the Gospel of Love.

So I ask the question: Do you believe in the Gospel, the God-spell, the good news brought to earth nearly two thousand years ago? Do you believe in love?

True, the word is much abused. It is employed to cover all sorts of emotions: sometimes mere physical desire; sometimes a pleasant sensation of approval and admiration for a scenic view, a fruit or a flower, a food or a sport; sometimes a mild amiability or even a neighborly feeling, a hearty friendliness. Perhaps the highest of such conceptions is friendship. Well, raise the temperature of a glowing friendship until it bursts into flame, and behold! you have love. Love lends color and joy to existence; it means romance, noble unselfishness, the submerging of one's smallness in a lofty devotion. Love between persons is more than infatuation, more than a passing sentiment. It is clean, with the purity not of snow but of fire. It finds its deepest satisfaction not in conquest but in surrender.

Now, do you believe that God is love? Is there One—the Eternal, the Unchangeable—whose mind is unfailingly one of kindness, whose purpose regarding man is always for good? How do you dare to believe in a God of love, when you remember the poisonous snakes and the ravening tigers and the burning deserts and the treacherous swamps and the terrifying earthquakes and the blasting volcanoes that make up a part of God's world of nature? How do you dare to believe that a God of love is dealing with this planet when you see the oppressions, the brutalities, the outrages, the distresses and tortures between men, after all these ages of our emergence from the beasts? Do you not cry sometimes in your heart, "How long, O Lord, how long?" If there is a good God on the throne of the universe, how can He permit such anguish to come even to the least or the lowest of His children? How can you believe that God is love, when you recall your own personal sorrows, disappointments, crippling accidents or disease, frustrations and losses? Do you still dare to cry with Job, "He knoweth the way that I take; when He hath tried me, I shall come forth as gold. . . . Though He slay me, yet will I trust in Him"? Do you still dare approach Him, not as one bows before a King, but as one who seeks a Father? Do you believe that Christ was right in calling Him always "Our Father"?

Can it be true, the grace He is declaring?
  Oh, let us trust Him, for His words are fair.
Man, what is this, and why art thou despairing?
  God shall forgive thee all but thy despair.

I say to thee, do thou repeat
To the first man thou mayest meet
In lane, highway, or open street—

That he, and we, and all men, move
Under a canopy of love
As broad as the blue sky above.

Yes, at the heart of the universe is not mere mystery, is not dark chance or fate, is more than sheer power. In the face of all the difficulties, you still have a right to sing with Whittier:

I see the wrong that round me lies,
  I feel the guilt within;
I hear, with groan and travail-cries,
  The world confess its sin.

Yet, in the maddening maze of things,
  And tossed by storm and flood,
To one fixed trust my spirit clings;
  I know that God is good!

I know not where His islands lift
  Their fronded palms in air;
I only know I cannot drift
  Beyond His love and care.

Is there not some undiscovered purpose of God which will explain the seeming contradictions of existence, which will convince us that "all things work together for good," that "the hidden heart beneath creation beating" is a heart of love? Are not those who put happiness first, just plain mistaken about God's plan? Why not all ease and pleasure and progress, why not all sunshine and calm?

Not long ago I heard a mountain preacher, speaking of the experience of the "timberline," the point at which trees have to fight for their very existence, tell how the wood grown in such trees has a toughness, a strength and a musical quality not found in wood grown lower down.

As a matter of fact, only out of struggle and pain can there come the kind of person and the kind of world that the wisdom and power and goodness of God are set to make. How could you teach any man courage if there were no fears to fight, no threats to overcome? How teach him kindness if there was no suffering to be relieved? How train him to endurance if there were no weary days and nights, no burdens to bear, no obstacles to surmount?

Ah, God is in it all! Not only the quiet shining of the stars tells of His presence, but the burning of the noon-day sun, the crash of thunder, the wildness of blizzard and tornado.

Often what is painful turns out to be best. Life was not planned to be a perpetual picnic for children but a school for adult education. The Chinese put it into a proverb: "The diamond cannot be polished without friction, nor man perfected without trials." There is no easy road to paradise or to perfection. Never has any man been "carried to the skies on flowery beds of ease." The silken saints have had to join the heroes and the martyrs and go to heaven in chariots of fire! "These are they who came out of great tribulation," is the heavenly word. Even the Captain of our Salvation was made perfect through suffering. Do not pray for easy times, as Phillips Brooks put it, but pray for strength, courage, grace enough to meet hard times and come off victorious.

The coming of the hard times—the disillusionment, the defeat, the loss, the pain—this is not evidence that God has forgotten or ceased to love. "Whom the Lord loveth He chasteneth." And the world has not slipped out of His hand. "He maketh the very wrath of man to praise Him and the remainder of wrath He will restrain." "God moves in a mysterious way, His wonders to perform." By green pastures and quiet waters, but equally through deep valleys and over rough roads, the Shepherd and Bishop of our souls leads on, toward the heights. He brings streams out of the rock and prepares a table in the wilderness. "He doeth all things well." God is wisdom, God is love.

Then two things follow: one, that the highest quality in man is that which is most like God—namely, love. And the truest measure of a man is not his ancestry, his native ability, the education and training which have developed that ability, his eloquence, his shrewdness, his success in surpassing his competitors and acquiring wealth, fame, power—but above even honesty, courage, purity, the question remains—is he a loving man? No creed, no ritual, no religious rite, no association with the men of goodwill can give the final answer. No mere orthodoxy or respectability can meet the test. "The Lord looketh on the heart."

In that matchless picture of the last judgment recorded in Matthew 25, our Lord has pointed out that the division between the approved and the condemned is determined by this simple, fundamental, inclusive virtue of love. Did you feed the hungry? Did you visit those in need of comfort? Did you perform "those little, nameless, unremembered acts of kindness and of love which are the better portion of a good man's life"? Are there visible those fruits of the Spirit which grow so largely from love—long-suffering, gentleness, understanding, sympathy, generosity, charity? Paul puts it thus: "If any man have not the spirit of Christ, he is none of His." John declares: "Love is of God, and he who loves is born of God and knows God. He who does not love does not know God; for God is love." Do you believe, with

Paul and John and Jesus, that love is the core and crown of character? Do you really believe in love?

Then, also, this other question: If love is at the heart of the universe, at the very center and seat of power, and if it is the supreme ingredient of human wholesomeness and wholeness, then is it not in this same love that we are to find the solution of all our problems?

"Science finally has discovered love," says a medical specialist of highest standing. "We doctors are learning how to prescribe human love."

Social experts who are students of youthful delinquency declare that they must have, as teachers, judges, probation workers and policemen, persons "who have a feeling for children." In other words, here, as elsewhere, money, education, law, are not enough. If society is to be saved, it must be by the inward compulsion of love.

Take the home, for example. Build a model house, with all the gadgets and conveniences that modern science can supply; give the family not simply an existence-wage but a comfort-income; enact better laws about marriage and divorce; provide for sex instruction that shall be at once scientific and romantic; bring in books and beauty, art and music; and if one thing is lacking, you'll have nothing but a handsome shell.

But if that one thing, that pure and passionate and devoted love, be present, a home, even with less comforts and privileges, may be not only a pattern for society but a foretaste of heaven! "Where love is, God is," as Tolstoi suggests.

Of course, the home is unique in its function as the training-school of love; but can we expect that the same spirit can be extended to the wider and more tempestuous areas of life? What about the working-day occupations of men—production and distribution, making and selling, banking and farming, mill-work and school-work, housekeeping and nursing, organization, transportation, teaching, law, medicine, scientific research, government, art, journalism, the writing of books and plays—the whole range of daily activities which rightfully absorb the major part of our time and thought?

Where does love fit in as an inspiring motive and as a pervading and controlling spirit? Have we a right to look for it in the N.A.M. and the great labor unions, in the realm of competition and cooperation, of profit and service? Is public life, in the city, the county, the state, the nation, the world, to be governed by love?

I am not saying that the use of force can at once be ruled out. If there is no force, how can there be law? Evil is not to be free to run riot, while love stands helpless by, wringing its hands in its impotence. Not so—we must have cooperation and union for the common welfare, and the general welfare may demand and justify the use of force to repress the evil, as well as to make possible the good. In the family itself, the absence of discipline (which involves compulsion of one sort or another) may weaken and disintegrate the home.

So beyond. City regulations, state statutes, national laws, world order, will not, humanity being what it now is, enforce themselves. Force, in some form, is inevitable. *But* the only adequate excuse for the use of force is good-will and not hatred—good-will for all concerned, for the victim as well as for the oppressor, for the present *and* for the future!

Good business calls for bargains that are profitable to both parties. Good government demands not only integrity and diligence, but respect and concern for all classes, and an overwhelming desire for the welfare of all men and all nationalities. "Do unto others as you would that they should do unto you" is a sound, constructive principle for business, for national and for international affairs.

Mussolini used to assert: "We wish to hear no more about brotherhood," and met the fate that the absence of brotherhood always brings. But long ago the voice of Jesus was heard: "Love your enemies, do good to them that hate you, and pray for them which persecute you." It is possible to love your friends, your competitors, and even your enemies. It is hard, bitterly hard, but there is a long distance between hard and impossible.

Dr. Helen Kim of Seoul, the president of the only college for women in Korea, found it possible to write to her friends as the Communists came to seize the city:

> There is overwhelming peace in my heart, and full confidence that everything is going to work out all right. With hope never-dying, faith that gives strength, and love that embraces even enemies,
>
> Yours always,
> HELEN

Generalissimo Chiang Kai-shek proved that it was possible to keep good-will, even in the midst of World War II, when he called his friends to prayer in his own home, and the burden of the prayer was that God would keep them from hating the Japanese.

Kagawa, the little stocky Japanese, who is one of the outstanding Christian personalities of our day, has proved that it is possible to love one's enemies. Booker T. Washington, years ago, cried: "I defy any man to make me hate him." The reign of love, and only that, can bring the end of war.

Dreams are they—but they are God's dreams!
Shall we decry them and scorn them?
That men shall love one another,
That white shall call black man brother,
That men shall cease from their hating,
That war shall soon be abating,
That the glory of kings and lords shall pale,
That the love of humanity shall prevail—
Dreams are they all;
  But shall we despise them—
    God's dreams!

Sermon on the Fifth Anniversary of the Victory of Monte Cassino

MOST REVEREND JOSEPH F. GAWLINA, D.D.
Bishop of Mariamme,
former Military Bishop of Poland, now resident in Rome, Italy

Bishop Joseph F. Gawlina was born of Polish parents in Upper Silesia in 1892. His philosophical and theological studies at Breslau University were interrupted by World War I. A German soldier by compulsion, he was constantly suspected of friendship with the Allies. He was a prisoner of war in Palestine from 1918 to 1920. Ordained to the priesthood in 1921, his first assignment was ministering to the working class. After the separation of Polish Silesia from Breslau, he became the secretary of the new diocese of Katowice and simultaneously editor of diocesan weeklies and organizer of Catholic Action for both Poles and Germans.

In 1927 he was designated by the Polish hierarchy to reorganize the Polish Catholic press, and thus he founded the press agency in Warsaw known as the "KAP" of which he remained manager till 1930. Bulletins were edited in Polish, English, French, German, and Italian. At this time Pope Pius XI elevated him to the rank of Monsignor and Canon of the Cathedral of Katowice, where he returned as Counsellor to his new Bishop. While in Warsaw he obtained the Doctor of Divinity degree at the University.

Longing for pastoral work rather than an administrative career, he resigned from all his dignities in 1931 and became a simple pastor of St. Barbara's in Chorozow, the greatest industrial center of Poland. There he labored among thirty-six thousand parishioners,, two-thirds of whom were Poles, the remainder Germans. It was a period of severe depression, but he proved to be a true friend and father to the workman.

In 1933, when the internal situation of Poland became very tense, Pope Pius XI appointed him Military Bishop, telling him: "You will be the buffer between Church and State." Bishop Gawlina, rated the most popular preacher in the Episcopacy at the time, was twice wounded during the last war, escaped from a P.O.W. camp and served his soldiers in Poland, France, and Great Britain. He was a member of the Polish National Council, president of the Commission on Foreign Affairs, and president of the Polish Red Cross.

Nineteen forty-two found him organizing religious work among the Polish soldiers in Russia. Dismissed after six months by the Soviets, he came to the U.S.A. where he exposed the truth about the religious situation in the

U.S.S.R. While here, he was received by the late President Roosevelt, whom he begged to aid some hundred thousand deported Polish children.

The end of the war saw him on the German front line. The only battle in which he did not participate was that of Narwick (1940). In 1946 the present Pope elevated him to be Assistant at the Pontifical Throne, and in 1949 also Protector of the Religious and Moral Life of the Poles abroad. In 1948 the Communistic government of Poland deprived him of the right to return to his native country and declared him a "malefactor." Suffering from the hardships of war, Bishop Gawlina now resides in Rome.

This sermon on the fifth anniversary of the Allied victory at Monte Cassino has some disturbing thoughts on war and life in our time. The Bishop's language and style are at the same time a rare combination of foreign and American.

Sermon Twenty-nine

TEXT: Is my hand shortened and become little, that I cannot redeem? ISAIAH 50:2

FIVE YEARS ago at this hour, we heard that, after a fierce battle, God had given us victory, and that the Polish flag was waving over the ruins of the Monastery at Monte Cassino.

Those were days of grandeur and heroism. For many months the Allies failed to seize this fortress; hundreds fell in the battle for its possession. The Monastery itself was turned into a pile of rubble. The enemy gloated over the idea that no one could repel him from this stronghold which he considered a symbol of his wartime honor—the eyes of the world were converged upon the Abbey of Monte Cassino, that Golgotha of the Nations.

In April, the Polish soldier, a pilgrim to a free homeland, stood at the foot of this mountain. He already had a long journey behind him. The war had pursued him to Russia. There, having at last gained freedom, he equipped himself for an engagement with the enemy in the tundras, the taigas, and the plains. He followed the trail of Alexander the Great through Samarcanda and Merve; he camped in the land of the deluge and in the country of the biblical patriarchs; he asked for strength at the tomb of the Saviour; and finally reached, by way of the land of the Pharaohs, the shores of delightful Italy. Others again, had previously overcome the forces of the enemy in the hot sands of Libya, emerging victorious at Alamein and Gazala.

Old and experienced was this soldier of the second Polish Corps—his age averaged thirty-eight years. There were fathers of families, whose dear ones had been deported to Germany or had not as yet been liberated from Russia; but there were also youths upon whom life smiled. But the same spirit pervaded all, the spirit of fighting for justice and freedom.

Prior to the battle, the entire Corps, from the leader to the last soldier, made their Easter duty and fortified themselves with the Bread of Angels. Thus, with God did we begin the battle of Monte Cassino.

On the 11th day of May, an hour before midnight, the hills resounded with the salvo of death. Two hours later, I began celebrating Holy Mass in the soldiers' tent. The Elevation of this Mass seems to have concurred with the moment of the march of the infantry for attack. Soon afterwards, the wounded were being brought in, and they repeatedly asked the same question, "Is the Monastery taken?" Alas! the following day we were driven back to our original positions. Reports of the death of many officers, leaders, began to pour in. We must begin anew. The battle continued.

There was something mystical about this gigantic struggle. Nowhere, neither in the field hospitals nor at the first aid centers, did I hear any moaning or complaining. The soldier lived with but one thought in mind—how to capture the Monastery. The wounded even struggled from the stretchers, eager to return to their platoons where death reaped mercilessly.

After five days a second crisis arose; victory, however, was in the offing. On the Feast of the Ascension, the news was flashed throughout the world that the Second Corps had taken Monte Cassino. God had granted us victory.

> On those who in battle, for freedom, struggled
> The Motherland the highest honors will bestow,
> As a tender wife, as a sister, as a mother—
> Not the survivors, but those whom death laid low.

There exists no military cemetery in which there does not lie among the brethren a shepherd, a faithful chaplain. Unusually high is the death toll of officers in this battle. These Polish soldiers from various parts of their homeland, from Warsaw, Galicia, Silesia, and prevalently from Eastern Poland, lay down to eternal slumber, having crossed one entrenchment after another, rampart after rampart.

Many of them were golden-haired youths, upon whose brows was engraved the name of the Lamb of God. You gave the world, dearest brethren, an example of how we comprehend freedom, and of the kind of protectors of justice we possess.

We dare repeat after St. Ambrose, "More fierce than death itself, which you despised, you have brought the burden of battle upon you. Oppressed by the defeat of the enemy, you have died in your own triumph."

You threatened the foe: "Thou shalt eat, but shalt not be filled: And thou shalt take hold, but shalt not save" (Mich. 6, 14).

We have offered today to Almighty God the Bloodless Sacrifice of Propitiation for them. May our hearts continue to offer prayers to the Lord of Hosts for those who have been faithful until death.

Their voices seem to cry out to the homeland: Verily, we have created of your name "a prayer that weeps, a lightning bolt that flashes."

"The bones that have been humbled shall rejoice" (Ps. 50, 10). Your mangled, desecrated bodies will rejoice on the day of Resurrection. Their bones will be joined, their bones will be covered with skin, and they will rise from their graves to praise God, Who kills but revives, and bestows upon His faithful life everlasting. "Exsultabunt ossa humiliata."

Monte Cassino has passed into history, but the battle continues. In truth, the principal fortress of Satan has fallen, but he who covers the scars of lightning, resulting from the fall from heaven with the palm of his hand—the Father of falsehood and exile from heaven—persists in leading forward to battle legions of Antichrist, though under various titles.

The struggle for faith and civilization goes on. Is it not significant, or perhaps even symbolic, that the cradle of western civilization, the Abbey of Monte Cassino, should in this war be completely destroyed? We are fortunate that we seized the Monastery, though we did not participate in its destruction. St. Benedict, then, was only a shield for St. Peter. Truly some memento is concealed in all this.

It may be that Europe, and with her, the world, unconsciously finds itself on the verge of the loss of culture, for modern civilization is the work of Europe. It may be that this transition is approaching at a faster pace than did the fall of the Roman Empire, of which St. Augustine wrote, "They have received their reward—vani vanam."

For every type of culture possesses an ideological subsoil from which it draws energy for constant social efforts.

Our own western civilization has its foundation in an eminent measure, on religious tradition. It contains, moreover, another element, namely the tradition of knowledge; nevertheless, not knowledge, but faith has always been the dynamic force of our culture. Learning, notwithstanding its preëminence, does not of itself furnish us with moral stimulus, or rather with spiritual technique.

Indifferent towards moral considerations—I speak not necessarily of theology—learning is comparable to the slave of the ring in "One Thousand and One Nights" who is prepared equally to "build a city or raze it, slay a king, excavate a river bed, or do anything else." Does not the modern discovery of Atomic Energy come to our minds?

Faith, then, stands as the groundwork of our western civilization, whose distinct trait, in contrast to that of the eastern search for knowledge exclusively, and profound research in soothing solitude, is charity—love, and combined with it, activity for the good of one's neighbor.

The most archaic document of western Christianity, the first Epistle of St. Clement, unfolds this practical ideal of social obligation.

The newly-founded Order of St. Benedict transformed the purely ascetical traditions of the desert monks into a social institution, in the service of the Universal Church. This is the significance of Monte Cassino.

Thus the newly-emerged culture, on whose tradition we draw, was eminently of religious origin. The apex (archimedal point) of our future, therefore, is religion; our saber and shield, the Gospel.

The more mature among us developed in an age of defense and apologetics when, in addition, man was handed one Fata Morgana after another, as a surrogate of religious decision. It was the age which did not favor religion, did not create martyrs, for they seemed superfluous. Today we stand before a decision. Its solution does not depend so much on laws, nor custom, as it does on the faith of the believers, the faith of those who live in this world. Victory is dependent on effort, and the fire and ardor of faith.

Defense and polemics do not suffice, for the adversary not only has intellectual claims, but apprehends life and its practices. The present generation, therefore, cannot evade this responsibility, for with it the future will stand or fall.

And we cannot return to the catacombs, there to muster new reinforcements for the world's conquest, though often we sorely need this subterranean solitude to heed the voice of God.

We must extract the last moral forces from the mystery of the Cross despite external weaknesses. Hence, we will search for leaders among men entirely devoted to Christ, just men who live by faith, practice asceticism, and desire to abide in the grace of God. The less we become attached to the concupiscence of the world the more energy shall we possess to transform the affairs of the world. If we do not live entirely by faith, if we are but listeners of the Word and not its doers, we would be likened to the biblical character who beholds himself in a mirror, and having left it, immediately forgets his countenance—likened to a builder who has lost his plans.

Were we to draw solely from cultural tradition, and not be concerned about the lively, decisive faith, a treasure gained through bloodshed by our forefathers, we then could be compared to a man who, gazing into a mirror, did not see his own face.

The so-called catholicism of culture does not hold precedence but rather the catholicism of faith, from which culture develops. When the walls of the fifth floor begin to crack, it does not suffice to cover them with wallpaper of church and national hues, but it behooves us to strengthen the foundations. The foundation is ardent, practical, and steadfast faith. "Seek ye therefore first the kingdom of God and His justice; and all these things shall be added unto you" (Matt. 6, 33).

This day we are commemorating the anniversary of a victorious battle. Our soldier marched on to his goal through seering flames and mined fields. Not only beneath this holy mount, but on all battlefronts the blood of our valiant men flowed freely. We wanted to serve God conscientiously with all our strength, to be of service to God through the service of our country. Wherever, O Lord, the cannon proclaimed Your Majesty with fire, and Your Name thundered in the hurricane of battle, we were present. Where day changes to night, where the heavens are ablaze with the break of morning, there proceeded the trail of our wanderings. Where "the voice of the Lord breaketh the cedars, divideth the flame of fire, shaketh the desert (Ps. 28, 5–7), there in battle we have cried to Thee. Amid the wonders of the deep and above the rainbows of the skies, we have wielded the "flaming sword, turning every way" (Gen. 3, 24).

When the forces of evil shouted, "Finis Poloniae," we sang out the more loudly, "Gaude Mater Polonia, prole foecunda nobili" (Hymn of St. Stanislaus).

But peace has not yet dawned—the peace that would be a triumphal hymn of the just man. Our homeland is still bleeding from a thousand wounds. The world has been to us both a flatterer and a tyrant. It matters not—our hope is in God. Let us first seek the Kingdom of God and its justice, and all else shall be added unto us. And let us remember the word of Jesus that "every plant which my heavenly Father hath not planted shall be rooted up" (Matt. 15, 13). The fallen angels also believe but tremble.

"Concussus, surgo." This is the inscription I recently read on one of the buildings of Rome. "Shattered, I rise again." Beside it another inscription, as if a commentary on the first, "Ardeo, nam credo." "I am afire, because I believe."

The ardor of faith burns in me. I know that not large numbers, but a more ardent faith will determine the outcome. We shall counteract the red conflagration, born of false imagination, with the fire of the Holy Ghost, particularly with that flame kindled in us by the Sacrament of Confirmation the Gift of Christian Fortitude.

And thou, St. Stanislaus, Patron of Poland, who dying, interrupted your song of life at the altar to complete it beyond at the throne of God, fill us with your spirit.

God has said, "I kill and I bring to life again." "Is my hand shortened and become little, that I cannot redeem?" (Isaiah 50, 2).

Sursum corda! In Te, Domine, speravi, non confundar in aeternum!

This Is America's Hour

Reverend Louis Hadley Evans, D.D., LL.D., D.H.L.
Minister, First Presbyterian Church, Hollywood, California

Dr. Louis Hadley Evans is the pastor of the First Presbyterian Church of Hollywood, California, a church of over six thousand members—now the largest Presbyterian church in the world. The church has the largest Sunday School enrollment of any Presbyterian church in the country, over four thousand.

Dr. Evans came to Hollywood in 1941 from the Third Presbyterian Church of Pittsburgh which has given two moderators to the General Assembly of the Presbyterian Church and which has been known for its strong philanthropies and benevolences. He studied at Occidental College and took his Divinity degree from McCormick Theological Seminary, Chicago. He also holds the honorary degrees of Doctor of Divinity, Doctor of Laws, and Doctor of Humane Letters.

While in college he was active in athletics, having been chosen by coaches as All-Conference and All-State in football and All-Conference and All-Southern center in basketball. He played a part of the season on a basketball team that won the national championship. Every year he spends at least three weeks on various college campuses and received invitations to lecture and hold conferences at some thirty different colleges and universities during the last twelve months. He was one of the main speakers at five international Christian Endeavor conventions, takes great interest in educational matters, spends considerable time lecturing at teachers' institutes and was a member of the special Citizens' Committee chosen to consult with the Los Angeles Board of Education on the securing of the Superintendent of Schools and other policies. His own church has a College Department of some six hundred members.

He has also visited colleges, universities and missions in Japan, China, Korea and India, visited Northern Africa, Syria, Palestine and Greece and traveled with representatives of the Near East Relief through these countries.

In 1936 he was elected to the Presbyterian Board of Foreign Missions, but in 1937 resigned this position to take the presidency of the Presbyterian Board of National Missions, which he held for three years, until moving to the west coast necessitated his resignation as president. He is now west coast representative on this board. He has addressed seven General Assemblies of the Presbyterian Church of the United States of America, is a member of the board of directors of the Protestant Film Commission, a trustee of the

San Francisco Theological Seminary, has served on the board of directors at Occidental College, World Alliance of Reformed Churches and others. He gave weeks to the government addressing our armed forces. He is author of Youth Seeks a Master.

"This Is America's Hour" was preached in the National Presbyterian Church, Washington, D. C., on February 19, 1950, caused much discussion in Washington among members of that church, and led the Honorable Edward H. Rees to call attention to it in the House of Representatives and to have it printed in the Congressional Record. Dr. Evans' discussion of God and national life shows America's need of a spiritual rebirth.

Sermon Thirty

TEXT: Blessed is that nation, whose God is the Lord and the people whom He hath chosen for his inheritance. PSALM 33: 12.

IT IS EVIDENT by the text that God has a philosophy of history—that He chooses what nations He will for His cosmic plan.

Deity has undoubtedly smiled at the many sacred anthropomorphisms, figures, similes and pictures that we have put together in our endeavor to describe Him and His attributes. We have pictured God as a Ruler upon a throne; as a Father in the heavens providing for His little brood; as the Lord of Hosts leading His forces to battle; as the Judge upon the universal bench ever doing righteously; in Genesis He is pictured as the Holy Spirit, as a Mother bending over the cradle of the universe and crooning over the young child humanity. Yet the most touching picture I have seen of God was one in which He was sitting at the table of the universe playing chess, and the nations and the men were His pawns. He was playing a great game of strategy and His aim was to build a Kingdom of Righteousness. The free will of man, as He chose to give it, seemed at times to allow such moves on the part of evil men and nations as to retard His progress and thwart His victory. Yet one was conscious always that God would win in the end and that the last maps would always be made in heaven. In the moves on the chessboard of history "careless seemed the Great Avenger, yet behind the great unknown stood God keeping watch o'er His own."

As nations served His divine purpose, they remained on the chessboard of history but when they got in His way, or in the way of the good of mankind, He removed them from the board. Nineteen civilizations had already been moved from the chessboard of history, for this strategic game was played with many men. A study of history has shown that nations have been His pawns and His pieces and this has been His strategy with knights and castles and kings and queens and bishops and pawns.

Benjamin Franklin at the framing of the Constitution, when he was eighty-one, spoke these words:

"I have lived, sir, a long time; and the longer I live the more convincing proofs I see of this truth, that God governs in the affairs of men. And, if a sparrow cannot fall to the ground without His notice, is it probable that an empire can rise without His aid? We have been assured, sir, in the sacred writings that 'except the Lord build the house, they labor in vain that build it.' I firmly believe this, and I also believe that, without His concurring aid, we shall succeed in this political building no better than the builders of Babel."

Let us glance now hastily at the way in which God has moved and used various nations in the over-all strategy in this game of the spirit and the kingdom.

Israel was to be used as His chosen people to convey His truth and His oracles to the nations with whom this people should come in contact. But first of all Israel must be tutored—she must learn obedience to God, humility and dependability upon Jehovah. This she must learn in the throes of captivity.

Thus Babylon is raised up and a part of these people is carried captive and under Babylonian yoke God's people learned obedience and a reaching out of hand and heart to Jehovah's guidance. The ministry of Babylon is done and she is soon to be removed from the chessboard of nations.

Persia now rises to power and by the prophets Cyrus is called the "tool of Jehovah" and God maneuvers this knight of history into such position as will enable him with his armies to knock with his mailed fists on the brass gates of Babylon and Babylon falls. God is done with Persia in its might.

Greece must now be born. A light of hope is soon to break. Immanuel, "God is with us," is soon to be born in Bethlehem and such speech will be His, such magnitude of thought and hope, that somewhere must be found a language that will rightly frame this message. It must contain fine shades of philology and meaning and be a fit vehicle for the expression of this new gospel. Thus Greece is born. God sets in the fanatical mind of young Alexander the determination that the whole world will think Greek, live Greek, and speak Greek. On the emery wheel of Greek education, Paul the Apostle will polish the facets of His mind and Philip the Evangelist, will receive his forensic powers. Much of the world will speak Greek and this will give God speech. Greece with her naked intellect will lack certain spiritual qualities and Greece will succumb.

The Romans now will come to power and they will make two definite contributions in preparation for the Christian era. Rome will give to the world a sense of world empire and kingdom sweep, until the expression "Jesus Imperator" will seem reasonable and the world will be ready for a cosmic king. Rome will also furnish means of communication and the Roman roads will make way for the dispersal of this new message to the far corners of the earth. Paul desires to "preach the gospel to them that are at Rome also," knowing that a gospel heard at Rome would soon be heard around the world. Having made these two contributions to God's strategy, Rome dies, as Papini, noted Italian historian and scholar says, for three reasons: "Because the temple, the bank, and the academy were against Jesus of Nazareth." Such triple inadequacy would mean the ruin and death of any nation.

The Goths and Huns now come to power. Beneath the barbarian skins of these Germanic peoples lie the potentials of philosophy and spiritual insight. The German mind builds up the German university and the Christian clergy swarm to the German classroom to receive their higher degrees in Christian philosophy. Then came the decay of the German mind, in the psyche of materialism. A great German philosopher, when asked why Germany collapsed so suddenly, replied, "Germany did not collapse suddenly, they pushed God out of the universities in Germany fifty years ago and that was the beginning of the death sentence for the empire." That could happen here. The great Germany died.

Now came the Anglo-Saxons and how vital they became. The Druids put aside their human sacrifices and sang hymns, and Gaelic peoples held high banners with crosses on them. Britain became the greatest colonizing empire in the world and with her often went the emissaries of the cross. Her most illustrious queen held in her hand a Bible and said, "England is the land of the Book." Now someone has said that approximately only fifteen out of every hundred Britons are active in any Christian church. We agree with Churchill that we would not wish to see the disintegration of the British Empire. Will England pass out of the picture in our day? May these warning words of Kipling, one of Britain's great sons, never come to fruition within the pale of your memory and mine:

> Far flung our navies melt away,
> On dune and headland sinks the fire;
> All of our pomp of yesterday
> Is one with Nineveh and Tyre
> Lord God of Hosts, be with us yet
> Lest we forget, lest we forget! [1]

Now America is placed on the board of history—so definitely held in the hand of God. There was no question in the minds of our early fathers that the express purpose of America's history must be to play the game in furthering this kingdom of God's righteousness.

[1] From "The Recessional," by Rudyard Kipling. Reprinted by permission of Doubleday & Company, Inc.

Our pilgrim fathers, by the light of the smoking lamp on the *Mayflower,* before landing at Plymouth, inserted these words into the Mayflower Compact. "We whose names are underwritten have undertaken for the glory of God to establish in Virginia the first colony for the advancement of the Christian faith."

Ten years later, in 1630, other Pilgrims said in the New England Federation Compact agreement, "We all have come into these parts of America with one and the same end, namely to advance the Kingdom of the Lord Jesus Christ." America was established as an experiment station to show what would happen to any people who cared and dared to follow Jesus Christ.

We received our Constitution from the church. John Witherspoon, a clergyman who signed the Declaration of Independence, was a Presbyterian elder and a professor in Princeton University. At his feet sat a young man named James Madison with a mind keen as a rapier, imbibing from his professor the principles of the Church of the old country—the sovereignty of God, the freedom of the individual, and the form of representative government borrowed from the Church of which he had been and was a part. Then, in the same era in which John Witherspoon drew up the constitution of the Presbyterian Church of the United States of America, James Madison, his pupil, was instrumental in drawing up the Constitution of the United States of America which was borrowed from the idealism and representative government propounded by his professor and the Church. So definite and great was this source and this contribution that Green, famous historian of the English people, said that the Constitution of the United States was the gift of the Presbyterian Church. Our history has been inseparable from Christianity.

The Pilgrims landed in Plymouth in the name of Christ. The Huguenots took New England in the name of God. William Penn settled Pennsylvania in the name of the Nazarene, and the padres landed on the shores of California with a crucifix in their hands. The kiss of God has been upon our faces as we have grown from national babyhood to what we trust is but our young manhood.

Scanning history we have never expected to do anything compelling without the leadership of God in Christ, and without the kingly guidance of the Nazarene, we seem to do nothing well. Washington is on his knees at Valley Forge and so vital is the contribution of the Church in revolutionary love for liberty that the Revolutionary War was called "a church rebellion." Lincoln rises from his knees, his eyes bloodshot with praying, and exclaims, "I know that liberty is right because Christ teaches it and Christ is God, and if there is a place for me to be offered I am ready now." How often did the pew of a certain church find Honest Abe sitting before the challenge of the Christian emancipation? Thus from the beginning of our free sovereignty we have had no human king. Ours has been a nation in which every man was royal but no man cared to wear a crown, yet conscious always that in the deeper chambers of our souls we had a king—

Our fathers' God to Thee, Author of liberty,
To Thee we sing. Long may our land be bright
With freedom's holy light, protect us by Thy might,
Great God, our King.

This is America's hour—an hour in which the world is watching her and demanding that she answer a question, the question is this, "Has your Hebraic Christian religion worked with you? Will your Christian ethics suffice? Is Christ sufficient? Or shall we look for another?" A Swedish prince gripping me by the hand said some time ago, "Dr. Evans, you Americans must not let God fail within your midst. He is the only One the world has left and the world is watching you. You must not fail us."

The world knows our background, and our history has forced them to demand this. Contend as you and I will that America has never been thoroughly Christian and that there have never been twenty minutes when all Americans have followed Jesus Christ, and yet they will say, "God has been there. We have seen his footprints on the pages of your history." They will not excuse our religious inadequacy; they will permit us no feeling of irresponsibility and will grant us no satisfaction or excuse for spiritual inferiority.

This question is hurled at us now. "What is the result of the cross in operation?" In this hour of fluid decision the world is getting ready to decide. This is a world of logic and the spirit of the times is pragmatic, experimental, practical, and utilitarian. Mankind will keep consistently asking, "Does it work for you?" This will be the genius of our world leadership—not what we say but what we are; it will not lie in our profession but in our performance. This is the logistics of the spirit—the foreign-missionary enterprise abroad will depend to a great degree upon the success of national missions at home. Our answer as to whether or not we will use God's program will have a great deal to do with whether or not America remains on the chessboard of His divine strategy.

America needs a rebirth of the spirit if she is to maintain her world leadership.

We could spend considerable time here reiterating the virtues and the greatness that have made us unwilling to trade her for any other land under the sun. "God bless America, land that we love."

Yet our consciousness of her inadequacies is not an antonym of love. The doctor's alertness to his patient's disease is a token of his concern. The nurse's sensitivity is a sign of her usefulness. The church's honest appraisal of a nation's need is loving to the end of its cure. J. B. Priestley once said: "We should behave toward our country as women behave toward the men they love. A loving wife will do anything for her husband except stop criticizing and trying to improve him.

"We should cast the same affectionate but sharp glance at our country. We should love it, but also insist on telling it all its faults. The noisy, empty patriot, not the critic, is the dangerous citizen."

No form of national betrayal might be worse than a blind, complete satisfaction with one's own land. Jesus loved Jerusalem, and wept over it.

In America we are facing a tragic malnutrition of the spirit. The head of one of our great national missions boards said that, "More people numerically (not percentagewise) are without God in America now than ever before in our history," and that "a greater number of Americans are out of touch with spirituality according to our standards than ever before in our national history." It is estimated that there are some 27,000,000 youths in America who have no definite religious instruction, and are as spiritually illiterate as though they had been born on the South Sea islands.

It has been estimated that 40 percent of our grammar school children have no connection with any religious organization; that 80 percent of our high school students have no definite affiliation with organized religion; that 90 percent of our university students "slept in" last Sunday morning.

What are the results of this spiritual malnutrition in our land? A crime problem, which is probably the worst in all our history, with a rape in America every fifty-two minutes, though we have applied to this problem our finest and our keenest brains. Insobriety and alcoholism strike at 4,000,000 persons annually, and 100,000 women and girls are in institutions for inebriates, sixty percent of them, the doctors say, as condemned to death as though they had cancer. Between one-third and one-fourth of our American homes are destroyed by divorce. And the newspapers reek with the records of our moral failures.

This is the hour for the Church. From her hearthside have come the missionary statesmen of the world. From her cupboard comes the spiritual food by which multitudes are fed. Her tables yield the vitamins of the spirit. From her wardrobe men must be clothed with the "garments of righteousness." Minds must be lighted by the spiritual fuel within her lamps, and the milk of her human kindness must feed a slowly starving flock of worldlings.

America is in the grip of a gnawing sense of loneliness that has seized the American people. Seventy million Americans have moved their residences since the war. Stephen Laird has pictured for us these people torn from home, toil, kindred, community church and community responsibility, and for the most part these ties have not been renewed. People are frustrated, not so much about taxes, the high cost of living, jobs, but about friendships. "Nobody seems to care." What has happened? America has become a nation of migrants. Going to camps, pouring in and out of war plants, fleeing from dust bowls, pouring into the city from the country and back into the country from the city. GI's are looking around trying to settle down, but the average veteran knows more about his bus ticket than he does about his Bible. Nobody seems to care what is happening to them. He goes on to relate that a certain writer in New York was asked the question, "Have you written anything since I last saw you?" "No, I just don't seem able to sit in an apartment and write. Almost every evening I find myself walking along the streets, dropping into bars, sitting opposite a man or girl my age in a restaurant, hoping to start a conversation, to make a friend or meet someone worth talking to." "Any luck?" "No," he said. "Everybody seems to be busy with himself. I get so lonesome I hate to go home any more."

Thus America, with plenty on her table, goes home hungry for companionship and God and for someone who cares. Along the crowded dogged streets, the lonely streets, men need that old hymn "What a fellowship, what a joy divine leaning on the everlasting arms."

The Church must invite these lonely souls to the firesides of the spirit. We must pursue these migrants of body and soul into every hamlet, Middlesex, village, farm, and metropolis, remembering that:

Where cross the crowded ways of life,
  Where sounds the cry of race and clan,
Above the noise of selfish strife
  They need Thy voice, O Son of Man.[2]

It is the spirit of God that we need in this hour—the upward pull of God. Should you attempt on a pool table, with ball and cue, to herd together in close proximity the other balls on the table you would probably end up with a total result of the several balls being farther away from each other than they were when you began your attempt. This horizontal effort would prove somewhat fruitless. If, on the other hand, you would with scotch tape affix to each ball separately a strand of string, and taking these several strings in your hand exert an upward pull, you would find that automatically the balls would nestle together in a compact, orderly way.

Our efforts at brotherhood and world fraternity have for the most part been unsuccessful because the effort has been on the horizontal. We have tried with the cue and ball of laws, treaties, wars, leagues, regimentation, and other mechanistic means to drive or herd men together into a fraternity. Over-all the result has been rather to separate them. If, on the other hand, we could subject individuals and groups to the upward tug of the spirit and allow God to exert the uplifting tug on the hearts of men we should find them nestling together in the collective fraternity that we all so earnestly desire. But this deeper fraternity and sorority must be of the Spirit.

It is quite evident that the horizontal force of war has failed to do this. A Boy Scout leader suggested that three members of his troop who had failed to do a good deed for that day be dismissed from the Scout meeting for a half hour in order to go out and perform it. They came back shortly and stood before him in a state of very bad repair. One had his face scratched, another had a bloody nose and a torn neckerchief, and the third Scout had lost both his hat and his equilibrium. The Scoutmaster asked, "What in the world have you boys been doing?" "We were doing our good turn, sir, and

[2] From "Where Cross the Crowded Ways of Life." Used by permission of Dr. Eric M. North.

helping an old lady across the street." "But why all the fuss and trouble?" asked the Scoutmaster. The spokesman said with due solemnity, "Well you see, sir, she didn't want to go." In the spirit of true Boy Scouts, you and I might wish to do our good turn by forcing humanity across the street from hatred to love and from war to brotherhood, but if humanity does not wish to go humanity will not go, and the sum total result will probably be bloody noses, torn neckerchiefs, lost hats and equilibrium, chaos and war.

In the realm of business, America needs the baptism of the spirit. A financial expert, this one not defined as "a man who has a Tau Kappa Alpha key at one end of his chain and no watch at the other," was asked to examine both personnel and program of a large factory to ascertain why they were going to pieces. After thorough investigation he called them together from the president to the board of directors to managers of personnel and labor on down to the last man who worked at the lathe. Calmly and decisively he said, "This factory needs conversion immediately." The president remonstrated, "But we are in the process of converting our machinery." The expert replied, "I am not talking about the conversion of machines. I am talking about the conversion of men. From the chairman of the board of directors to the last man who works at the lathe, you are all rotten with selfishness. Unless you convert the hearts of this firm from selfishness to unselfishness you will suffer nothing but disaster." Why not get at the root of the matter —self? You and I know, in our best moments, that the passing of the Taft-Hartley bill or the repealing of it will be little more than treating the pimples on the skin, when deeper down the blood stream is bad.

Sociologically America needs a new birth. It is one thing to have good building plans as the sociologists do for a new society, but where will they find the new materials? One clergyman has stated, "No clever arrangement of bad eggs will ever give you a good omelet." No clever sociological collective arrangement of bad persons is going to give you a good society, as much as we believe in organization. The soul of improvement will always be in the improvement of the soul. You cannot have a new deal with the same old people nor a fair deal with unfair individuals. "If any man be in Christ, he is a new creature," then come these words, "Behold all things are becoming new." It is difficult to change things externally without changing people internally. Social reconstruction will rest upon individual regeneration.

So far as our racial differences and prejudices are concerned, we deal here with an equilateral triangle. Our God-man relationship is the altitude which determines our man-to-man relationship which is the base. It is an adequate sense of the fatherhood of God on the perpendicular that will force us to an adequate brotherhood of man on the horizontal and it has been proven that it is difficult to have an adequate sociology and love for man unless you have an adequate theology and love for God. Until then, our nation will be broken up into brittle little bits on the basis of blood and ethnological conceits.

American education needs a baptism of the spirit. When Horace Mann claimed that if we would send people to school, we would cut down crime nine-tenths, it was a hundred years ago in the State of Massachusetts. We have tried the experiment and greatly increased our educational facilities but in that time crime has increased eight hundred percent. It is quite evident that all our problems do not revolve around a cranium, and naked intellect alone is not enough to see us through. Educators are not so much interested now in the size as in the sort of our pupils. Education is no longer merely about information but about formation of character. It is no longer merely a matter of what our young people know but, more important still, of how they feel.

Speaking to one of our great American writers an alumnus of Yale said, "You know, sir, Yale has added another branch of research to its curriculum. Yale now has all the branches." The great scholar and writer wistfully replied, "Yes, Yale has all the branches, but Yale has none of the roots." That is an unfair thing to say about Yale. I lunched with its great professor, William Lyon Phelps, shortly before his death and that conversation was along the line of a previous statement of his: "Jesus knew more about political economy than all the professors in all of the colleges of the world. He knew more about the human heart than Shakespeare. While I am not sure of many things, I am as sure as I stand here that Jesus is the greatest leader, the most absolutely right person the world has ever known. The leadership of Jesus Christ in our lives ought to be made constant in everything. . . . But I am not out of my mind when I say that the only way the world can be saved today is by following Jesus Christ." What the commentator on that university wished to portend and to say was probably this—that if we build up a colossal system of secular education as a gigantic tree with all the branches of research, the twigs of knowledge and a massive trunk of secular content, and fail to ground education in the roots of spiritual purpose, merciful motivation, and religious aim, we shall see this thing topple over on us and crush us in a scientific age.

A great nostalgia for God is sweeping the universities of America along with the realization of the insufficiency of humanism and the mere factual. Both faculty and students are beginning to realize that unless the "soul catches up with the brain" the degree B.A. will again, in the next era of chaos, stand for "Builder of Alibis."

Our homes need saving and the sides of the triangle—husband and wife—will be best held together by the extrovert base, God and His Kingdom and their unselfish purposes. While one out of every three or four homes is broken by divorce in America, someone has stated that only one out of every fifty-seven church homes suffer this decay. There is evidently something, then, in Christianity and the presence of God at the hearthside which gives the home fifteen times the chance of standing and surviving with religion than it would have without its benefits.

America is not leading the nations in law-abidingness. We shall never attain that leadership until in the realm of law men feel the guilt more than the punishment. J. Edgar Hoover, head of the FBI, has reminded us of how difficult it is to reform the adult criminal and how vital it is that youth receive the teaching of the Sermon on the Mount and the Ten Commandments. He reminded us not to forget that the Sunday school and the church were the greatest moral forces in our nation. Until we set up in the tribunal of the human heart a love for the laws of God and society we shall go on for another generation "making new laws for old sinners to cheat by."

America has won the war—but only on paper. We do not conquer the Japanese when we merely take the guns out of their hands. Not until we have taken wrong goals out of their hearts have we won a lasting victory. A personal friend of mine reported that MacArthur had made a request for 18,000,000 Bibles for Japan and wanted 3,000 missionaries in order that Japan might be properly exposed to Christian teachings. He felt that this war was "a theological war" and that unless the bad theology of Amatarasu, the Sun goddess, loving only the slant-eyed people, could be changed to the universal fatherhood of God through Jesus Christ and that high concept, Japan would never have a sociology and a brotherhood that would enable her to make her contribution to a lasting peace. As one GI said, "America with her rich spiritual heritage should remember that it will never do for her merely to give her enemies 'a licking,' she must also give her enemies 'a light.' "

Communism is challenging us to outlive it with Christianity.

John Foster Dulles claims that communism enters a nation as a Trojan horse. In clever disguise it enters the gates of a city or a nation, and only after its entrance do people realize what lurks inside. Communism thrives on the failures of others. It makes its garments out of the holes and frets in others. Communism builds its ship of state out of the driftwood of the wreckage of other systems that lies along the shores of contemporary failures. Communism does not work. It only thrives on the supposition that other ideologies have not worked. It is the question of the cross as over against the sickle. They will whisper to the African in Africa, "How do you like the way black men are treated by Christians in America?" They will say to those in Mexico, "Are you pleased with the treatment received by Mexicans in the American Southwest?" To those of China they may make the query, "Are the 28,000 Chinese of the San Francisco area impressed by the Americans who live under the sign of the cross?" Of labor they inquire, "Has Christianity taken the crown of thorns from your brow?" Of those of great possessions, they do ask, "Along with your property do you possess peace?" The line is down now and the question is inescapable. Will the cross outdemonstrate the sickle? Our own type of democracy at its best was born in the Christian concept and democracy will not "democ" without Christ, as someone contended. America will not be able to stand, any more than Rome was able to stand, if her banks—her economic system; her academy—her educational system; and her temple—her religious system, are against Jesus of Nazareth.

It will be difficult to match the propaganda of communism but we must outmatch her performance. We must not only "out-promise" them, we must also "out-passion" them. We must teach our youth the concepts of the Kingdom of God with greater earnestness and thoroughness than they teach their youth of the Kremlin. The open hand of democracy must prove itself more potent than the clenched fist of communism. We will find it difficult to outlaw communism; we must outlive communism. And this is the hour of dynamic demonstration.

I am not advocating a narrow nationalism when I say, "God bless America and give us success." In this I am pleading for a supranationalism. I want to see her continued on the chessboard of history because she makes her contribution to this strategic game of righteousness among the nations that God is playing. I want to see her continue to be a tool in the hand of Jehovah.

This is a personal equation. U. S. spells "us." America fulfills her spiritual Christian destiny only as God possesses our individual purses, powers, and persons. Each must do his and her part for the collective righteousness.

A student at Penn State College said to me some time ago, "Dr. Evans, I prayed a great prayer last night—the bravest I have prayed." "What was it?" I said. He replied, "I prayed, 'O God, may I always be honest enough never to ask Thee to do for the world collectively what I have not first allowed Thee to do for me personally.' "

Let us not expect of the nations any more ready forgiveness of their enemies collectively than our forgiveness of our enemies personally. Let us never expect the nations of the UN to be any less selfish than we are in our personal relationships.

> We who are Americans must be too big for shame.
> We are the guardians of a faith from which our glory came.
> Ours is a splendor truly won, in us all people trust.
> We face this world as men who are great hearted, kind, and just.
> And he who fails to play his part in high or little place
> Shames not alone himself but brings disgrace upon his race.[3]

The light that shines brightest at home shines farthest abroad. The light will shine farthest when we wash the lamp chimney or cleanse the lens through which it shines.

We do pray "America first"—first in usefulness to God. First in her availability to the spiritual kingdom enterprise. First in power that she may bend over the weak and suffering. First in wealth that she may share it with others in mercy. First in light if she be a lamp in the darkness. First in strength that she may bear the infirmities of the weak. First in purity

[3] From "We Who Are Americans" by Edgar A. Guest. Used by permission.

that others may sup from our cleansed wells. Let us couch our national concern for our nation and our flag, in this strategic hour, in the words of G. Ashton Oldham:

America first, not only in things material,
But in things of the spirit.
Not merely in science, invention, motors, skyscrapers,
But also in ideals, principles, character.
Not merely in the calm assertion of rights,
But in the glad assumption of duties.
Not flouting your strength as a giant,
But bending in helpfulness over a sick and wounded world like a good Samaritan.
Not in splendid isolation,
But in courageous cooperation.
Not in pride-arrogance, and disdain of other races and peoples,
But in sympathy, love and understanding.
Not in treading again the old, worn bloody pathway
Which ends inevitably in chaos and disaster,
But blazing a new trail, along which, please God,
Other nations will follow into the new Jerusalem
Where war shall be no more.
Some day, some nation must take that path—
Unless we are to lapse into utter barbarism—
And that honor I covet for my beloved America.
And so in that spirit, and with these hopes,
I say with all my heart and soul, "America first." [4]

NATIONAL AND INTERNATIONAL

A Christian View of the International Crisis

REVEREND O. FREDERICK NOLDE, PH.D., D.D., LL.D., L.H.D.
A Minister of the United Lutheran Church
Director of the Commission of the Churches on International Affairs of the World Council of Churches; Dean, the Graduate School, Lutheran Theological Seminary, Philadelphia, Pennsylvania

Born in Philadelphia, Pennsylvania, in 1899, he attended Muhlenberg College and Lutheran Theological Seminary and did graduate study at the University of Pennsylvania where he earned his Ph.D. in 1928. He taught religious education at the Lutheran Theological Seminary, Philadelphia, and became Dean of the Graduate School in 1931. At the University of Pennsylvania, he was assistant professor and lecturer in religious education. He

[4] Used by permission of Bishop G. Ashton Oldham.

was Executive Secretary of the Joint Committee on Religious Liberty, constituted by the Federal Council of the Churches of Christ in America and the Foreign Missions Conference of North America, 1944–49, and is now director of the Commission of the Churches on International Affairs, established by the World Council of Churches and the International Missionary Council and Associate General Secretary of the World Council of Churches with portfolio on international affairs.

He was a member of the former Commission on a Just and Durable Peace of the Federal Council of the Churches of Christ in America and is currently a member of the Department of International Justice and Goodwill of the National Council of the Churches of Christ in the U.S.A.; Associate Consultant to the United States Delegation at the United Nations Conference on International Organizations, San Francisco, 1945, representing the Federal Council of Churches; consultant, representing the C.C.I.A., the United Nations Economic and Social Council since 1947; accredited representative since 1947 to the United Nations Department of Public Information, Member of the Committee on Human Rights of the Commission to Study the Organization of Peace; research affiliate of the American Association for the United Nations, Inc.

In 1932 Muhlenberg College conferred the D.D. upon him and an LL.D. in 1946, and in 1951 Wittenberg College recognized his achievements in the cause of international justice and peace with an L.H.D. He is a contributor to religious and educational journals and is the author of several books, including Yesterday, Today, Tomorrow (1933); Truth and Life, (1937); Christian World Action (1942); Christian Messages to the Peoples of the World (1943); Power for Peace (1946); Freedom of Religion and Related Human Rights (1948), *Chapter V of* The Church and International Disorder, *Study Volume IV, of the First Assembly of the World Council of Churches;* Freedom's Charter (1949), *Foreign Policy Association Headline Series No. 76.*

"A Christian View of the (present) International Crisis," an able address, was delivered at Cleveland, Ohio, November 30, 1950. In this field it is a real contribution to current religious thinking and speaking. Technically not a sermon, it is, nevertheless, included for its penetrating analysis of the present world situation and the road to an enduring peace and a solution of many problems facing men today. His last paragraph should be copied, quoted, and reprinted in thousands of church bulletins to remind Christians of their personal part in helping to make a peace under God in our time.

Sermon Thirty-one

YOU will not misunderstand me when I say at the outset that I approach my task tonight with fear and trembling. My disturbance does not arise from the limited time available for preparation but rather from the subject which I have agreed to propose for your consideration. I would be at measurable ease in describing for you the program of the Churches' Commission on International Affairs, a joint agency of the World Council of Churches and the International Missionary Council—the manner in which the Commission is developing a structure for continuing Christian testimony to the world of nations and the representation it is making on world problems currently before the United Nations and related bodies. That easy course is not open to me tonight. In face of tangled evidence and obvious ignorance of details, I am forced by request and circumstances to attempt an appraisal of the present critical situation, and to suggest lines of procedure which will reflect Christian conviction. In this effort, I do not speak officially as Director of the C.C.I.A. I approach my task as an individual Christian and a citizen of the United States, remembering that my recent activities have permitted exposure to points of view held by Christian leaders in many parts of the world.

My approach to the present situation is on the background of two presuppositions. The first is my conviction that God in his goodness makes available strength in proportion to the needs of the hour. Christianity was born and bred in adversity, and Christians have risen to greatest heights when confronted by the most critical tests. God will provide strength at this moment of history if we will only use what he places at our disposal. In the second place, I hold to the belief that even in this trying hour a third world war is not inevitable. In accepting that view, we must not delude ourselves. Events of the last few days have considerably diminished the margin of possibility. We are on the brink of a precipice. To avoid being plunged into the abyss, we must exploit to the full every possibility which that thin margin affords us.

The broad outlines of the world's dilemma are generally known. With full recognition that we are not without guilt with regard to the total situation that has arisen, we are convinced that an act of aggression has been committed. The United Nations authorized an international police action which has, on varying grounds and in various ways, been opposed and resisted by North Korea, by the Peoples' Government of China, and by the U.S.S.R. In the United States, as well as in other countries, a far-flung program of military preparation is under way. While there are some who believe that force should never be used, the preponderant majority reluctantly accepts the necessity of military strength to serve as a deterrent to aggression and, wherever aggression occurs, as a means of opposing it. The reality of the situation is that the world is divided into two armed camps and that the United States and other governments are making every effort to have a military power sufficient for any emergency. In face of these circumstances, the definition of our problem is not too difficult. We must find those affirmative steps which will decrease the need for using military measures now under way and, if further need does arise, to keep such military measures at every point to the absolute minimum which circumstances will allow.

In seeking to determine these affirmative steps, we must ask ourselves what can be done by the people to whom we can speak and who will listen to us. We cannot at this time hope to get sympathetic audience from Soviet controlled areas. What we can do is so to govern our own immediate actions by justice and goodwill as progressively to extend the area of our influence and ultimately win a hearing from those who now oppose us.

Before proposing concrete steps, let me stress the importance of building into the fabric of our individual and community life the two presuppositions which I previously mentioned. Our people need the comfort and inspiration which come from the conviction that God does make available the strength which the emergency requires. The knowledge that there still remains a marginal possibility of averting a global war should intensify their efforts in this critical time.

In suggesting eight possible lines of procedure, I shall first of all indicate the danger against which we should guard and then explain or illustrate the positive means by which this may be done. Needless to say, isolation of any one proposal from the others may lead to a faulty conception of what is intended.

I

Guard against hysteria which robs action of moral perspective and political wisdom. Confronted by imminent danger, people are disposed to cast reason and judgment to the wind. In alarming proportions, there is, for example, the demand that we should now drop atomic bombs on Communist China and promptly invade Manchuria. I unhesitatingly express my personal opinion that these are the very things which we should *not* at this moment do. In the first place, until all other means have been explored and found impossible, the dropping of atomic bombs would not be morally justifiable. In the second place, we would be veritably playing into the hands of those who want to pin upon us the tangible responsibility for starting a world war. Be it noted that in saying this, I speak in the light of conditions as they exist today and allow that the conditions of tomorrow may be different. I would not give to those who are resisting the decisions of the United Nations the comfort and assurance of a pledge not to resort to drastic measures if their actions leave us no reasonable alternative. By calmness and confidence, born of the conviction that God will ultimately prevail, we must meet our responsibility in a way that reflects moral principles to the fullest possible extent permitted by a tragically evil situation.

II

Guard against self-righteousness and hatred which give impulse to the monstrosity of a holy war or a preventive war. We cannot overlook evils in others; in fact, we must condemn them. But the conviction of our implication in world guilt and the knowledge of what total global war means rule out any possibility of war ever being holy or of a preventive war ever being justified. If war should come, and God forbid that it may, it must be irrevocably thrust upon us by others than ourselves and every alternative recourse which our interpretation of God's will for man can entertain must first have been honestly explored.

III

Guard against unilateral action which rejects a moral judgment as reflected by majority opinion in the United Nations. One hopeful factor in the United Nations' decision to recommend military measures against aggression by North Korea was the emergence for the first time in history of international police action, imperfect though that recognizedly was. That new factor must be retained and improved. The General Assembly, by action on November 3, has invited each member of the United Nations to maintain in its national armed forces elements for service as a United Nations unit or units. If military forces stronger than those now operating in Korea are needed, they should be provided in accordance with this action of the United Nations. If when all such designated forces have been made available even more are needed, additional forces should be provided within that same framework. The United States, as a government of preponderant power, must submit its actions to international consideration and decision. The day when any one country has the right by itself to declare war on another is past.

IV

Guard against false pride and face-saving tendencies which close the door to open-minded and effective negotiation. We must be constantly willing to review our policies, especially as they apply to particular situations. For example, an original commitment may have called for military action embracing all Korea. Alternate possibilities should not be arbitrarily discarded, such as the creation of a buffer strip to be controlled by the United Nations for a cooling-off period. Great importance must in this connection be attached to the offers by intermediate or small governments to negotiate differences between the major powers. Moreover, other nations respect us when we respond to such proposals by the U.S.S.R. as hold promise of offering effective solutions. The mood of the United States Congress will be a crucial factor in determining what our foreign policy can be, and public opinion should make unmistakably clear that critical international issues must be settled on merit and not become the football of party politics.

V

Guard against complacency which comes with increasing military strength and which may bring a disposition to risk an incident or even to create an incident for inciting conflict on a world scale. The United States must indicate, and has expressed, readiness to expose its action to impartial scrutiny. The United Nations has provided a medium for this by setting up a Peace Observation Commission to be located at every potential danger spot throughout the world, a plan which was originally proposed by the Churches' Commission on International Affairs, and presented by the United States Government with the support of six other Governments. With the accusations and counter-accusations on rearmament by occupying powers in the Eastern and Western zones, Germany is in a critical position so long as the tensions between the Soviet Government and the Western Powers continue. It offers a logical point at which to inaugurate the Peace Observation Commission's operations. Informal conversations with church and political leaders in Western Germany reveal that such a Commission would be welcome. It should be located in both the Eastern and Western zones but could serve a useful purpose even if its establishment were limited to the Western zone. The United States can demonstrate readiness to expose its actions in Germany to international scrutiny by taking the leadership in calling for a Peace Observation Commission in both the Eastern and Western zones and by agreeing to the operation of such a Commission in the Western zone, no matter what the decision of the U.S.S.R. may be. Once the start has been made, similar commissions can be located in other areas.

VI

Guard against making our economy so dependent upon military production as to give the impression that we are unprepared to recognize the importance of economic health throughout the world. The President of the United States has expressed before the United Nations our desire to achieve a real disarmament which will include *all kinds* of weapons, be based on *unanimous* agreement, and be *foolproof*. He has further held out the hope that our armaments would be transferred into food, medicine, tools for use in underdeveloped areas, and into other aids for human advancement. Numerous leaders in the United States have publicly professed that our war preparations are designed in the interest of peace.

So long as the marginal possibility of averting war remains, these commitments must be kept prominently before the world. We must let it be known that we have not forgotten our promises and that we continue to study ways whereby the goal of diverting our resources to constructive enterprises may be achieved. In so doing we would provide much needed assurance to our own people and to the people and governments of every country throughout the world. At all times we must indicate our readiness to beat swords into plowshares at any moment when reconciliation has reached the point of providing convincing evidences that aggression need no longer be feared.

VII

Guard against the prejudice which refuses to see the evils in situations which are predominantly good and thus deserving of support. Rarely if ever is one party to a dispute entirely right and the other entirely wrong. Few situations may be described as completely "black and white." The mixture of good and the evil which characterizes virtually every international issue is a substantial obstacle to effective solutions. Christians must be particularly vigilant in making sharp distinctions so that, in supporting a predominantly just cause, they will not condone the injustices which are inherent or which emerge in the situation.

Many illustrations could be cited. I mention only one. When the churches commended the United Nations' action in Korea they felt justified because two new factors had appeared—the presence on the scene of a neutral international commission and the effort to take action by international police measures. This original commendation may in no sense be taken as a cloak to cover subsequent mistakes or injustices. Christians must now insist that (1) the benefits of the land reforms which were effected in North Korea should be conserved, allowing that injustices worked in the process of instituting those reforms cannot go unheeded; (2) the principles of justice and law as related to reported atrocities must be equally applied in North and in South Korea; and (3) provisions must be made for North Koreans freely to exercise their electoral right in determining the government of a free and unified Korea. Such impartiality is incumbent upon Christians if they seek a just settlement of world issues. Clear-cut political planning along these lines would tend to ease many fears. The pressure of military needs has too often obscured the importance of political preparations.

VIII

Guard against impatience in seeking a sound conclusion of the world's present plight. Americans by tradition want to get a job done quickly and to have it out of the way. If we are bent upon creating a world atmosphere in which conflicting ideologies can compete peacefully, we cannot expect a speedy conclusion. It is in large measure this lack of patience that makes some people contend that communism can be successfully combatted by military action. The result of a global war, even though we were victorious, would probably be to spread communism far more quickly and widely than has been possible through Soviet tactics of coercion, infiltration, revolution, and deception.

We must be prepared in patience to persevere in what may continue to be a long drawn-out struggle. In this connection two lines of activity are imperative. Because of their intrinsic merit they would be imperative even if there were no need to oppose communism. The program of economic assistance to underdeveloped areas of Africa, Asia and the Far East is the only just and effective answer to the upsurging demand for independence and for an acceptable standard of living. Equally important is the universal observance of human rights and fundamental freedoms. We must guard against the totalitarian methods which we condemn in others, set an example of true democracy in our domestic practice, and speed the completion of an adequate International Covenant on Human Rights.

Measures such as these require patience and their fruits are not always readily apparent. Nevertheless their inherent rightness calls for patient continuance in well-doing.

Conclusion

I have sought to outline for you certain measures which it is within our power to undertake. If time permitted, the illustrations I have cited could be further developed and other measures added. I do not presume to say that these steps will prove effective. I recognize that full catastrophe may break upon the world at any time. Nevertheless, I make bold to contend for my fundamental thesis. So long as there remains even a marginal possibility of averting total global war, we must utilize every means which will not betray conviction or offend conscience. In this process—and even if war should come—we must penetrate the artificial curtains by which we are momentarily separated and experience the bonds of humanity and faith which unite men of different nations and races. Particularly must the churches throughout the world stay together.

The struggle for peace must go forward unremittingly and men must be driven by the conviction that the peace can yet be won. In these trying days and always, we pray God, through Jesus Christ our Lord, for the faith which will enable us to stand in face of principalities and powers of darkness. From Him alone can come strength to make the expression of our faith adequate for the needs of our day.

Be Not Afraid

Reverend Theron Chastain, Ph.D.
Minister of the First Baptist Church of Phoenix, Arizona

Some sermons are made in the study. This one grew out of the struggles of the preacher and emerged full grown when the occasion called it out. When first delivered as the annual sermon at the American Baptist Convention meeting at Boston in 1950 it is said to have provoked more newspaper comment than any similar sermon had ever received.

Dr. Chastain is the son of the turbulent West with its insecurity both secular and ecclesiastical. He grew up on theological controversy. Planning as he did from earliest years to enter the Baptist ministry, he felt the full force of the storms which have troubled this communion in the past three decades. In an earnest search for solid ground on which to build his Christian faith and ministry, his period of schooling was extended. To Bible Institute training he added the College of the Pacific and six years of intensive study at Southern Baptist Theological Seminary at Louisville, Kentucky, where he earned his doctorate in the field of Christian Sociology.

After Seminary he returned to the West Coast as state director of Christian Education, later taking the pastorate of the First Baptist Church of Santa Barbara, and in 1945 became pastor at the important First Church in Phoenix. He now is President of the Arizona Council of Churches, President of the Arizona Convention of American Baptist Churches, and is on many school and church boards including the Berkeley Baptist Divinity School, Redlands University, Cook Christian Training School, and the Board of Education and Publications of the American Baptist Convention.

As pastor of the largest American Baptist Church in Arizona, he has had to give himself almost wholly to the work of the practical churchman. The membership of the church has been increased to 2,200, the finances have more than doubled, a new building has been erected, and several mission projects have been sponsored. The denominational and cooperative church situation has required unusual efforts in reorganizing both the Baptist work and the State Council of Churches. In his schedule of innumerable administrative activities, community work, and five to eight addresses and sermons each week he places considerable importance on his sermon preparation. This sermon illustrates his concern for current problems and the application of the Gospel of Christ to each need.

Sermon Thirty-two

TEXT: Be of good cheer; it is I; be not afraid. MATTHEW 14:27

THE DISCIPLES were with Jesus when he fed the five thousand. They heard him teach. They were impressed by his miracles. Like the others, they were caught up in the enthusiasm of the crowd which sought to make Jesus king. Political ambition swept them into veritable clouds of glory. Even after Jesus had slipped away from the multitude and sent the disciples off in the boat without him and had disappeared into the mountains alone, they were still afire with visions of greatness as they started to row toward the other side of the Sea of Galilee.

In their preoccupation they did not notice the storm that was brewing. Suddenly, with terrifying fury, the storm broke. Dreams of conquest and royal splendor were blown out of their minds as they set about the grim struggle for mere physical survival. Almost instantly they passed from bright hopes to dark despair, from brave imaginings to dull terror.

Again they were startled. Jesus came to them in the storm, walking upon the water. Unruffled by the wind and unworried by the threat of the sea he spoke quiet words of assurance, "Be of good cheer; it is I; be not afraid."

The terror that strikes out of bright dreams is no unusual experience. Like the disciples, we forget the perils of life when all goes well and we are thrown into utter panic by the very unexpectedness of disaster. Not long ago the lush days of progress smiled upon our nation until the dream of the Kingdom of God on earth seemed very real. Today the dream is all but gone and there are few who dare hope for more than mere escape from the worst of our fears. Wars, hot and cold, are recurring with terrifying frequency. Liberties, once thought secure for all time, have been betrayed and crucified. Even God seems a long way from helping us.

The Christian Church has been called to speak words of hope and peace to mankind. Christ did not leave his disciples to perish alone in the sea, nor does he leave the world without hope today. Amid all the fear of modern life, Christ is speaking again "Be of good cheer; it is I; be not afraid." The trouble is, our world is not hearing the words of Christ. As men gaze about them apprehensively, they see only the fear-ridden stares of other terrified seamen. Unless the hope of deliverance comes to them in a voice they can understand, our generation is doomed. The Christian Church has been sent into our world to be that voice, the only voice the Master of storms can use to reach this troubled generation. And the great tragedy of this moment is: the Church has all but lost its voice!

The Christian Church is speaking feebly for Christ today because it is sometimes so concerned about saving itself. The Church is caught in the same tumult that is frightening the nations today. War has drained mankind dry of blood, sweat and tears. The agonizing cry of humanity may be heard, calling for God in a great wail of fear. Just when we could speak peace to a war-torn world, we lose our voices. Christ has commissioned us to speak for him, "As the Father sent me even so send I you." We have been given the message of the peace of God. But instead of speaking out in tones of confidence and faith, we whisper so low that few hear us and fewer still are convinced.

Political and economic confusion have taken from men everywhere their old sense of confidence and destiny. Democracy is being beaten down, not by arms but by fear and uncertainty. Only Communism and other brands of totalitarian statism seem to be sure of where they are going. Everywhere men are questioning the worth of the individual. Some agency must supply him with the necessities of life. Like helpless children, people run to some earthly father to protect them from the uncertainties of life. The world is too much with us and God is far away to most people.

Why should the Church choose this time in the world's history to carry on one of the greatest theological controversies of all time? Just when the world needs certainty and assurance, the Church loses confidence in its own faith. Just when a war-torn world cries for peace, the Church girds for the wars, and in the clash of theological big guns is all but deaf to the cry of a lost humanity. The reason is not hard to find. The same forces of confusion which have destroyed international peace are at work in the Church. We are at war, both politically and ecclesiastically, because our faith has been disturbed. Old patterns of thought have been torn to pieces and we have not made new adjustments to a more complicated civilization. Like sailors in a storm who no longer trust their ship, we are terrified.

To many the past looks like a peaceful haven of rest, and so they flee to some conservative refuge. A famous British poet deserted his Church for what he describes as a firmer foundation on which to place his feet in this changing world. Plaintively he complained that his Church no longer spoke with authority. It raised far more questions than it answered. His timid soul was in panic flight into a lost past which he imagined as security.

Conservatism is solidly planted in the past. Communism is confidently out to conquer the future. Must we be slaves either of a dead past or a mad future? God have mercy upon us if we settle for either one!

If all we desire is a sure footing in this troubled time, we probably will be driven to one of these extremes. If we are concerned only to save our own lives we have no hope. But let us remember that our Lord assured us that he who would save his life shall lose it but that he who will lose his life for Christ shall find it. Christ did not send his disciples into our world to save themselves but to bring assurance and salvation to others. Words of peace are not needed in a dead calm. No one listens to a messenger of peace unless there is war. Dare we lose our lives for Christ or are we merely to run like frightened rabbits to the nearest shelter?

One thing is certain: we cannot spend our energies trying to save our little private systems if we are to speak for Christ now. "Seek ye first the Kingdom of God and His righteousness" must come ahead of our own favorite schemes. Think how long God managed to get along before my way of life ever saw the light of day! God does not depend upon any particular organization of state or church. He can get along without any of the familiar trappings of modern life, and so can we. The one thing He must have is our unswerving loyalty and love and our willingness to sacrifice ourselves for Him. We cannot speak for God today unless we are willing to have the dearest associations and the most secure ties broken for the sake of God's Kingdom. If we have spoken with too feeble a voice before, we must fashion a new medium for making the Word heard by all the world, no matter what it may cost.

Nor may we indulge in fearfully protecting the Gospel we have been sent out to proclaim. Too many of us have been like peacetime sailors who have spent so much time shining up the bright-work of our little vessels that we are unwilling to risk the uncertainties of battle. The Gospel of Christ was born in conflict and has little meaning except as it meets men where the battle rages. It was made for the storm and will take care of itself. The Gospel has suffered many indignities at the hands of irreverent men. The greatest indignity, however, is for us to try to protect it with our feeble hands. Defending the faith is a useless impertinence. The faith must defend *us*. Christ is *our* Saviour. We do not protect Him as idolaters protect their helpless idols. The God of storms is not afraid of the wind! Is not our concern for the safety of the Gospel only a reflection of our own insecurity? We rely upon ourselves all too much and upon God too little.

Then, we must be willing to live by the Gospel we preach. If we rejoice that "God so loved the world that he gave his only begotten Son" for us, let us also be willing to take up our cross and follow him. This is indeed dangerous. It is a dangerous thing to love one's enemies. Modern total wars are not won on that basis. Nor is modern cold war supposed to prosper on love. But win or lose, survive or perish, we are called upon to believe in God enough to obey Christ whatever the risk involved. The timid soul who will follow Christ only through familiar paths on sunny days knows nothing of faith. We need to plunge into the storms with Him where we may not always see Him in the confusion but where we will hear His voice saying to us, "Be of good cheer; it is I; be not afraid."

The most significant fact about the storm was that Jesus was in it, calm and confident. Peter became so disturbed about the tempest that he forgot about the Lord of the tempest. Like him, we also forget that God is in the midst of storms. Indeed, out of storms have come many of God's greatest blessings. Somehow the voice of God has been recognized more often in the storm than when he has spoken in quiet times.

The Prophets were not ascetics in ivory towers separated from the conflicts of life. Far from it. They were hard-pressed men living amidst suffering and sin. When the nation was destroying itself, Jeremiah spoke from God in tones so thunderous that we still hear his cry. In the tragic year that King Uzziah died, Isaiah saw the Lord. Amos left the peace of the Judean wilderness for the stench of the back alleys of Bethel before he spoke from God. Out of the clash of falling empires and the agony of dying civilizations God has always spoken through prophet, poet, and seer. Thank God for the green pastures and still waters, but let us not be afraid of the storm. It is out of just such days as ours that God has led his people into new life and hope. The storm is important today but far more important is the presence of Christ. He has the answer and the power which we need. Listen and you will hear him speak peace. Look to him and the storm will lose its terror.

When Jesus was led to Calvary, the disciples thought the end had come. It was a dark and stormy day. But out of that darkness came the hope of the world. God brought our salvation out of storm.

The stoning of Stephen scattered the disciples and deepened the darkness of the primitive church. But God spoke to the heart of a young Pharisee out of that dark storm and the Apostle Paul was born into the Christian life to bring the light of his gospel to a waiting world.

When the wild German tribes destroyed Rome and civil government broke down completely, it looked like the end of the world. Indeed, it was the end of the world that had been, but a new world was born out of that storm. Christ became the Master of that storm as He has always been. To be sure, in the calm that followed, Christ was forgotten through dark ages. The storm was not as dangerous as the calm. The disciples listened and heard Him when they were in jeopardy though they forgot Him when the danger was past.

Christians should be the most ready to accept the storms of life, and our Baptist Church was born in storms and strengthened in adversity. Roger Williams did not discover the truths of religious liberty in the calm of his study. Nor did his message get a hearing in the academic atmosphere of the classroom. New England whipped up a storm around his head out of which the voice of Roger Williams became the voice of God to the world. Those who see only the storms of controversy miss the meaning of all these events. The storms became only a platform on which the Son of God walked before a listening world. When the storm begins, look for Christ in it. Be sure he has a word to say to us then.

We should have learned this truth more quickly in our own small circle of experience. Has it not been in the midst of personal confusion and suffering that we most often heard the voice of God?

One day I went into a sanitarium to see a man dying of tuberculosis. His wife was in as serious a condition as he. He was undergoing severe surgery in a desperate effort to save his life. I found him confused, bitter, almost in despair. Only one thing seemed hopeful. He asked me to bring him a Bible he could read. A few days later when I visited him again I saw a new thing. A kind of excitement lighted his face. He had been reading the little Book. There he had found truths which answered the cry of his heart. He wanted to talk about the Christian faith. Week after week he grew in physical strength as he also grew in spiritual hope and understanding. The miracle had happened. Out of the storm and confusion of his illness, he had caught a glimpse of Christ. Neither he nor his wife will ever be physically well. But he never ceases to thank God that in the storms of his sickness he met Christ, for he is sure he never would have seen him in fair weather.

Why should we doubt that Christ is just as much in the storms that beat around nations and cultures? Is Christ so small that He is not able to master the worldwide problems of men? He is no less in the storm that rocks our Ship of State. It is not likely that the paganized world will see him. They are too concerned with the storm. But the Church must see Him and hear His voice and it is for us in the Church not only to hear but to *be* His voice in this troubled day. "Be of good cheer; it is I; be not afraid."

The strange truth about this situation was that the disciples feared the very source of their salvation. "And when the disciples saw him walking on the sea, they were troubled, saying, It is a ghost; and they cried out for fear."

Christ often comes to his people in unfamiliar guise. Pain has often been the agency of God.

But fear is destroying us. All the world's prejudices, suspicions and hatreds are born of the fear of ghosts that walk through the darkness of our life's storms. Because a race has a different color of skin, or a national group speaks a different language, or a culture has strange customs, we imagine fearful evil is hidden in them. The Chinese often spoke of the Christian missionary as a "foreign devil." But we are little better. When I was a boy there was a Chinese village in our town in California. It was an old-fashioned block of houses with their backs toward the street and all the buildings facing into a court. There was a narrow gate off the street into the courtyard. My friends and I would pass fearfully by that gate and hastily peer in to see what terrible things went on in there. Of course we never saw anything, but we were sure they were there. We had been told that the Chinese ate the ears of little boys. We knew they dressed differently, for it was before they cut off their cues. And they would walk down the street single file—not like we do, two or three abreast. They talked a strange language. Surely they were a fearsome people! We American boys would run out bravely within hearing distance, but not too close, and shout insults at them.

Today it is more serious than boys hurling insults at a few Chinese. It is now war. The Russians see all kinds of fearful ghosts when they look across at us, and we see ghosts behind the Iron Curtain. In our mad witch-hunt for Communists let us beware that we do not destroy our own freedoms. There is as much danger in our fears as there is in the storm itself. The sad truth is that the Church has excelled in finding ghosts instead of Christ. To some of us, every new thing has been a fearful evil to avoid and destroy.

Back in the days when the youth movements were just beginning in the churches, and Christian Endeavor was leading the way, Baptist youth started the B. Y. P. U. One of America's leading Baptist pastors set out to put a stop to this evil heresy. Why, he said, if these young people get together and start leading their own meetings, some of the girls will learn to speak in public. If the girls do it, the women soon will be speaking in our churches and what will become of the Church if that should ever happen! With the immense prestige of his position and the blazing oratory of his day he set out to save the Church from this spector. The women of the churches began to speak and they are still speaking. But instead of it being the death of the Church it has been the renewal of life. Christ was in the Christian youth movement and those who fought it fought the will of God.

So it was earlier with the Sunday School movement. Before that the Church was horrified at the prospect of printing the Bible in the language of the people. What would happen to the sacred Scriptures if they were permitted to fall into the hands of the common people!

In view of our present-day ecclesiastical fears, let us look a little more carefully at the ghosts that cause such concern lest we find ourselves fighting God!

Peter responded to fear just as any young fanatic might. Swinging quickly from fear to bravado he asked permission to walk on the sea. Climbing out of the boat he started to meet Jesus halfway. It was a silly thing to do, but it was just what we often do when we are frightened. In our effort to cover our fear we lash out in some extreme action to prove to others and to ourselves how brave we are! "And Peter answered him and said, Lord, if it be thou, bid me come unto thee upon the waters. And he said, Come. And Peter went down from the boat, and walked upon the waters to come to Jesus. But when he saw the wind, he was afraid; and beginning to sink, he cried out, saying, Lord, save me."

Note that Jesus did not commend Peter for climbing out of the boat. To the contrary, he reproved him for his lack of faith. The exaggerated bravery which springs out of fear does not last long. Jesus literally had to drag Peter out of the water to keep him from drowning.

Let us be careful of protesting our faith too loudly. Those who shout the most about defending the faith often have the least faith to defend. We take for granted the things we believe most deeply and spend very little time arguing about them. Why should I get all disturbed about whether tomorrow will come or not? If you do not think it will, all right. I shall live as though tomorrow will come. But if there is a real doubt in my mind and I am frightened, I may get quite heated about the matter and spend considerable time refuting any who dare express doubt about it. I may even hate the man who sets my doubts going. If he really bothers me, I may try to silence him. In silencing him I am only trying to quiet my own fears.

No, it is not the man who shouts his faith the loudest that enjoys the most powerful convictions. The only real proof of the power of God on earth is to demonstrate it. Peter would have been far more convincing had he relaxed and invited Jesus into the boat.

The Gospel is today in danger from only one source. If we do not proclaim it, it cannot save men. No denial can take away its power. The logic of philosophers and the armies of nations cannot defeat the Christian Gospel unless we can be silenced. So long as there are any who will proclaim the message it will go on demonstrating itself as the power of God unto Salvation.

Out of the stress and terror of the storm Christ emerges to bring peace into our hearts and to restore our faith. Our text ends with the significant statement that "they that were in the boat worshipped Him, saying, Of a truth thou art the Son of God." Those who come through the storm with Christ need no other proof of who he is. All the fine arguments of theology fade into the background, and the blazing experience of finding Christ in the midst of life's great storm will answer all questions.

I believe we are about to enter into a new day of faith. We have not come through these years of uncertainty and trial without purpose. Christ has been in this vast hurricane. We are hearing as we have not heard for many decades the confession of faith—"Surely this is the Son of God." Uncertainties are beginning to fade. Let us take Christ into the boat with us here, and then we can expect the storm to subside and faith to be justified. Then we can walk about the earth fearlessly declaring through word and life that Jesus Christ is indeed the Son of God, our Saviour, and the Master of every storm.

Why Do One's Best When the World Is Going to Hell?

REVEREND LOWELL RUSSELL DITZEN, D.D.
Minister of the Reformed Church, Bronxville, New York

Lowell Ditzen is one of the brilliant younger ministers of the Reformed Church whose preaching is attracting considerable attention for its freshness and directness. He attended school in Kansas City, Kansas; Park College; William Jewell College; McCormick Theological Seminary; the Divinity School of the University of Chicago; and Union Theological Seminary. Park College conferred the honorary degree of Doctor of Divinity upon him in 1943.

Ordained by Topeka Presbytery (1936) he has risen rapidly during the last fifteen years. His pastorates include: The Pullman Presbyterian Church, Chicago, 1934–38; the South Shore Presbyterian Church, Chicago, 1938–42; assistant minister of the Brick Church, New York 1942–43; First Presbyterian Church, Utica, New York, 1943–50; the Reformed Church, Bronxville, 1950–. His present church of two thousand members includes many leaders in national and international life. Two Sunday morning services are required to accommodate worshippers.

Early promise was shown in his winning first place in sixteen collegiate oratorical contests; participation in international debating with the University of Mexico, 1932; honors in history; Theological Seminary prizes for research essays in Old Testament, New Testament, church history, theology; and in his being granted the "Nettie F. McCormick Fellowship in Old Testament Hebrew" (1936) for two years' graduate study.

His sermon on the occasion of the 300th anniversary of the Westminster Assembly won first place in a national contest sponsored by the Presbyterian Church, U.S.A., 1944. In 1951 another sermon was given second national honors by Freedoms Foundation.

In interdenominational activity he has represented his Church at "The Alliance of Reformed Churches Holding to the Presbyterian System" meeting at Geneva, Switzerland, 1948; the World Council of Churches, Amsterdam, Holland, 1948; and as one of two Americans to attend the Pastor's Conference, Ecumenical Institute, Céligny, Switzerland, 1949.

In radio and television he has been minister on "Gems for Thought" (A.B.C.), "Faith in Our Time" (Mutual), WOR Sunday Radio Chapel, and on WABD, Dumont Television Chapel.

"Why Do One's Best When the World Is Going to Hell?" was delivered in the Reformed Church of Bronxville, December 31, 1950, and was selected for its aliveness, sane, constructive thought and relevance to life today.

Sermon Thirty-three

TEXT: If the foundations be destroyed, what can the righteous do?
PSALM 11:3

SOME DAYS ago, at a banquet in the city, I sat next to a gentleman of distinguished ability and character. A man of affairs, he had dealt successfully with hard business and legal problems over a period of years. In the course of the evening he said, "I have to answer the most difficult letter I ever received. It came two days ago from my son, who is a sophomore in college. In substance it said: 'Dad: Why should I dig into history and literature, the physical sciences and sociology, when it looks as if I would be pulled into the Army and probably spend the next important years of my life in a war? The fellows here are discouraged—there doesn't seem to be much chance for a normal life for us. Some are saying, 'Why try to do our best when the world is going to hell?' "

The problem of that troubled son and concerned father is a problem that has come to your attention, I know, as it has repeatedly to me, during these past weeks. For the sake of the superb group of young men and women of this church who must deal with it—and for the sake of the same concern that broods over the youth in the heart of each of us—let's try to give some answer to that question: Why do one's best when the world is going to hell?

To begin with let us realize that we are not the only generation which has faced an immediate future that seemed shrouded in uncertainty. Five hundred years before Christ, a little band of people faced the destruction of their homeland, the tearing apart of families and a slavery no less horrible than that practiced by Hitler or Stalin. They asked the same question, though in different words, that the college sophomore asked his father: "When the foundations are destroyed, what can the righteous do?"

When, in the sixteenth century, the Tartars invaded Europe a Polish student at the University of Krakow might well have said, "Why continue the discipline of my studies? I might as well raise hell now—because I'll be dragged into it tomorrow in battle."

Try, down the corridors of time, to find a set of young people who have entered adulthood with skies all untroubled and the way ahead full of light.

Look and you will look in vain! Somewhere for each generation there has been the threat of war, or pestilence, or famine, economic or political turmoil, enemies within or without. Planning has always been uncertain and the future hazardous. Tune in on any period of human history and you will hear people say, "Why hold on? Why keep trying for what is right when everything is going to pot?"

Now I know that no previous generation has had the atom bomb, nor an Einstein to suggest that its use might so deplete life and resources that a future war would be fought with clubs. Yet I find no difficulty in believing that young people of the fourteenth century faced a more teeter-totter and terrifying future than do we. Then a cycle of epidemics, known as the Black Death, decimated whole villages and cities, with the best estimates ranging from one-fourth to three-quarters of the total population of Europe taken by death. Can't you almost hear heart-sick young people in London or Amsterdam or Paris saying then, "As for me I'm going to take my pleasure where I find it. Eat, drink and be merry, for tomorrow it will be my turn to die. What's the good of being good—when everything seems bad?"

But recognize that in spite of the despotism and destruction caused by the Horsemen of the Apocalypse, life *has* gone on. In spite of war, famine and plague, values of significance have survived. It is well to have some such perspective as this when we face our immediate mood and problem: Every generation has had to face a hard future! For every generation the world has seemed to be going to hell!

With that as a general back-drop to keep us from the unwarranted and hysterical conclusion that our situation is unique in human history—come now to closer grips with our question, why hold on to the best we know when the pulls all seem to be toward the worst?

One answer of conservative Christianity is that you are an immortal soul, and that your life here on earth is but a fragment of your total possible life. You are made for eternity. How you live here determines the form of eternity you will enter—toward heaven, or toward hell. That's a potent reason for doing your best—you have "forever" to face.

If this is hard for you to accept, don't brush it lightly aside. Think of it carefully. Consider it thoughtfully. It is a faith that has been pivotal for centuries of Christians, giving a new dimension for existence and a powerful force for successful living. It may become pivotal some day for you even though it may exert only a modest appeal now. Only the other day a daughter whose fine father had just died said: "I didn't know heaven and eternity could be so close." It comes close in a moment like that when we are faced with the finality of bodily death and yet sense, too, the living presence and sweetness of the spirit which death cannot claim. Eternity is close indeed! It is close in the Bible; in the spirit which enables us to get outside and above ourselves; to have an influence beyond our physical life. We all have "intimations of immortality."

One reason to live at one's best is that we have a "citizenship in heaven," and while in this finite world we must see the foundations of what we cherished or hoped for changed and destroyed, there is a part of us, too, that is infinite.

Various interpretations have been made as to what happens at death, but none is more profound than that of the Old Testament when, "—the silver cord is loosed, or the golden bowl is broken—the dust returneth to the earth as it was, and the spirit returneth unto God who gave it." The New Testament intensifies and personalizes that exalted idea when it affirms that those who live at their best, in spite of everything, will, at the end, hear the Father say: "Well done, good and faithful servant, enter thou into the joy of thy Lord." To grow into such a faith gives a potent reason for living at one's best though one must experience hell itself.

Now, on the purely pragmatic side let us ask, "If worse comes to worst, *what will be needed*?" Suppose history's most devastating war became an actuality, and Einstein is proved to be right, the few who are left find themselves back at the cave-man stage—what would remain in the remnant? What would have priority? Psychology, sociology and anthropology would agree that though but a hundred people be left in the world, the indestructible will to live would be present. The report of the officer who asked for twelve men to volunteer for a dangerous mission in the last war is illuminating. "Only one of you may expect to come back," the officer said. Yet twelve stepped forward and each, looking around at his fellows, inwardly said, "I'm surely going to miss the boys!" That ineradicable desire and hope to live would surely be present even if civilization came to the end of its rope.

And to live, some elements would be needed. What are they? Wouldn't intelligence be sought for? Wisdom? Sound judgment? The ability to solve problems? To work? To be faithful and thorough and dependable? These would be the things to make life continue and to make it continue well.

There is no denying that such qualities are needed now. They have always been valued. But if the worst came they would be more needed then than ever.

There was a Lincoln whose first name was Thomas. He was, I presume, an agreeable person. But he was shiftless—a putterer, trying this and that. The pastures for him always looked greener across the fence. Something of a vagabond, he would be a complete nonentity, except for a son whom he sired named Abraham. In that son was a different quality. He engaged in the discipline of reading and the most difficult discipline of all, learning to think. He observed. He evaluated. He exposed himself to the wisdom and exaltation of the Bible. He schooled himself in hard tasks. He sought rather than evaded responsibility. He learned to consider his fellows and to act helpfully toward them. When the judgment thunders split this nation, and civil war threatened to end the American experiment, what of the Thomas Lincolns? They were a dime a dozen. But, thank God, there was an Abraham Lincoln—a man who had lived at his best. It was the stuff of which he was made that held us up and carried us through when the American experiment seemed on the skids to destruction.

"If the foundations be destroyed what can the righteous do?" They can go on being righteous. *If never before* they'll get top billing then, when things have really gone to smash.

Once again, come closer home still as we deal with our question, and see that no matter what the world comes to, you and I still have to live with ourselves. I wish it weren't so. But it is so. As the old Negro put it, "I's always got myself on my hands."

For a time you and I can seem to lose ourselves in an idea. We can be caught up into the exaltation of music, or poetry. We can forget a part of ourselves for a time in the concentration of a game, or in some intriguing work. But even then you have yourself from which you can't get completely away: with memories—good or bad; skills—excellent or mediocre; temperament—unrestrained or controlled; knowledge—sharp or fuzzy. How terrible to have this "you" that you have to drag around—nothing more than a bundle of regrets and mediocrity! How sad not to measure up when the time of testing and opportunity comes! Get on a ship or a plane, hide in a palace or a jungle, and you take yourself with you.

I've used the word hell several times in this sermon. One of the most interesting definitions of that word is this: "Hell begins on that day when we look back over our lives and see our wasted opportunities." So it is!

> The saddest word of tongue or pen
> Is this sad word, "It might have been."

If not for God! If not for your fellows! Then for *yourself alone*—save yourself from the bleak remorse of having had the chance to develop and live up to your best—but of having muffed and messed it up.

There must be no higher satisfaction than that which comes from a backward look across one's life and when a man may inwardly whisper: "Things really hit bottom. But I can honestly say 'I tried to do my best even in the worst times.'"

Two summers ago I preached in London's St. Andrew's Church. Following the service the senior Elder, well in his eighties, told me of the bombings in the last war. He said that, on one of the worst nights, he saw from a distance the whole center of London aflame. As I wrote them down immediately afterward, these are his words, "It seemed all was lost, the war, England, the values of our civilization. As I saw the docks and the center of London all burning I wept like a child. But then a gust of wind cleared the smoke for a moment, and I saw the gold cross of Christ still standing on the dome of St. Paul's. I stopped crying because then I knew there was a power stronger than the swastika—a power that would carry us through and which would live on."

It is symbolic that today St. Paul's still stands in London. The ground around it is leveled. But the miracle of the cross is there at the apex of the great dome. It still catches the first rays of the morning sun, and reflects the last light at the day's end. Silently it affirms the old man's insight that the "best" has greater power than the "worst." It remains through storm and calm to bless the children of men as it continues to remind them of a power able to carry them through.

O youth, in every heart here today—no matter what may happen, some things abide: God; and Jesus Christ; faith and hope; truth and honor; loyalty and high purpose. These are the *best*. Touched with heaven they survive the fires of hell itself.

Hold to them! Live by them! And they will give us a victory over the powers of darkness.

Prayer:

Lord take these wooden words—
And by Thy grace set them aflame—
So that the youth in some heart here—
Now cold with discouragement
May be warmed with hope—
Some mind darkened with uncertainty—
And the temptation to compromise—
May have anew the light of honor,
Diligence and faithfulness—
And may face into this New Year—
Determined to do The Best—
For Jesus' sake. Amen.

The Proper Study of Mankind

REVEREND LYNN HAROLD HOUGH, TH.D., D.D., LITT. D., LL.D., J.U.D.
A Minister of the Methodist Church
Sometime Dean, Drew Theological Seminary, Madison, New Jersey

Lynn Harold Hough is a distinguished Methodist minister, author and teacher. As dean and professor of homiletics and the Christian criticism of life at Drew Theological Seminary, Madison, New Jersey, he has influenced the future career of hundreds of men who attended the Seminary.

Dr. Hough laid a sound educational foundation for his brilliant career, took his A.B. at Scio College, his B.D. at Drew Theological Seminary, did postgraduate work at New York University, and took his Th.D. at Drew in 1919. He received the D.D. from Mount Union-Scio College, Garrett Biblical Institute, Wesleyan University, the Litt.D. from Allegheny College, the LL.D. from Albion College, the University of Detroit and the University of Pittsburgh; the L.H.D. from the University of Vermont, and the J.U.D. from Boston University.

Beginning his pastorates in 1898 in a small church in New Jersey, he rose through his preaching ability to be the pastor of churches in Brooklyn and Baltimore. Then from 1914 to 1919 he was professor of historic theology at Garrett Biblical Institute, Evanston, Illinois, and in 1919 he became president of Northwestern University.

In the following year he was called to be pastor of Central Methodist Church, Detroit, where his preaching attracted wide attention. In 1928 he was called to the American Presbyterian Church in Montreal. In 1930 he took the chair of homiletics at Drew and was dean from 1934 to 1947. In the last forty years he has written forty books, including In the Valley of Decision, The Significance of the Protestant Reformation, The Civilized Mind, The Christian Criticism of Life, Patterns of the Mind, The Dignity of Man, *and* Christian Humanism and the Modern World.

He gave the Cole Lectures at Vanderbilt University (1919), the Merrick Lectures at Ohio Wesleyan University (1923), the Fernley Lecture, Lincoln, England (1925), the Fred J. Cato Lecture, General Conference of the Methodist Church in Australasia, Brisbane, Australia (1941), and half a dozen other famous courses.

In 1918 he was sent to England by the Lindgren Foundation of Northwestern University to interpret the moral and spiritual aims of the first World War. At the invitation of the British Ministry of Information, he spent eleven weeks in England during the summer of 1942, preaching to the congregation of the City Temple, London, and making addresses in army camps and to the general public. He is as popular in the British Isles as he is in America.

Since his own freedom from administrative duties at Drew, he has preached widely in Canada, England, Scotland, and the various cities of the United States, and has given several important series of lectures. His sermons show his concept of true freedom in our world and the need for disciplined living and thinking. Dean Hough's fine mind penetrates to the rich ore of spiritual wealth and his language brings his message clearly and convincingly to every audience. In the Summer of 1951 he preached at St. Giles' Cathedral, Edinburgh, Scotland, at Richmond Hill, Bournemouth and at St. George's, West, Edinburgh. He also addressed the Methodist Ecumenical Conference at Oxford, England.

In 1952 he will give the Ezra Squirer Tipple Lectures for Drew University on "Five Great Humanists."

There were two sermons by Dean Hough which the editor would have liked to include in this volume, "The Importance of the Individual," and "Horses Among the Myrtle Trees," but "The Proper Study of Mankind" was chosen for inclusion for the intellectual stimulus it may afford many younger ministers everywhere. It was given as the first address in the Dana Lecture Series at Carleton College, and was delivered at an All-College Convocation in Skinner Memorial Chapel on December 8, 1950. It is used by permission of Dr. L. M. Gould of Carleton College.

Sermons must have substance as well as form. "The Proper Study of Mankind" represents the substance of the thinking of a man who has given his life to preaching and to training other young preachers.

Sermon Thirty-four

THE GREEK writer Isocrates was a contemporary of Plato. He was very sure that he knew what is the proper study of mankind. The proper study of mankind is speech. He established a tradition of writing which profoundly influenced Cicero and is still deeply felt in the best modern prose. To him speech and writing were the clear and noble expression of true thoughts. He sharply separated himself from those sophists who quite without moral inhibition used rhetoric to further any cause in which they were interested. He was the remorseless foe of the conscienceless use of speech. The man who by wise speech influenced the thought and action of his fellow men was a person of genuine power. And Isocrates wanted to have the definitive share in making this person. Quintilian in his *Institutio Oratoria* gave the tradition an ample setting. To him the truly educated man was one who had full access to the materials which are necessary for good thought and knew how to use these materials with complete competence and to express the result in clear and exact and persuasive speech. Isocrates wrote with much practical sagacity about the conduct of life. And he endeavored to set forth an adequate policy for the Greek states of his time. He was proud of the leaders he had trained. He really believed that he was a master of good words to express good thoughts about the good man and the good state.

That the study of speech is a most important matter is illustrated by the speeches of Winston Churchill. He finds juices in words which other men have never suspected. And he can turn words into bullets and send them singing a wicked song as they go straight for the heart of his foe. Could Isocrates have listened to his words, he would have responded with delight. And he would have read his books had this been possible with a sense that he had just the power over words which he so much prized.

Since those days in the fifth century B.C. when Isocrates won the admiration of Athens, there have been almost countless men who knew the secrets of the magic of words. To have the music of their speech in one's mind is to be on the way to making such music oneself. When words decay and sentences languish and paragraphs sink into a kind of literary coma, bad days have come to the men who are busy about the adventure of living. When we cease to believe in words and to have power over them, democracy itself is losing its reason for being. As long as men are men, words will be a part of the serious study of those who have any sort of appreciation of what it is to achieve and to maintain a civilized mind.

So we can imagine a man impressed by these things turning his conviction into rhyme and writing:

> Great secrets only words can teach;
> The proper study of mankind is speech.

The first Greek philosophers thought that they knew what is the proper study of mankind. They would have declared that the proper study of mankind is the physical world. The student of genuinely awakened mind and of true intellectual curiosity in any century looks upon this world which gives him a home with the deepest interest. How much can he learn about the structure of the world? How do things fit together and to what degree, if he learns the secrets of the world, can he turn them to his own uses? Thales noticed that water could be soft and flowing. It could be hard as ice. It could vanish as vapor. In other words, as we would say, it was a solid and a liquid and a gas. Surely it was the fundamental substance of which everything was composed. Empedocles thought that there were four substances: earth, air, fire, and water. The Eleatics emphasized substance. Heraclitus emphasized change. Democritus thought of elements whose character gave unity and whose combinations accounted for the variety of the world. All this was the beginning of the tradition of man as a detective following illusive facts to their most hidden lair. It revealed man as a creature of infinite intellectual thirst eagerly desiring to drink from the cup of knowledge about the physical world. It is a tradition with a long and glorious history of which we shall have more to say as we go on.

We can easily imagine a man who, impressed by these things, turns his conviction into rhyme and writes:

> The knowledge of the world makes men its kings;
> The proper study of mankind is things.

Socrates was all the while busy with the workings of the endlessly active mind of man. He was perpetually calling for the examined life. And he was all the while busy using his own mind and wisely guiding other people to use their minds to get at the real meaning of thoughts and actions. By endless conversations in which he forced men to face the logical implications of their own assertions, he furthered the cause of clear thinking and of wise action. Always, implicitly or explicitly, he brings you into the presence of man, the controller of his thoughts and the controller of his deeds. And it is the mind of man which he sees sitting on the throne of control. The tradition of the examined life is the profoundest intellectual tradition of civilized man. It was many centuries later that Alexander Pope, impressed by the possibilities of man's study of his own life, wrote the lines:

> Presume not God to scan;
> The proper study of mankind is man.

By the time the middle centuries had arrived the old civilizations had decayed and Christianity had become the most powerful force in the world. The great Greeks were remembered and gradually what were believed to be their wisest thoughts were used as an aid in constructing a great system of Christian thought about existence and experience. St. Thomas Aquinas found in Aristotle a frame for all his manifold and farreaching thoughts about reality. The great Christian thinkers of the Middle Ages were sure that they knew what is the proper study of mankind. The proper study of mankind is God. Life had come upon much frustration. Many men and nations walked in ways of disillusionment. The old cultures had decayed. The old political structures had fallen down. But from the eastern shores of the Great Sea the moral and spiritual insights of the great Old Testament people had spread about the world. And then the shining light of Christianity had blazed forth upon mankind. From the first, there had been men to whom Christianity was a mental as well as an emotional and moral and spiritual experience. Paul had found it necessary to be a Christian with his mind. The great Greek fathers had made notable attempts to combine Greek insights with Christian loyalties. The great Latin fathers, notably Augustine who has been called the schoolmaster of the Middle Ages, were very busy about the task of making the mind Christian. Then came the synthesis which the Middle Ages produced. It was dialectic in St. Thomas Aquinas. It was poetry in Dante. It was architecture in the great Gothic cathedrals. Here man really attempted to think God's thoughts after Him. A genuine attempt was made to see all human experience in the bright light which came from the face of God and with the clear thoughts which it was believed came from the mind of God. God was indeed the Alpha and the Omega. He was the creator and redeemer of men. He had made them for Himself and their minds as well as their souls were restless until they found rest in Him. All ultimate thought was thought about God. Theology was indeed the Queen of the Sciences. In God man lived and moved and had his being. You understood man because you had seen a vision of God.

It is easy to imagine a man, impressed by these things, turning to rhyme and writing:

> All else decays beneath the sod;
> The proper study of mankind is God.

This may seem like the great consummation. And there are those to whom the thirteenth seems the most wonderful of the centuries. But the human story is long and many-sided. And so after the Middle Ages came the Renaissance. It was, most of all, a retreat in order to go forward. The great Greek thinkers and writers seemed to rise from their graves in order that they might haunt the imagination and inspire the thoughts of living men. And once more man himself came to the center of the picture. In a sense the characteristic note of the period was sounded by Pico della Mirandola in his famous tractate on the Dignity of Man. Man was seen between two worlds, one above him and one below him, by the sentence of his intellect choosing to ally himself with one or the other. Sometimes the Renaissance sunk into a world of impulse from which all discipline was lost. Sometimes, as when it was guided by a noble Platonism, it kept its eyes fixed upon the heights. But in its characteristic moods it was full of the wonder of life, full of a happy confidence in the strength of the mind and the powers of man. In its most characteristic experiences it could repeat the words: "The proper study of mankind is man."

In the Greek world the physical basis of life was studied first and the mind of man later came in for its own. In the modern world the process was reversed. The Renaissance was based upon the dignity of man. And modern science, when it came later, was based upon a study of the physical world. The thought that there was a mathematics explicit in nature which corresponded to the mathematics implicit in the mind of man was the basis of a new and productive study of the material universe. The Saga which tells the tale of it is the story of the most characteristic achievements of man since the middle of the nineteenth century. The sentence, "the proper study of mankind is things," may well be transformed into the other sentence, "the proper study of mankind is science," when one attempts to describe the life and the achievements of the last century. The name of Darwin has a place all its own in the story and various theories of evolution come in for an important place. In a sense the world ceased to be thought-minded and became thing-minded. Such control over the forces of nature as had never been dreamed of was actually achieved. The one great faith was faith in the reign of law. And men could discover the laws which controlled the world in which they lived and could use their knowledge to remake the world after the fashion of the mind's and heart's desire. The face of the world was changed. The habits of men's lives were changed. The steam age came and went. The age of electricity came and remained. Nuclear activity submitted itself to the control of man.

At first the new age seemed to be an age of the happiest expectation. A brilliant Englishman wrote a book called *The Century of Hope* to describe the direction in which the process was moving. Men thought more and more in terms of materialistic utopias. Physical comfort was to become universal and automatically all other good things were to follow in its train. But somehow things did not turn out well. The control of the forces of nature was not paralleled by control of the human forces. Standards were slipping and morals were disintegrating. One great war followed another. And vast sections of the world were left in a state of industrial and political confusion with signs of physical destruction everywhere. Brilliant inventing went on apace. And soon it was evident that the power, the preoccupation, which science had put into man's possession might lead to the destruction of civilized life and the reduction of life on this planet to the lowest possible levels. The mastery of the forces of nature had made it possible for the race most brilliantly and successfully to commit suicide.

When all this was analyzed by the most penetrating thinkers it was discovered that the preoccupation with things had been accompanied by a loss of the sense of values. The preoccupation with mathematics had led men to forget moral principles. The study of impersonal forces had left men without the compelling sense of controlling personality. Somehow the sense of values must be restored if civilization itself was to be saved. The sense of the moral structure of the world must dominate the understanding of its physical structure. The sense of the controlling person must once more be put upon the throne or life would fall completely apart intellectually and morally and physically.

Most happily new insights in respect of the philosophy of science came to support the new insights in respect of the moral and spiritual ongoing of life. Sir James Jeans came to see that the very structure of the universe suggested the presence not of an impersonal mathematical formula but of a great mathematician. Men like Whitehead began to refer to the schemes of complacent materialism with words of ironic scorn. Sir Arthur Eddington saw and said that there are elements of reality not reflected by those instruments of precision whose measurements were the be all and the end all of the immediate concern of the physical sciences. The new vistas were made visible in a brilliant History of Science and its relations with philosophy and religion by Sir William Dampier.

Serious and critical thinkers began to get a new look at all the problems with eminently good results. It was seen that in a curious way the men who worked out a philosophy of existence based upon the achievements wrought out in the laboratory forgot that you must always include the free and exploring mind of the scientist in an interpretation of the work of the laboratory. He sets the form of the experiments. He draws conclusions. He does not go into the laboratory in the morning and find that by some inscrutable process formulas have been worked out and discoveries have registered themselves while he slept. Free intelligence gets into the process of scientific investigation at the start and remains all the while. If uncritical minds do not see the significance of this fact they must be encouraged to become more critical. But where there is a free mind choosing among alternatives, as is always the case in the work of the scientist, there must also be erected standards of judgment, and so we begin to live once more in a world where standards are not only possible but are inevitable.

But there is much more to be said. This logic implicit in man's mind, which is answered to by a logic which is found to be written in the physical universe itself, is the very basis of all the achievements of the physical scientists. Does it not suggest and, indeed, make necessary the thought of a supreme intelligence which is the author of this correspondence, which made nature to be understood and mastered by the mind of man and made the mind of man for the understanding and mastery of nature? This coherence is the foundation of science. It is also the very basis of the intellectual security of religion. The belief in the great Intelligence which puts a logic in man's mind which answers to the logic of nature itself has the security which belongs to the key which fits the most important lock in the universe. A great thinker once said, "God cannot die while man lives." He meant by this that the very structure of man's life makes it inevitable that he shall look beyond himself to the great mind which created and sustains his own mind. The more you study man, the more it becomes necessary to think of God. From this point of view you can transform Pope's lines almost out of recognition with an improvement of their insight if not of their poetic grace and say:

> If God you'd scan,
> The best beginning is the mind of man.

The truth is that there is no conflict between the truths of science philosophically apprehended and the truths of religion critically understood. There is often a battle between the scientist who has not really inspected the philosophical relationships of his own positions and the religious thinker who is characterized by more piety than intelligence. The truths involved in the validity of the scientific process are just those which are essential to the ethical and the religious life.

A true humanism is based upon the belief in a free man choosing among alternatives in the light of standards. This belief is reached by a direct inspection of the processes of human life themselves. And it leads on by a deep and inevitable logic to the belief in a free ultimate intelligence perpetually choosing in the light of the permanent standards which grow out of his own perfect nature. And these very principles make secure the deepest necessities of the moral and spiritual life.

When, after inspecting these great matters, we turn to the Christian religion in its classical form, we find that the positions we have reached fit that religion as a hand fits a glove. The Scriptures begin with the conception of man made in the divine image. They bring us into the presence of a great Person who is perfect in knowledge, in power, and in goodness. They see men always confronting the necessity of moral choices. They see God at work in history making His will felt through the thought and the actions of men. And in Christ they see that God is made available to men in human experience. And in him they see perfect goodness revealing itself as perfect and sacrificial love. They see the men who have made bad choices summoned to change their minds and to enter once more upon the way of good choices. They see man's life in the light of God's character and God's will as these are revealed in Christ. They see the world of things made the servant of moral and spiritual meanings. They see the physical wearing the livery of the spiritual. They see man as a creature whose life cannot be made complete in time and who can be satisfied only with eternity. So the Scriptures bring us to that which will fulfill the requirements of man's deepest intelligence and of his most profound moral and spiritual desires. They see men over nature and under God finding marvelous richness in a life of fellowship with each other. They see the strength of God made available for the needs of men and men so made that something within them is constantly reaching out after the strength of God. They see the signature of the mind of God everywhere in the world which He has made for man's understanding and for man's use. And all of this fits together in a perfectly coherent view of life which reflects man's deepest experience with nature and with other men and with God. And the thought about it and the understanding of it are made possible by a net of clear and vital words themselves the expression of the profoundest qualities of man's intelligence and of his humanity. The proper study of mankind is man, and Nature and God and the implements of that study are found in those words whose study makes them sharp and decisive instruments for the subtlest work of the intelligence.

It is when we realize these things that we are ready to make the most of the processes of education and to choose wisely in the ways of learning and in the ways of life. It is when we have understood these things that we are ready to make the most understanding use of the long story of man's thought and feeling and action. It is when we have understood these things that we are able to conserve and to use the best of the past, to make golden the meaning of the present, and to fill our thought of the future with that expectation which will not come upon disillusionment or defeat.

PHILOSOPHY AND RELIGION

The Jewish Thesis

RABBI LOUIS FINKELSTEIN, PH.D., S.T.D.

Rabbi and Chancellor, The Jewish Theological Seminary of America, New York

Dr. Finkelstein is one of the truly great Jewish leaders today. Since his appointment as president in 1940 his leadership has been the most important force in the development of the Jewish Theological Seminary and the Institute for Religious and Social Studies. To meet the needs of the nation at war, he instituted the accelerated program of the Rabbinical School which made it possible for the Seminary to give one hundred of its graduates to the chaplaincy of the Army, Navy and Maritime Services of the United States and Canada. He is also the directing genius behind the annual Conference on Science, Philosophy and Religion.

Educated at the College of the City of New York, he took his Ph.D. at Columbia in 1918, and the degree of Rabbi from the Jewish Theological Seminary of America in 1919. He has been the Solomon Schechter professor of theology at the Seminary since 1931, a lecturer at Johns Hopkins, Oberlin, Duke and Harvard, and is adviser to the committee on Judaica research at Yale University. Dr. Finkelstein was elected an alumnus member of Phi Beta Kappa, and in 1944, at a special convocation, Columbia University conferred upon him the degree of Doctor of Sacred Theology "in recognition of many years of work as religious teacher, scholar and author."

Among the many organizations with which he is affiliated are the Rabbinical Assembly of America; United Synagogue of America; the Executive Committee of the Jewish Publication Society; the Executive Committee of the Joint Distribution Committee; National Conference of Christians and Jews; the Board of Directors of the American Friends of Hebrew University; the National Council of the League for Religious Labor in Palestine; fellow and member of Executive Committee of the American Academy for Jewish

Research; member of the Religious Book Commission of the American Library Association; and the New York Committee of the National War Fund.

He is the author of Jewish Self-government in the Middle Ages, The Pharisees, Akiba, *and other books and articles. He is co-author of* The Religions of Democracy and Faith for Today, *and the co-editor of the eleven symposia of the Conference on Science, Philosophy and Religion. In 1940 he was awarded the Townsend Harris Medal, and was appointed by President Roosevelt to succeed Dr. Cyrus Adler as the Representative of Judaism advising the President on steps toward world peace.*

Once again Dr. Finkelstein proves himself the pulpit master with this great religious address, on Judaism, the Talmud, the Bible, the Halachah or Way of Life. He shows the relation of religion, war and peace; discusses the mind of man, the Mosaic Tradition, Torah, education, and the teachers of a hundred generations. This is a brilliant study of Jewish tradition with a look ahead to the world of tomorrow in wisdom and love.

Sermon Thirty-five

IN THE Jewish tradition, we pay respect to those who are dear to us by devoting time to study in their memory; and we like to select for study the fields in which they excelled. It is, therefore, fitting that part of this evening, dedicated to the memory of Milton Steinberg, one of the great theologians and saints of our time, be given over to a discussion of Jewish theology. Therefore, the few moments at my disposal will be devoted to such a discussion, which would, I believe, have pleased him.

A hundred generations divide Moses, proclaiming God in the wastes of Sinai to a horde of former slaves, from the present conclave of his disciples re-affirming his faith, in the mightiest city of history before all mankind. Each of these generations has added its contribution, sometimes small, sometimes great, to the evidence for his teachings; and yet in each age, many have questioned them. In our time, the evidence for the Mosaic thesis is stronger than ever and the challenges fiercer than ever.

Stripped to its bare essentials, the underlying idea of the Mosaic tradition is that the Universe and man have a history. Time is not a succession of chance events, but an ordered affair. Life is not an accident, but the product of thought. The universal process which includes all matter and all living things has meaning, purpose, direction. God is not only the beginning of all beginnings, He is also the goal of all goals. The spirit gave birth to matter, the world process is matter striving to become spirit.

To us, children of religious traditions, this may seem obvious common sense. But in its own setting, it was a revolutionary doctrine. The people to whom the insight was revealed inevitably became the fathers of historical writing. Instead of recording the events of their day as independent achievements of great conquerors, they described each as part of a Divine drama, which begins with the Creation, and leads to the indefinite and infinite perfection of man.

A large part of the five books of Moses, all the earlier Prophets, and almost all the later ones, are devoted to a discussion of this Divine drama. The stories of the Creation, of Adam and Eve, of the Patriarchs, of Moses himself, and of the kingdoms of Israel and Judah, are related as the unfolding of the great theme, namely, God's relation to the world. The Midrash is a commentary on this theme, and carries it forward to the end of time.

But the ancient teachers of our people knew that man does not think only, or even primarily, in words and sentences. They noticed what most of us discover only when we have grandchildren—that human beings think before they talk; and that it is possible to think in visual symbols, as well as in words. A baby looks to its mother for permission to play with a new toy, to accept a new friend, before it can formulate its problem in words, and perhaps before it even understands them.

It is even probable that in adult life our decisions are more often made in symbols of action than in words. We dream in symbols; and many of us express our innermost ideas far more effectively in gestures and in deeds, than we do in words and propositions. We learn to read our neighbor's face and eyes, and to judge from them, more than from his words, what he really thinks. More is expressed by a cordial handshake, and far more is implied, than by the words which accompany the gesture.

The faith of Moses is unique among those of the world in its emphasis on the place of visible action in human life. It seeks to express its underlying ideas in action symbols rather than in any other type of symbolism. To say, "I believe," does not in the Mosaic tradition confer any benefit on the speaker; the question is whether the belief influences his behavior. Nothing is gained by announcing that all men are made in God's image, unless that truth is reflected in the treatment of every man as the child of God, who bears His image.

In the tradition of Moses, therefore, moral action is not something we owe to one another, but something we owe to ourselves as heroes in a Divine drama; something we owe to the Universe, of which we are a part; something we owe to God, as the Author of this drama. When we pay respect to our fellow-man, we are asserting the kinship of both himself and ourselves to God; when we act unrighteously, we deny that kinship.

But the tradition of Moses is not satisfied with demanding of us righteousness toward one another. It also maintains that we can express our belief in God and in the meaning of the world-process through a series of purely symbolic actions. These are the many ceremonies and rituals which are at the very heart of Judaism; for it is in them that Judaism expresses most clearly and unmistakably its basic ideas. In the Jewish religion, life as a whole is a pageant, consisting of ritual forms and moral actions, all of which express the Divine theme. The observance of the Day of Atonement or the Passover is mandatory, not only because the rituals are beautiful in themselves, or because they are habitual with us. We observe these ceremonies to assert truths which they express.

Perhaps the most difficult idea which Judaism seeks to transmit is this: a life dedicated to express in action the basic principle that the world has meaning, is in itself a fulfilment of that meaning. When we in our moral and ritual behavior depict the idea that God exists, that life has purpose, and that this purpose is Divine, we are acting in accordance with that purpose. The world reaches the beginning of its full expression in men who, free to choose between good and evil, turn toward good. It is in such men that the material begins to be the spiritual.

So the ancient rabbi said of the man who studies Torah, the whole world is not more worthy than he; in devoting himself to the study and fulfillment of the Torah, he is giving meaning to the whole process which has led from aboriginal chaos to man.

There are, therefore, two parts in Judaism. One is the description of the world order, such as the narratives of the Bible and the homilies of the Talmud provide—we call that Aggadah. The other is the portrayal of the same ideas in moral action, legal system, and communal institutions. That we call the Halachah, the way of life, prescribed for us in the Jewish religion. To the extent to which the words that I am now speaking are simply a passive description of the world in which we live, they belong in the realm of Aggadah; if they involve any stimulus to action, and relate that action to the basic concepts of Judaism, they are part of the Halachah of our people and our faith.

As there were chronicles long before the ancient Hebrews wrote history, so there were ritual, ethical, and legal systems millennia before the emergence of the Halachah. But the Halachah differs from all other systems of ritual, morals, and law, in its determination to transform the whole of life into one consistent expression of ideas. In fact, under the Halachah, the whole of life becomes a martyrdom, not, however, a martyrdom in the Christian and derivative sense of suffering at the hands of tyrants, but in its original Hebraic sense of bearing witness to the truth of God and of His revelation in the world of matter and men.

"Ye are My witnesses," says Scripture; translated into Greek, this reads, "Ye are My *martyrs*"; and the well-known Rabbinic comment adds, "If ye are My witnesses, I am God, but if ye refuse to be My witnesses, then—so to speak—I am not God."

Under the Halachah, therefore, the whole human family becomes a kingdom of priests and a holy nation; "from the rising of the sun even unto its setting, the name of God is great among the nations," or at least ultimately will be; and everywhere offerings will be presented unto His name when history is fulfilled. There is none so humble that he cannot contribute greatly to the fulfilment of God's will in the world; none so great that he can have any other significant mission in life.

So profound and so comprehensive is the Mosaic concept that, once accepted, it transforms every act of man, and every one of his endeavors. In a sense it may be claimed that the Mosaic revelation has exalted man's sentimentality into humanitarianism and philanthropy; his love of ritual into theistic religion.

In a degree, Moses is no longer the teacher of Israel alone, but of all mankind. So far as profession of faith and intellectual conviction are concerned, the concept that the Universe is a process and has a meaning, has now been accepted by more than half the world.

Yet obviously something is wrong; for almost all men seem to proceed on the theory that life is a plaything. Those who take life seriously, and weigh their actions with anything like the care of actors in a pageant, are few and far between. We agree that we are acting out the most important drama of all, that of which God is both Author and Audience; yet we live as though there were nothing to life but child's play.

The Mosaic concept of the world is far too profound to be accepted easily. Forty days after the revelation on Mt. Sinai we are told that the ancient Israelites made a golden calf. And even after that calf was destroyed, the tendency toward easy-going paganism did not disappear. When the Mosaic idea penetrated the whole Western world, it was accepted in words, but not implemented in deeds; and even less in social institutions. As our ancestors retained such institutions as slavery and concubinage, though neither was consistent with the basic concepts of Moses, so European Christendom retained and developed a feudal system which was pagan, rather than Mosaic. The Roman Empire after its conversion to Christianity was not noticeably more peace-loving or democratic than it had been before. Superficially everything had changed; yet underneath, almost all remained the same. It was as hard for men in the early centuries as it is for us today to accept in daily life and action the principle that all men are brothers. They were intellectually persuaded that life derives its meaning from the fulfilment and expression of the meaning which God has given it; but it was not easy to accept such a truth emotionally, and so to translate it from the realm of abstract ideas, into that of practical affairs.

For these many centuries, therefore, the whole Western world has been the scene of bitter struggle within each man of his mind against his passion. His mind moves him to peace; but his passion to war. His mind moves him to serve; but his passion to power. His mind calls on him to create; but his passion to possess. His mind demands that he live with a purpose; his passion that he forget that he has a purpose.

Religion itself has been infected with paganism. Its great teachers announce that it has come into the world to unite men; actually it has often divided them. There is no scandal in history greater than that of religious persecution, when men slew one another in the name of God, re-enacting the tragedy of Cain and Abel, but adding to the sin of homicide the sin of blasphemy.

The confusion which made men identify religion with their particular denomination, and regard themselves as justified in hating, persecuting, and even killing those outside it, was repeated on the whole scene of human life.

The impact of the Mosaic tradition had aroused men's desire to serve; but it had not always been able to arouse them to service of God. Often they stopped half way, serving their particular group and their particular skill. The creators of art, science, and letters, did not generally give their talents to the advancement of their particular forms of expression and failed to attain, like the foremost among them, Maimonides and Judah Ha-Levi, the heights of giving all their talents to God. Habits of creation were fostered; not for the sake of explaining God's ways to man, but as outlets for self-expression, each with a ritual of its own and for its own sake, and each protesting that *it* constituted the goal of all civilization.

In the ritual of scholarship, the production of books is an end in itself. In the ritual of science, the accumulation of data is permitted to stifle wisdom and even truth. In the ritual of art, it is considered heretical and philistine to inquire whether any ideas at all are being expressed. The ritual of social improvement is often held to justify the sacrifice of human life. The ritual of increased production is impoverishing the earth, and robbing our descendants.

For many, ritual alone is taking the place of religion. Obviously, it is not the concentration on skill which is harmful, but the failure to develop the skill for the sake of God. The creative spirit who loves God will labor even more zealously to perfect his technique than does he who knows not God. But the results of such labor will reflect spiritual insights.

And so the modern age has emerged—misshapen offspring of the vision of Moses and the confusion of paganism; able greatly to produce, yet producing not for man's advancement or for the service of God. It is an age when man, drawing on the spiritual resources of his ancestors, performs miracles of creation, but unaware of the purpose of his creativeness, achieves only unhappiness and discontent. He is building a tower of Babel, which may reach the heavens, but can only divide mankind. No science, but mutually unintelligible fields of research; no philosophy, but mutually hostile schools of thought; no art but techniques; no great books, but oceans of clichés; no great orators, but skillfully polished phrases and orations. An age whose statesmen feel little responsibility for the fate of man and for his real problems, and who make no effort to translate insights into action which might help him on his hard road. An age, in which religious leaders themselves often forget their common duty to the Author of their work, and strike at one another, instead of evil.

The hundredth generation since Moses, a generation which has seen him vindicated as never before, finds his teachings in greater danger of defeat than ever before. Again, as in the last months of his life, he seems to stand on Mt. Nebo, a promised land before him, but with the threat of death hanging over him, barring his entrance into the land itself.

No one has a clearer duty in this world crisis than those who believe themselves to be the direct spiritual descendants of Moses. If the tradition he founded has been preserved with particular care by our fathers, we ought to consider whether we have done what we can to make it effective in the lives of men.

Obviously, the Mosaic conception of life as a Sanctification of the Name of God, is inconsistent with much of modern thinking. The test of success in the Mosaic tradition is neither happiness nor power, but service to God and fulfilment of His purpose. The tradition of Moses offers us, indeed, a simple test of our loyalty. Given the choice for our children of captivity and slave labor in a concentration camp, and the opportunity of building one for others, which would we prefer? And unless we can say wholeheartedly, without reservation, that we would prefer to see our children suffer the torment of prisoners of injustice, rather than partake of the wickedness of the perpetrators, we have betrayed the tradition of Moses. Between suffering wrong and inflicting it, Moses gives man no alternative. If life's martyrdom involves pain, there can be no repining. In the creation of so magnificent a work as a good life, incidental pain and labor are irrelevant. Pain and suffering must be lessened and eradicated, when possible; but when unavoidable, are to be accepted as part of man's assignment, with joy, and the knowledge that it is part of his role in the greatest of all causes. So Rabbi Akiba spoke with the utmost sincerity and precision, when, incredibly tormented in his last hours, he said that all his life had been a preparation for that supreme moment when he might at last discover the measure of his love for God. Having discovered that he loved God literally with all his life, Rabbi Akiba's joy in the triumph of the spirit offset his physical agony.

Only in a world in which men cease to pursue happiness and self-expression will they attain them. Submission to God is freedom. Dependence on Torah is independence of men. New prospects of happiness and achievement will open to men when all of us, servants of ritual, finally become servants of God. A new statesmanship, nobler in vision, profounder in insight, and endlessly more practical than any now existent, will appear among men who serve not parties and states, but mankind and God. New wisdom awaits philosophers who can forget their place in the history of philosophy, and their stake in their own fragments of knowledge, and can seek to join others in the search after truth. New vistas of achievement, beyond our present imagination, are open to the artist, who selflessly can dedicate his gifts to the interpretation of God to man. Religion itself, including our own Jewish religion, and the faith of our particular gathering here, will attain new vision and opportunities for service, to the extent we forget our immediate objectives, and think of how we can as individuals, as institutions, and as groups, bring about greater love for God among all men.

Of all Sanctification of the Name such self-dedication may be for modern man the most difficult to achieve. Bred in cultures that stimulate the hunger for fame and power, emphasize the role and memory of the individual, and permit him, at best, only to sublimate his strivings for personal gain in institutional advancement, we are captives of a pagan idol-worship, engaged in mortal conflict with the Mosaic tradition.

Each of us thus contributes to the vast paganism of our time; and our rituals and institutions become not the means to realize God, but to blind us to His truth. As in the days of Jeremiah, so in ours, the Ark of the Covenant, intended as symbol of men's nearness to God, has become a fetish.

Those who contribute to the Sanctification of the Name can measure their gain by its very hardship. The times call on us to rise to a dedication attained only by the greatest of our forefathers. The period of grace in which to decide between God and Baal is running out.

Fortunately for us, the truth is not far to seek. In the Torah of Moses and its commentaries, it has been spelled out so that it cannot be mistaken. There is a miracle to be performed; but our teachers have shown how to perform it. "The least of you," said the Roman emperor to Rabbi Judah the Patriarch, "resurrects the dead." Many here present have breathed life into lifeless bones, have given vitalizing life to young and old, preserving their physical and mental, as well as their spiritual health.

Your achievements as individuals can now become the basis for greater achievement on the national and world scene. Your contribution to the advancement of Torah can be extended to new areas, so that our combined martyrdom may bring about a re-orientation of our own people, of America, and of the world. It is for us to re-affirm, in our own actions and decisions, the basic truths, bequeathed to us, that the world process has meaning, and that the supreme task of life is to express and clarify that meaning. A rare opportunity for magnificent service is open to the small band of pioneers who will move and take others with them from the world of today into that which Moses and the Prophets envisaged.

Members of the Rabbinical Assembly, by your training and by your calling, you are eminently qualified to be such pioneers. The teachings to which we are dedicated, make no allowance for convenience or comfort. We regard the land of Israel as the Holy Land, not because it is more fruitful or materially more promising than any other, but because it was the cradle of Prophecy. We insist on reading the Scriptures and prayers in the very tongue of the ancient Prophets, though that requires us to be bi-lingual. We observe the Sabbath on the seventh day, and follow the food laws of Scripture and Talmud, though both practices become increasingly difficult in a standardized and mechanized world. We insist that our children be instructed in their faith, though that requires hours of devotion which might be given to play. We choose to be a minority, often a minority in the midst of a minority, because we believe that is our particular task in pursuit of Torah.

Having been trained in love of Torah, let us follow the example of our predecessors, the Men of the Great Synagogue, and perform an even more difficult service, merging our diverse wisdom and experience, in order to deal effectively with the common problems of all men.

Let us not seek escape from our task in a false humility. This conclave, measured by gifts, training, experience, and dedication, is no ordinary assembly, but a University. The University of Judaism, envisioned by our own Professor Kaplan, exists in the combination of our four hundred and fifty congregations, and the Seminary, from which they draw inspiration. You and the staffs of the Seminary constitute its faculty. Fortunately for us the faculty includes our revered masters, Professors Louis Ginzberg, Alexander Marx, Mordecai M. Kaplan; and now there have been added to it such great teachers as Professor Saul Lieberman and the group of younger scholars, together constituting the foremost gathering of Jewish scholars in the present world. It also includes among you some of the outstanding interpreters of Judaism produced in many generations. The establishment of such a University is no mean achievement. But a greater one beckons, if Professor Kaplan's dream is to be realized.

This University can succeed only if it achieves the team thinking that has thus far eluded all others; and takes for the subject of its research not separate segments of Jewish and world affairs, but their totality. All the wisdom we can muster, all the piety we can attain, all the grace that may be granted us, will be needed; and we must not fail. With the leadership our Assembly possesses, with the increasing insight of our laity, we have the personnel needed for the reconstitution not of a Sanhedrin to deal with rituals, but of a Keneset Ha-Gedolah, to deal with life.

If we can surmount the divisions among ourselves, we may become the instrument to draw together men of all faiths and philosophies for the establishment of a human family, a diverse yet unified human culture, an enduring and creative peace. We can seek, and I believe find, the help and cooperation of our brothers in the Roman and Eastern Catholic Churches, in the Protestant denominations, in the Moslem and Eastern religions. Together with them, we may yet wrest the world and mankind from destruction. Undoubtedly the first step toward that end will be an understanding among the Western faiths. (For that purpose, we should above all, seek the understanding and cooperation of the Christian and Mohammedan, as well as our sister Jewish groups.) Perhaps through our joint efforts a forum may emerge where scholars of the various faiths may discuss their separate symbolisms, their diverse and common problems, overcome religious antagonism, take a measure of the world's spirit, and achieve religious creativeness and understanding.

A hundred generations of teachers look down upon us, and many more generations of their disciples will follow us. Ten, twenty, a hundred thousand years hence, the Torah will still be studied; men will still grapple with the problems of faith and life. May it be given us that future generations, even if they forget us, may look on our work with love and gratitude. May the Jews of that distant time see in our effort an expression of Divine love, and be drawn by it to greater love of God. And may man today so live that when each day those men of the future come to the prayer, "We thank Thee, O Lord our God," they may regard our generation as we do that of Hillel and Rabbi Akiba, and of Moses himself, as among the foremost of the blessings for which man is indebted to the Almighty Creator.

PHILOSOPHY AND RELIGION

On Christian Certainty

REVEREND DAVID E. ROBERTS, PH.D., D.D.

A Minister of the Presbyterian Church; Associate Professor of the Philosophy of Religion and Systematic Theology, Union Theological Seminary, New York

Most of Dr. Roberts' preaching is in colleges and universities: Princeton, Wesleyan, Hamilton, Sweetbriar, Connecticut College for Women, Smith, Union (Schenectady) and Wooster (Ohio).

As a member of the Joint Committee on Graduate Instruction of Columbia University and Union Theological Seminary, he influences the intellectual life of many of the graduates of Union.

He was educated at Occidental College, Union Theological Seminary, Edinburgh University, Marburg, Göttingen, and Oxford. He is a member of the Presbytery of New York, and is associate professor of systematic theology and the philosophy of religion at Union Theological Seminary, New York.

He is editor of the "Union Seminary Quarterly Review" and wrote, in cooperation with H. P. Van Dusen, Liberal Theology.

When Dr. Roberts was first included in Best Sermons, *several years ago, so many people expressed appreciation of his "Christmas Story," that we decided to include him again and it is a pleasure to present this philosophical statement of his. Where, after all, can man find certainty? What is the relationship between philosophy and religion? (The reader should see Dr. Douglas Horton's sermon on this theme, in more popular vein, earlier in this volume). Dr. Roberts' sermon is a real contribution to current religious thinking.*

Sermon Thirty-six

IN THE ELEVENTH chapter of Romans St. Paul offers an explanation of God's dealings with men. He shows how the rejection of Christ by his own people has, in a sense, made the faith of the Gentiles possible. God has so used the unbelief of the Jews as to make it contribute to the universality of the Gospel. Therefore the Apostle urges that the universality of the Gospel will ultimately embrace those who now reject it. Since God can use lack of faith to produce faith, Christians should include the Jews in their hopes instead of excluding them.

But the explanation concludes with an outburst which seems to contradict it. "O the depth of the riches both of the wisdom and knowledge of God! how unsearchable are his judgments, and his ways past finding out!" (Romans 11:33). How can God's judgments be unsearchable if the Apostle has just succeeded in delineating them? Which side shall we choose? Shall we believe St. Paul when he tells us that he has reached insight into God's wisdom, or shall we believe him when he suggests that no man can know the mind of the Lord (verse 34)?

Actually we must accept both sides. From beginning to end, the Bible tells us that we can know God and that God is unfathomable. As a consequence, Christians can always go astray in either of two directions. First, they can become so confident about their doctrines, creeds, systems and proofs that the mystery of God is forgotten. Secondly, they can represent divine truth as so utterly inaccessible that many people give up the quest altogether, or decide that one man's religious hunch is as good as another's because nobody really knows.

But how can we possibly accept both sides? How can we say, with the thirteenth chapter of First Corinthians, that "we know in part?" How can we admit that this partial knowledge is always imperfect, always subject to human distortion, always in need of correction—and at the same time declare that it is knowledge instead of sheer ignorance, error and illusion?

I do not propose to discuss this problem as an intellectual puzzle. I raise it, rather, because all of our most urgent practical difficulties come back to this one. How can we, as Christians, stand firm against fanaticism without falling into fanaticism? How can we believe in the universality of the Gospel without falling into arrogance toward non-Christians? How can we effectively preach a mystery to men who clamor for definite, simple answers? They can be persuaded or dragooned into laying down their lives, but only in response to tangible promises of power or security or revenge. How can we ask them to give their lives, their whole selves, to the service of God, if we must tell them in the next breath that no one knows fully what the will of God is?

Some humane and potentially religious people turn away from the Church because they fail to find satisfactory answers to such questions. They feel, to put it bluntly, that knowledge of God is not a major issue. They observe that a man can strive for freedom, truth and brotherhood, whether he believes in God or not. They also observe that a man may be enslaved to ambition, hatred and lies, whether he believes in God or not. Indeed, it might be more accurate to say that every individual is himself a battle-ground where these opposite sets of forces struggle for supremacy. Therefore the real issue lies within man himself. The real issue is whether humility, justice and love can release us from bondage to regimentation, materialism and war. As Erich Fromm has put it recently: "Is it not time to cease to argue about God and instead to unite in the unmasking of contemporary forms of idolatry?" [1]

We must take this question seriously. Sometimes we *have* been guilty of talking about the existence of God in such a way that it has no discernible bearing upon the decisive events of human life. Whenever God becomes simply an idea that we want to defend against competing ideas, then it is quite right to say that we have lost touch with the real issue.

Let us start, then, with man, no matter how much such a proposal may horrify some theologians. Let us start with his hopes and fears, his assets and liabilities, his power and weakness. Let us start with his inner battle between slavery and freedom. In that instant the problem is not how we can bring God into connection with our theme, but how we can possibly avoid Him. For I cannot take a step toward deeper self-knowledge without discovering that the answer as to who I am, and what I should be, does not depend exclusively on me. I have my own ideas about what freedom, justice and love mean; but they are precisely as defective as I am. They reflect the limits of my experience, my insight, my character. If they are ever to be corrected and improved, then the change must come in part from beyond what I am.

You may say that they can be corrected by means of what other people teach me; and this is true. But how do I decide what to accept and what to reject among the influences which come to me from other people? Ad-

[1] *Psychoanalysis and Religion,* p. 118.

mittedly my ideas are defective; but so are theirs. Many things about myself I am quite incapable of seeing. But certainly no one else sees me in such a way that there is no distortion, no exaggeration, no blindness. Thus we must say of every human being: "The truth about him is there all the time; but no man knows it fully." Starting with man, we cannot move a step without encountering the fact that a judgment is true only insofar as it participates in something which transcends us.

At times, when we are asked about God we reply as though we were trying to hand out an item of information. But if God is real, then knowledge of Him is not like acquiring information. Instead, it is like the inner upheaval which occurs when we face a new crisis or a new friendship. The basic requirement is not so much that I shall apprehend something as that I shall be able to stand being apprehended.

Picture a man who has organized his life around a stubborn effort to justify himself. Every cause he supports is *ipso facto* noble, and all those opposed to his cause are, as he sees it, obviously vicious. What he demands of life seems reasonable to him. If at times he becomes unfair or unreasonable it is always because others have taxed him beyond what flesh can be expected to bear. Every criticism directed against him is a malicious attack. Every criticism he directs against others is a piece of commendable candor. This man is, in short, a kind of godlet who judges others by how they fit into his scheme of values.

I say: "Picture such a man." But with minor variations such a man comes very close to being every man. Every man is tempted to justify himself beyond the point which is really justifiable. Every man views the worth of others in the light of how they affect his own security, his desires and ambitions. And whenever conflict breaks out between two such human beings, neither is completely fair to the other. That is why legal systems have to settle disputes by introducing a judge who is not directly implicated.

But now picture such a man brought into a personal relationship where the Other is completely free from the prejudices and distortions to which all men are subject. Imagine that this Other is able to reveal exactly those truths the man has always managed to evade. All his grandiose notions are shattered. The evil he has done, and the evil that he is, stand out in stark clarity, with no retouching, no possibility of shifting the blame. At the same time, all of his redeeming qualities are recognized. He is not allowed to take credit for more than he is worth; but everything worthy in him is given full weight. Yet the man is more enraged than satisfied by this perfectly fair appraisal. And when he tries to expostulate in self-defense, when he tries to fight back by accusing the Other of ulterior motives, there is no retaliation, no argument. The Other simply looks at this man with eyes which see right through him—and waits.

Accept this situation as a parable, and you will see why the chief obstacle to knowledge of God is not lack of information. What is called for here is not that we should have brains enough to grasp some irrefutable philosophical theory. What is called for is that we should have courage enough to look at ourselves, and to be looked at with the eyes of Eternity. *This* is the major issue; and it cannot even be formulated without bringing us face to face with God, whether we use the word or not.

The man we are talking about is now in the presence of the kind of truth, the kind of love, that can save him; but he can also reject it. He can dig in stubbornly and refuse to see what the Other sees. Indeed, if the Other were interested merely in humiliating or in crushing him, one would have a perfect right to fight back. The most humiliating thing about standing in the presence of God is that He has no interest in humiliating us for its own sake. The most crushing thing is that He does not ask us to wallow in remorse. His presence simply unveils what we are.

This is one of the points at which mystery enters. We can be saved by accepting the truth which comes to us from beyond our own distortions. But in order to be saved by it there has to be something already accessible within us. You and I have seen people fight desperately to hang onto what might be called "the indispensable lie." We have seen them ward off every appeal to reason and fairness and sympathy. And then sometimes, when they have been driven absolutely to the wall, it is as though they were able to let go. They have relinquished their insistent claims; they have stretched out a hand to the enemy; they have turned from bitterness to forgiveness. And we say of them: "Well, those fine qualities must have been in them all along. Why did they fight so stubbornly against letting them come to the surface? Why did they have to be cornered, and almost smashed, before they were able to give in?"

Here is the mystery. We do not save ourselves. We do not conquer our own egotism. It is only because of the harsh inescapability and the loving inescapability of the truth from beyond us that we are ever saved. And yet we participate. The transformation takes place within us. It is ourselves, our human potentialities, which are tapped in fulfilling the conditions. The truth from beyond becomes the truth which we "produce" within.

And there is another mystery. The man we were talking about is free, because he can accept God's truth or reject it. But notice the odd feature of his rejection. When he holds out, when he will not capitulate in the presence of perfect justice and perfect love, it is as though some demon were driving him. It might even be more accurate to say that he cannot give in, he cannot let down his defences, he cannot have a change of heart. So the point at which man's freedom becomes most obvious is also the point at which his bondage becomes most obvious. Mankind is free to enslave itself to illusions, to egotism, to conflict. It is free to do so indefinitely—to the end of time. But man is not free to change God. He is not free to change the nature of justice and love. He is not free to turn lies into truth. He is not free to make his own blindness honesty.

If, then, we return to our initial question, asking how God can be both known and unfathomable, the answer is that His presence is both inescapable and eternal. What a man brings into this presence is simply himself. For all of us that means: one who is neither completely inaccessible to the truth nor completely open to it.

From such a standpoint it is God who is supremely important. Our ideas, our doctrines, our lives are important only insofar as they point to Him and participate in what He is. And because they always incorporate our limitations, we must always be ready to alter them whenever they prevent us from entering more deeply into what He offers. In other words, our certainty of God goes hand in hand with a *lack* of certainty about the finality of our own formulae. Genuine faith continually breaks the bonds of any concepts, any symbols, any words which try to hold it captive.

Thus the Christian message is, indeed, at a disadvantage wherever men want simple, dogmatic assurance that *their* answers are the right answers. And, more precisely, this is a perennial disadvantage which Protestantism should expect to suffer in comparison with Roman Catholicism. But one does not really succeed in disposing of mystery by demanding something he can control. Nor does one capture mystery by organizing an institution to be its custodian. It surrounds our lives unpredictably at every point. Who understands his own compulsions and his own failure to do what he knows is right? Who understands his ability to enter deeply into communion with some people and his inability to find any common ground with others? Who understands why every assertive step that we take toward bolstering security leads to a fateful increase of insecurity? Who understands why man should have appeared on the scene at all? Who understands the birth of any person or the death of any person?

It is in the midst of such uncertainty about ourselves, and our own righteousness, and our own destiny, that we point beyond ourselves to the Other, as to the constancy of truth itself. Some people reject Christianity because they think it requires pretending to be sure where one is not sure. They look upon the Church as filled with men who claim that their beliefs are absolutely right while everybody else's are wrong. They reject Christianity because the information which we furnish about God does not seem very convincing. And all I can say is that we as ministers and laymen have done a rotten job. Otherwise such people would realize that honesty, humility and a large dose of skepticism toward neat theories, do not exclude a person from faith. On the contrary, they are pre-requisites for reaching faith.

This is not quite the last word on the matter, however. For Christian faith rests upon the singular claim of one historical life. Surely at this point, one might think, there can be no tentativeness about our doctrines, our creeds, our formulae. And yet, if what is true of God is true of Christ, then our certainty attaches to Him, not to ourselves. The same human fallibility, the same distortions, the same blindness affect our interpretations of what Christ means, precisely insofar as they interfere with our fellowship with God. Sometimes we overlook this fact. We talk as though the ultimate nature of God were concealed from us, but the historicity of Christ makes certain truths as plain as this morning's headlines. But what we ought to say is that God is supremely accessible to us in Christ *and* supremely mysterious to us in Christ. Let us be thankful for the accessibility. The message can be told as a story about events upon this earth. It breaks through the barriers of sin, reaching the wise man and the simple man right where they are. For those who want something tangible and definite, here it is. But in the next breath we must say that it is unfathomable. What can possibly be less obvious, less to-be-taken-for-granted, than that God was made man and took upon Himself the burden of our guilt? Here, as nowhere else in history, a human life, concrete facts and events are linked to inexhaustible meaning. Christ is, for us, both the mystery and the knowledge of God.

And it is as though the Epistle to the Colossians were echoing the eleventh chapter of Romans when it speaks of "the full assurance of understanding . . . the acknowledgment of the mystery of God, and of the Father, and of Christ; in whom are hid all the treasures of wisdom and knowledge" (Col. 2:2f).

PHILOSOPHY AND RELIGION

The Sanctuary of Life

RABBI ISRAEL BETTAN, D.D.[1]

Rabbi and Professor of Homiletics, Hebrew Union College, Cincinnati, Ohio

Rabbi Bettan is one of the preachers who molds the young rabbis of Reform Judaism, and in his own preaching he has force, intellectual power and spiritual perception. Born in Lithuania in 1889, he completed his studies at the University of Cincinnati and Hebrew Union College. He was rabbi of B'nai Israel Congregation, Charleston, West Virginia, from 1912 to 1922, was president of the Charleston Federated Charities, chairman of the City Survey Commission, member of the West Virginia Child Welfare Commission and member of the West Virginia School Code Commission. In 1915 Hebrew Union College conferred the D.D. upon him.

During the first World War, he was chaplain in the United States Army. He has been professor of homiletics and Midrash at Hebrew Union College since 1922, is chairman of the Committee on Responsa, Central

[1] Sermons by members of the Advisory Committee were contributed at the request of the editor and are included on his responsibility.

Conference of American Rabbis; member Committee on Liturgy C.C.A.R.; member Tract Commission; member Committee on Ceremonies. He is the author of a number of monographs on homiletical and historical subjects; chief work: Studies in Jewish Preaching, 1939; The Five Scrolls *(a monumental work on The Song of Songs, Ruth, Lamentations, Ecclesiastes, and Esther).*

He gave a series of lectures at Garrett Biblical Institute, the Divinity School of the University of Chicago and Duke Divinity School. At Hebrew Union College he helps to mold the minds of hundreds of young rabbis of the Reform tradition, and is modern in his homiletic methods, while following sound interpretations of religious matters in the Reform Jewish pattern. His personal courage in his preaching and thinking, his wise teaching, his guidance of his students in their serious problems, and his gentle way of life make him a man worth knowing.

In "The Sanctuary of Life," Rabbi Bettan has a message on God, the all-pervading spirit of the universe, and shows how doubt and unbelief upset man's spiritual peace. His discussion of the faith of the Jewish people from the Reform viewpoint is at once interesting and stimulating.

Sermon Thirty-seven

TEXT: And let them make Me a sanctuary, that I may dwell among them. EXODUS 25:8

THE POSITION of the Bible in human thought must prove baffling to the modern mind. Although a considerable portion of its content has been relegated by Biblical scholars to the realm of poetic fancy; although many of its ancient laws and statutes have long ceased to operate with any real force in the life of man, the Book is yet cherished and revered by the great bulk of mankind.

The fact, so often overlooked, is that the Bible comprises in large part a body of universal principles specifically applied. The specific instances, determined in most cases by local conditions and requirements, may have but temporary validity; the general truths underlying them are immutable and eternal. Only its perishable elements have yielded to the ravages of time; its fundamental principles, expressed or implied, are as true and potent today as when first enunciated. What, for example, if science discredits the Biblical account of creation; what if it questions the authenticity of the story of the Flood—the great affirmations of faith that there is an all-powerful spirit back of visible phenomena, pervading and controlling all, and that virtue and not vice can permanently endure are thereby neither disproved nor impaired.

The special case to which a general principle is applied may be but local and transient; the principle itself is as lasting as it is universal. The luster it sheds through the interminable ages may at times grow dim and faint; it can never wholly fade into darkness.

One of those luminous truths which, though clothed in a garb of local coloring, bears the clear stamp of permanence is implicitly expressed in the Scriptural text read to us. The children of Israel are ordered by divine command to erect out of the material voluntarily offered by each of their members a sanctuary as the habitation of the Lord, that His spirit might dwell in their midst. "And let them make Me a sanctuary, that I may dwell among them."

Few there are among us whose conception of the deity is such as would permit the restriction of His presence to any one circumscribed place. We no longer think it possible to define the illimitable and set bounds to the infinite. We discountenance any attempt to localize the presence of God. We rather echo the sentiment voiced in Solomon's prayer "But will God in very truth dwell on the earth? behold, heaven and the heaven of heavens cannot contain Thee; how much less the house that I have builded" (1 K. 8:27). But our text is clearly the particularization of a general truth, a specific instance beneath whose surface lies hidden one of the most profound truths which Judaism has contributed to the spiritual riches of the world. God, the all-pervading Spirit of the universe, manifests His presence and reveals His glory in the pure atmosphere of holiness; and if man is to discern the divine essence that permeates the visible world, if he is to fathom life's mysteries and find himself face to face with God, he must raise his being to the plane of the spirit, weaving the golden threads of lofty ideals and aspiration into the ethereal texture of his soul and gathering out of the ordinary everyday experiences fit materials for the sanctuary of his life.

The world as constituted is dual in nature. There is mind and matter, soul and body, spirit and substance. Both nature and man are products of these component parts. God alone, according to the teachings of Judaism, while manifesting part of His glory in perceptible forms and objects, is in essence pure spirit. He lives and works within us, about us, and above us. But the material form in which our being is set checks the free inflow of His holy spirit. To surmount this obstacle, to enable our souls to pierce through the solid crust of the material and approach the throbbing heart of the universe, feel its pulsations and mingle with its essence, we must spiritualize our lives. We must invest our surroundings with holiness, transforming the whole of life into a vast sanctuary in which every mode of action, every line of conduct, every human relation is purified and sanctified. "And let them make Me a sanctuary, that I may dwell among them."

We live in an age of doubt and widespread unbelief. In our day men question, without actually denying, the religious certainties in which their fathers found comfort and inspiration. They demand that the fundamental truths of religion shall be submitted to the tests of the laboratory. They will accept nothing as true that cannot be verified by scientific demonstration, nothing that cannot be proved by straight lines, by balanced scales, by critical analysis. They are honest and sincere in their unbelief and hardly merit the derision and rebuke they often get. Yet this tendency to apply to spiritual things the evidences that belong to the realm of things material, this clamoring for a God whose image is limned on a tangible canvas, betrays a confusion of thought which is not at all compatible with the new insights of which our generation seems so proud. Higher truths can be perceived only by higher faculties. Spiritual truth will be discerned by the spirit of man made sensitive to its reality. We come to truth in different ways and apply to it various modes of evidence. We do not measure the fragrance of the rose by a yardstick; we do not gauge the power of the sunbeam by its weight in the scales. Nor can we grasp by the use of the senses anything that does not lie within the scope of their power. And doubt it as we may, there is a spirit in man which when properly trained and directed, when nurtured on the sanctifying influences of a consecrated life, will burst its narrow confines and soar aloft to the unexplored regions of the unknown. The recorded experiences of prophet and seer, the authentic visions of saintly men everywhere, bear indisputable testimony to this fact.

Man's spiritual faculty, only too often in a dormant state, will be quickened to perception by a life that embodies and expresses the divine principle of the world. In the sphere of higher truth, perception always follows in the wake of performance. We live by love first and then apprehend its nature and the full measure of its power. We gain a clear comprehension of the character and significance of justice and truth only by first being just and true. "Thou shalt" and "Thou shalt not" were the impelling forces that first put the mighty engine of morality in motion; and only after centuries of ceaseless endeavor to bring our lives under the controlling influence of these moral truths have we learned to grasp their true meaning and import. For the deeper mysteries of life will be explained and the deeper problems solved not by thinking but by living. And in a certain sense it is true that while we ourselves were created by God, every one of us in compassing the higher life creates God for himself. "And let them make Me a sanctuary, that I may dwell among them."

This general principle, of which our text is only a special application, is the chief corner-stone on which Judaism rests. It is the fashion of the day to talk disparagingly of old practices and observances. Religious forms and usages, which are no longer in vogue, evoke but rarely a sense of reverence. They are petrified rites, we say, gritty relics of a hardened faith. We forget, however, that they were once volatile and fluid, alive with the spirit that animated them. Taking for its basis the profound truth that only by building a sanctuary will man reach unto God, that only through the spiritualization of life will he attain unto a conception of higher realities, Judaism has formulated a system of conduct which for comprehensiveness and loftiness of aim has no parallel in the world's religions.

The purpose of the elaborate system was to bring the Jew to a fuller realization of the meaning and value of life. All things reveal God; all right activities and forms of life disclose His presence. The world is sacred; man's life and relations are sacred. Every act of his, every step and movement must be forged into a chain linking him to his God. He rises from his sleep—he is required to express his gratitude for the life restored to him. He sits down to his meal—he is soon reminded of the solemn nature of the act. The table is an altar and he a priest bringing an offering to Him who is the source of life and all its comforts. He is successful in some enterprise; he escapes misfortune or injury; he recovers from illness; he dons a new garment; he sets out on a journey; he arrives home safely—whatever experience he undergoes is by some symbolic act or utterance sanctified. He is always to be conscious of the Being from whom all things emanate, who rules over all, possesses all. He is to love truth, justice, benevolence and purity because God is true, just, merciful and holy.

Such, in essence, is the faith of the Jew as expounded by the ancient rabbis. It was a sound system of belief and practice, superbly conceived and productive of most fruitful results. Religion became the vocation of the Jew, his distinctive pursuit and endowment. But in the course of time the natural limitation of all outward expressions could not but make its appearance and lay bare a flaw of perilous nature. The invisible breath of an ideal, once infused into a visible substance, will sooner or later begin to congeal and harden. External forms, meant to give expression to an inner faith, will often assume an independent role of unwarranted prominence. The means first outgrows the end in importance and then seeks to supersede it altogether. We cannot make our formularies fixed and unalterable without running the risk that these rites and beliefs, even when no longer alive with the original spirit and purpose, will continue to exact our utmost devotion. When law, even divine law, hardens into an unchangeable legal code, it ceases to be a vital factor in the life of the spirit. Blind obedience may mold a group into uniformity; it will bring no enrichment to the individual personality.

Liberal Judaism aims to steer clear of the baneful effects of externalism. Not that it would have us sever the ties that bind us to Torah. Ours is still a religion of law. An ethereal faith, like a disembodied spirit, can have no permanent abode in our lives. We shall never dispense with the traditional disciplines of faith. But we shall ever test their validity and value for us. We are concerned with such means as rites, forms, customs, only in so far as they help us achieve our larger purposes. The burden of Judaism's message to us today is the same as it was yesterday: Make a sanctuary of your life, that God may dwell in your midst. Live a higher life, the life of the spirit. Live inwardly rather than outwardly. Sanctify your life. Learn to see in everything you do and encounter its inner essence, its higher relations. How? By what means? Methods in abundance are at hand. If among the old ways you can discover a path leading to this goal, follow it. If in the ashes on your ancient altars you can discern a glint of light shooting upwards, cherish it. If from the anvil of your own experience a spark flashes forth that can lighten your way, use it. If the experiences of others who, hard by your side, are struggling for a richer life in their own way and in their own light can afford you safe guidance, accept it. Search for all that is good and helpful and appropriate the best, for the best leads Godward.

This is the old message in its newer form. As in centuries past, it urgently summons us to a higher life. Never before has the call been so distinct and insistent. And what has been our response? What have we done to make God dwell in our midst? We who deem ourselves in the van of religious progress, we who pride ourselves on marching under the banner of a living Judaism, do we in our everyday life show the influence of a purer faith? Do we live the life of the spirit? This is the real challenge to the liberal Jew of our time. Have we escaped the shoals of externalism only to be wrecked on the craggy shores of secularism? Have we emancipated ourselves from a rigid legalism only to become enslaved to a sordid materialism?

If religion is to be a real force in our lives, we must strive to live in harmony with its true spirit and rise above the shallowness and superficiality of a life alienated from God. With God in our hearts there is no obstacle we shall not overcome, no difficulty we shall not surmount. The seemingly insoluble problems facing us will become surcharged with menace only if we forget the source of our strength and hope, only if we lose consciousness of the God who made us and shapes our destiny. As Jews, dowered with a rich spiritual heritage, it behooves us to remember that the place of high distinction we hold among the peoples of the earth we owe not to Jewish blood or power, not to Jewish skill or possession, but to Jewish idealism and Jewish spirituality. The Jewish vision of God and the Holy Bible of the Jew, these have glorified the name Jew.

This solemn day is a time for deep meditation and earnest self-searching. We know in our hearts how far we have fallen short of our possibilities. We also know that the task to which this sacred hour calls us is not one that can be undertaken by us as a group. Each one among us must heed the summons and meet the challenge of the hour. Each one of us has a life to live; and it is this single life that must be purified and sanctified if the vision of God is to break upon its horizon and glorify it. Then, and only then, shall we merge again into "a kingdom of priests, and a holy nation." Then, and only then, shall we lay rightful claim to the title Israel, the champion of God. *Amen!*

PHILOSOPHY AND RELIGION

Religion's Answer to a Troubled World

RABBI DAVID DE SOLA POOL, PH.D.[1]

Rabbi, Spanish and Portuguese Synagogue, Shearith Israel, New York
(Orthodox)

Dr. Pool is the leader of the famous Spanish and Portuguese Synagogue, Shearith Israel, which was founded in 1655, when New York was a village and the western boundary of what is now the United States was east of the Delaware River. This great synagogue has had a distinguished history, and Dr. Pool has shown himself to be a capable leader of the Orthodox faith, a scholarly preacher, and a good pastor. He is active in interfaith work and has the respect of Catholics and Protestants.

He was born in London, England, in 1885, and was educated in universities of Europe, being graduated with honors from the University of London in 1905. Later he studied at the universities of Berlin and Heidelberg, receiving his Ph.D., summa cum laude, *from Heidelberg.*

Since 1907 he has been minister of the Spanish and Portuguese Synagogue. He is the representative of Jewish army and navy chaplains to the Chiefs of Chaplains.

In 1917 he was one of three Jewish representatives appointed to serve on Herbert Hoover's food conservation staff; in 1919 he was one of three American representatives on the Zionist Commission to Palestine; and from 1938–40, he was president of the Synagogue Council of America.

Dr. Pool has written numerous prayer books, pamphlets and reviews, and is the author of The Kaddish, Hebrew Learning Among the Puritans of New England, *and* Capital Punishment in Jewish Literature.

This distinguished sermon asks several startling questions concerning war and peace, religion and man, government and citizens. One of the great rabbis coöperating with the Chaplains' Corps and the Army, Navy, and Air

[1] Sermons by members of the Advisory Committee were contributed at the request of the editor and are included on his responsibility.

Corps on chaplains' matters, Dr. Pool still shows his hatred for war and its brutalities. When will men learn to devote their creative abilities and statesmanship to enduring peace and the welfare of all men and all races?

Sermon Thirty-eight

TEXT: The work of righteousness shall be peace and the effect of righteousness quietness and confidence forever.

WHAT IS religion's answer to a troubled world? Before we attempt to suggest an answer to this question let us first consider some of the troubles of our world.

The greatest of them is war. We are setting up vast organizations of military offense and civilian defense. We disclaim the inevitability of war; yet we are planning and acting as if there is little that our leaders in Washington or we as individuals everywhere can do except feverishly to prepare for war's carnage and destruction. Is the building of air raid "shelters" in which to cower from the blast and burns of atomic bombs the summit of our practical wisdom at this moment of supreme challenge to mankind's purpose on earth?

In our day the traditional madness, cruelty, bloodiness, and vileness of war have been intensified in unprecedented measure and with a scientific thoroughness and efficiency unknown to the most sanguinary chapters of history. It is war which is today mankind's supreme catastrophe and supreme blasphemy.

Practical men of affairs as well as realistic philosophers have looked upon this supreme challenge of war. One after another they find the ultimate hope of a warless world only in a new religious education of the souls of men everywhere. Nations are as the men who compose them. Only when nations are made up of men with a martyr religious resolution that they will never go to war any more, will they beat their swords into plowshares and not lift up sword against nation, neither will they learn war any more, shall we have peace.

The next great evil threatening us is presented by man's failure to control his new powers. We have achieved a fabulous control over physical matter. We can create for our children an undreamed-of wealth of living. Man's brain has gone from triumph to triumph in building machinery of incredible creativeness, in attaining magical new skills, in mass production that gives a wide diffusion of rich comfort and ease, in scientific agriculture, in destroying distance, in moving with supersonic swiftness, and in a thousand ways making possible a hitherto unimagined enrichment and broadening and deepening of living. Yet we look out on our world of today with an almost paralyzing sinking of the heart when we see the terrifying inadequacy of the human material into whose hands is entrusted this world of marvelous, potential power and beauty. The climactic crisis is caused by failure in the soul of man. There can be no real happiness and no security for ourselves and our children except in the measure that we are ruled by what men of all denominations call religion.

Another of the great evils shaking the world is class warfare. Economists and social thinkers have attempted to analyze the racial antagonisms, the crippling strikes, the class strife and the economic nationalisms that are shaking the world to its foundations. They also declare that ultimately legislative enactments or solemn covenants can never assure peace and a just economic order so long as men are unworthy. We can never have security so long as the power groups of nations and within nations are moved by a conscienceless, greedy self-interest without regard for the rights and welfare of others. Mankind can have an assurance of peaceful coöperative living in brotherhood only through a rebirth of social justice. Such a passionate stirring of social conscience can arise only out of a religious recognition of the moral law and religion's basic social teaching of love of neighbor.

And what shall we say about certain new concepts of the state? The new technique of the all-assuming dictator state has frankly and avowedly torn down the age-old moral law which religion has been laboriously building up and fortifying through the weary ages. In our generation every method of sadistic violence has been made legitimate and acclaimed as desirable by the autocratic power-state if it subserves the end of concentrating power in the leaders of the state and eradicating every focus of difference or opposition. Every lie is justified, every promise forsworn, every solemn treaty and covenant torn up, to serve what is declared to be the interests of the man or men heading the state. Every form of freedom for which religion has struggled since the days of the exodus from Egypt can be denied the citizen. His physical freedom is made to yield to the physical or economic corvee. His freedom of thought, speech and spirit yields to loyalty hunts, ruthless propaganda or fear of the concentration camp or bloody purges. The human personality which religion regards as the image of God is deliberately enslaved to the soulless state. The very concept of justice which stands at the center of Old Testament ethical teaching is warped and perverted into any action or policy that may be considered to serve the interest of the state, however subversive it may be of religious concepts of justice. The love, kindness, mercy, tenderness, humility called for by the Bible—these are denounced as the decadent virtues of inferior peoples, the sign of weakness unworthy of the super-race and super-state that are destined to dominate the future. Strength is built on foundations of hate, not of love. "Walking humbly with thy God"—that is cynically mocked by the thrasonic claims of the master race and master state. The new commandments declare as a positive command, "Thou shalt steal." If it seems advantageous, then it is right to rob individuals, groups, and nations in order to subserve the purpose of one man or one clique or party dominating the state. Thou shalt murder men, women, children and gentle nations at peace under their vine and fig tree. Thou shalt covet their lands, their crops, their resources. Thou shalt bear false witness against anyone and everyone who ventures to question the rightness of any word or action of the bloody tyrant astride thy nation.

So runs the new code of moral law given to the world not from the heights of Sinai but from the heights of the bombing plane, not to the accompaniment of awesome thunder and lightning, but to that of the tank, the machine gun and bursting bomb, not from the mouth of God, but from the mouth of cannon and torpedo. So rings out the challenge to all that we have held sacred and to all that Church, Synagogue, and Temple the world over, have struggled to make into a reality in the lives of men, peoples and states.

The misery and suffering which in our lifetime have been heaped on untold millions of mankind are the direct and inevitable outcome of all these violations of religion's fundamental law. There is no escaping the results of evil doing. Religion is not the opiate of the people, but the repudiation of religion has been the murderer of peoples. It is not war which is the disciplinary test of the worth of men and nations, but war which is the corrupter and destroyer of men and nations, victor and vanquished alike. As never before we have been made to see that evil and wrong-doing destroy the wicked and the good alike, but "the work of righteousness shall be peace, and the effect of righteousness quietness and confidence for ever." There has been borne in on us the inescapable truth that only when "all thy children shall be taught of the Lord, great shall be the peace of thy children." Only so shall the world find peace.

How then can religion make itself felt? What is called for is something stronger and more vibrant than a tepid ethicism. If mankind is to be saved, we need a revival of religious goodness among men. We need the birth of a new and passionate spiritual conviction. If man is not to succumb to the triumphs of a soulless mind, he must be fired by a religious vision of the possibilities of the human spirit and of spiritual purpose in life. Only religion can regenerate the bewildered scientific monstrosity that twentieth-century man has become.

No utilitarian considerations of self-interest alone can stir and inflame man to withstand man's own threatened sadistic onslaught on his very existence. The hideous tragic experience of the last two decades has taught us only too well that a nicely calculated morality did not and could not stand against a flaming Hitlerian or Stalinistic fanaticism. Conventional plebeian virtue did not and could not resist organized perfervid nationalism or unscrupulous economic imperialist aggression. Well-dressed conventional propriety has been ruthlessly overwhelmed by the blitzkrieg of soulless intellectualism and atheism. It is such cyclonic forces as these that have swept men from their frail moralistic moorings, and that threaten to obliterate the cities, the culture and the soul of man in their violent hurricane strength. Man needs stronger anchorage than lukewarm lip-service to morality if he is to resist and survive today's and tomorrow's onslaught of power politics played by Titans of class hatred, racial arrogance and imperialist greed armed with the world-shattering robots of the technological triumphs of tomorrow.

Yet, despite the need, why is it that all too often Church and Synagogue fail to engage the active and enthusiastic allegiance of so many of our youth and some of the fine spirits among them who are devoted to the cause of human betterment? Perhaps they see religion's earnest echoing of the ten commandments and the golden rule of neighbor love often only as an intellectual assent. The mere iteration of Biblical ideals, which for centuries have commanded universal recognition, fails to touch men to action and raise them to higher standards of living. Can Church and Synagogue create new heavens and a new earth by mere enunciation of superb ethical and social principles, and denunciation of manifest evils? Are they fulfilling their sacred trust through inactivity, however eloquent and prayerful be that inactivity, in the face of humanity's burning problems of race conflict, imperialist aggression, international prejudice and animosity, war, governmental inadequacy and corruption, industrial waste and oppression, social injustice and exploitation, and the spread of corroding, cynical materialism?

The one imperative forward step which our Churches and Synagogues must take if they are to shake themselves free from a numbing aloofness from the more vigorous and serious relations of life, is to adopt a policy of participation in broad, progressive movements of hope. "And the Lord said unto Moses, 'Wherefore dost thou cry unto me? Speak to the children of Israel that they go forward.' "

An ideal of prayer, piety, saintliness and millenarian ethics alone too often provides not a remedy for but an escape from the problems of organized social living. Religion must set as its goal something more than a self-centered quest of individual human salvation leading to a world-negating lure of an afterlife, or of Nirvana. It must demand neither a passive resignation of unquestioning faith nor a devout fatalistic acceptance of man-made evil. It cannot permit almsgiving to compound for social evils. Religion must sound a ringing call to refashion the tottering structure of civilization by building an ideal commonwealth organized on the social foundations laid down in the Bible. Religion must offer more than words; it must offer a program of mankind's weal.

Is there overproduction, too much toil for some, too little work for others? The Bible's social principles limiting hours of labor hold out a promise of hope. Is wealth distributed with irrational, heartless inequality? The Old

Testament principles of compulsory taxation show a way for modern society to distribute God's blessings so that all may share them. Are problems of land tenure basic to our chaotic living? Apply the teachings of the Levitical law and you can cure this age-old evil. Does the political tyranny of dictatorship threaten to sweep away human freedom? And are the relations of capital and labor breaking down? Turn to the pivotal breaking of Egyptian bondage and the derivative laws of the Old Testament regulating both civic and industrial freedom, and you will find principles of social healing. In short, the Bible offers to a thwarted and baffled world not bland and beautiful words of general exhortation or solace but the outlines of a social code with a practical program. To be potent and effective, religion must face social realities with a vision of organized justice and spiritual sensitiveness.

Increasingly men are demanding that their religion be interested in their welfare as a whole. They ask that it fearlessly battle poverty, injustice, oppression and the curse of war. They ask from it more than the glow of personal mysticism that they can attain without religion, through beauty, through human love, through the pageantry of the passions. Thinking men, crushed by the enormities of our social system, expect from their religion something more constructively helpful than soothing words of an acquiescent, unreasoning faith, consecration of needless suffering, and nebulous promises of reward hereafter.

The greatest menace to Church and Synagogue today is the extent to which they may fail to bring to the coming generation the social message for which it is hungering. Church and Synagogue are in danger of becoming secluded in a world of wordy unreality unless, cutting through the mumpsimus and sumpsimus of theological sectarianism, they give themselves once more with magnificent and passionate fervor to those eternal social verities which formed so large a part of religion's original message.

For how often in history was it the social message of religions that first aroused the enthusiasm and the adhesion of their followers, and thereby opened a way for spiritual truth? Moses began his career as a social reformer summoning his people to liberty. The Jewish people came into being through the politico-social triumph of the Exodus. The religious message of Sinai gave divine sanction and spiritual support to this call to freedom, law and an ideal commonwealth on earth. Psalmist and Prophet could not have arisen except on the foundations laid by Moses, the lawgiver and social reformer.

First free the body and spirit of man by giving him a world fit to live in; then you may talk to him of more abstract religion. The other worldliness of the Middle Ages is gone. A religion which speaks only of spiritual solace to men who feel themselves to be living in a world of cruelty, violence and injustice may see them seek their social and spiritual freedom in Communism or outside the Church, venomously declaring that religion has been the opiate of the masses.

The social salvation of man is the antecedent of his spiritual salvation.

"He hath told thee, O man, what does the Lord require of thee, but to do justice, to love mercy and walk humbly with thy God." First we must achieve justice and humanity in our social relations; only then may we be deemed worthy of walking humbly with our God.

Religion must prove to men that it is not a meaningless survival from an outlived order of society, but that it identifies itself with the collective struggle of idealistic mankind by caring for the plight of men, constructively, inspiringly, and with vision. It must show that it has a message which can resolve the outer chaos in the world of men as well as the inner chaos in their soul.

All men are one. The unity of mankind is but the terrestrial obverse of the supernal truth of the divine unity. That God is one is the central teaching of Judaism. The Jew has always argued from the unity of God to the unity of mankind—if there be but one universal Father, then all mankind, His children, must be brothers. Our modern world is being forced into the position of arguing that since biologically, economically, sociologically, and in every way demonstrably we are all brothers, therefore, if we grant the God concept at all, there must be only one God.

The God of the modern world is the God of all mankind. We must come to see God not in a blinding maze of national and sectarian mirrors, but in one universal mirror in which God can be seen as one, supreme, universal. Then religon will have its message not only for some but for all men in this troubled world.

Prayer—Source of Spiritual Energy

Reverend Lewis L. Dunnington, D.D.
Minister, First Methodist Church, Iowa City, Iowa

Dr. Dunnington was in Moscow when the Revolution broke on November 7, 1917. He lived there for a year at that time, attended the First All Russian Congress of Soviets, met many of the early leaders and heard Lenin speak for two hours at the Congress. He has returned to Russia on four subsequent occasions at five-year intervals, to travel widely and study the unfolding of the strange experiment of the Russians.

In 1939 he was sent to Europe as a newspaper man by the Duluth "Herald News-Tribune" to write twelve articles on "The Danger Spots in Europe." He succeeded in getting into Czechoslovakia after that little country had been under Hitler's iron heel for six months, and brought out the first story of the tragedy being enacted there. Though followed and shadowed by the Gestapo for ten days in Prague, he held secret interviews and gathered invaluable material on the designs of the Nazis toward subject peoples. In the fall of 1948 he took a flying tour of seven countries of Europe, including Czechoslovakia and Greece in order to gather firsthand information on the standing of Communism in Europe.

*Dr. Dunnington holds degrees from four colleges and universities. He is the author of four books that are best sellers in the religious field—*Handles of Power, More Handles of Power, Start Where You Are, *and* Something to Stand On. *His fifth book,* Keys to Richer Living, *will be published by Macmillan in the Spring of 1952.*

He has been the pastor of the Methodist Church in Iowa City for the past ten years. During that time the attendance at the Sunday morning service has grown from an average of four hundred to about two thousand, necessitating two identical worship services each Sunday at 9:30 and 11:00 A.M. *Two-thirds of the attendants are students of the University of Iowa.*

This sermon on prayer has a moving story about Roland Hayes and illustrates a sermon technique for getting and holding attention so that the minister can make the real points of his sermon.

Sermon Thirty-nine

THE WORLD TODAY is full of baffled and confused people. Many of us have lived through two world wars with their unbelievable destruction of life and property only to find ourselves helplessly drifting toward a third and worse holocaust. What is it all about? Does life have any ultimate meaning or significance? Does God care? Is there a way to find integration and power for the living of days like these?

Millions are turning to alcohol because it furnishes momentary forgetfulness. Nine billion dollars is being spent this year for a thing that may give a fleeting sense of importance but that leaves a ghastly trail of human wreckage in its wake.

Prayer is the only answer. Properly understood and utilized, it is the most powerful form of energy that one can generate. Its results can be measured in terms of increased physical buoyancy, greater intellectual vigor, moral stamina, and a deeper understanding of all human relationships.

Habitually indulged, prayer will profoundly alter any life and stamp our actions and demeanor with its indelible mark. Observing friends will not fail to note in us a tranquility of bearing, a facial and bodily repose and an emanation of radiant spirit that we never before possessed.

Within the deep places of the human heart a flame kindles that throws a searching light into our true character. We see our selfishness, our silly pride, our greed and our baseless fears silhouetted against the radiant background of Christ's life and thought. Comes then repentance, humility, and the beginning of the journey of the soul toward full-orbed personality. Only in prayer do we achieve that complete and harmonious assembly of body, mind and spirit which transforms weakness and confusion into unshakable strength. Prayer, like radium, is thus the one source of luminous, self-generating energy.

Such prayer is not begging God for this thing or that. It is not a slick and easy way of cajoling a cosmic Errand Boy into giving us money and gadgets to make life easy. It is not primarily asking for things. "Seek ye *first* the Kingdom of God, and his righteousness; and all these things shall be added unto you," was the way Jesus put it.

But how? Use prayer as a supreme opportunity for the soul's communion with the Creator and Source of all energy, all beauty and all power. When a Russian peasant was found sitting quietly before an altar gazing intently at a painting of Christ and someone asked him what he was doing he replied, "I am looking at Him and He is looking at me." Just so! No one can look steadily at the pure, selfless, forgiving, understanding Christ over a period of time without experiencing a mysterious transformation in his own life.

When we pray, we link ourselves with the inexhaustible motive power that spins the universe. We humbly ask that a little of this power be apportioned to our needs. Through faith we boldly affirm that it is ours. As we arise, quieted and strengthened, we go forth to meet people and a variety of complicated life-situations and we find it is true. Instead of confusion and despair we show forth serenity and quiet power. We have been helped. We are better people and happier people. Our health improves and our affairs straighten out.

"Yes," you say, "that is true sometimes, but why do I have so many bad days? Why is it not always true?" The reason, my friend, is because we do not pray enough. Most of us pray a little, morning and evening perhaps, and then try to go it alone the rest of the time. Paul said to pray "without ceasing." Said Epictetus, "Think of God more often than you breathe." True prayer is constant communion with the Great Companion. In order to mould character and personality, prayer must become a habit.

Let me illustrate. Some years ago Roland Hayes came to Duluth to give a concert. Dr. A. Raymond Grant and I were both pastors there then and we decided to pay the distinguished Negro tenor a little visit.

When my friend and I knocked at the door of his hotel room he was busy telephoning, but he cheerfully called out to us to come in and find seats. He flashed us a smile and waved his hand in friendly greeting as he continued his telephone conversation, and we felt immediately the warm, courteous, engaging atmosphere that his very presence creates.

I noticed that he had just finished eating his breakfast in his room, for the dishes were still there. So I began the conversation by remarking that I as a Christian minister was exceedingly sorry for a condition in our social order that made it necessary for a colored man of his sensitive nature to stay away from the public dining rooms and to enter hotels by side doors. "There is nothing that you or any white man can do to alter that," said Mr. Hayes. "That is a job for me and my people. I am trying to live every moment with such consciousness of the Divine Presence without any trace of bitterness in my heart that that condition of prejudice and racial antipathy shall disappear. And I am trying to get my people to do likewise."

He paused and looked about the room and smiled as he resumed. "I am perfectly happy here by myself, and nobody in all the world can hurt me except myself." Then he told of the marvelous old soul that taught him as a boy. This colored master of the art of singing told him that as a black artist he would suffer many things if he allowed the barbs to get inside. "But always remember that if your heart is right and your spirit divinely disciplined, nobody in all the world can hurt you."

I then asked him if he had certain hours of communion and spiritual preparation for his concerts. He surprised us by saying that he did not. Years ago he had had certain hours. "Now every breath I draw and every moment of the day is a communion with Him that is my preparation." Then he asked if we would like to know how he captures his audience during his first minute on the stage. We were eager to know. "I stand there perfectly quiet with hands clasped before me and pray—that Roland Hayes may be blotted entirely out of the picture—that the people sitting there may feel only the Spirit of God flowing through melody and rhythm—that racial prejudice may be forgotten. The audience instinctively feels what is happening as I commune with my Father—and I capture them that moment and never let them go until I am done."

The radiance and joy that emanated from this quiet-spoken little black man was a thing that electrified the whole room. "What a time I have had this winter!" he said. "I have given up my expensive managers and the high-priced tickets they used to insist upon. Now I have no manager and I am free to insist that the prices be kept low so that the poor who long for my songs may come and hear them. The color line disappears—rich and poor, high and low, forget the lines that ordinarily divided them, and we all become sons and daughters of a common Father, hushed and quieted by the haunting power of the message of melody and rhythm and song."

Then he told us a very interesting story—only typical, he said, of what is happening to him constantly. In a town in New Jersey a Southern family decided to attend one of his concerts in order to show the fifteen-year-old son "what a horrible mess a 'nigger' makes out of life when he thinks he possesses talents that should be possessed only by white people." The parents had filled this lad with their prejudices ever since he had been able to talk.

He came in that spirit. Mr. Hayes used his usual technique. He blotted himself out of the picture. The rhythm and beauty of God shone through. The atmosphere now of harmony and peace and now of pathos and sadness was not marred by the ego of the artist breaking through. After the concert, this young man sought out his black brother and threw his arms about his neck. Through his tears he confessed that Mr. Hayes had done in two hours what all the books and orators in the world never would have accomplished. His lifelong prejudice was gone. "And now every time I see him," said Mr. Hayes, "he assures me that he is devoting his life in an effort to eradicate some of the prejudice in others similar to what his parents had fostered in him through the years."

"If I thought I could relate it without breaking down, I would tell you about the most remarkable thing that ever happened in my life," Mr. Hayes went on. He finally did do it, however, and what a story it was!

"A few years ago I gave a concert before a large and enthusiastic audience down in Alabama. The next day I journeyed to the old plantation not many miles away where my mother had been a slave. The old master and his wife were still alive, but what a change the years had wrought! The affluence of the old days was gone. The plantation itself had gone to ruin and had been sold for debt. The old gentleman and his wife, upwards of ninety, were existing in a little shanty. I introduced myself and asked them if they remembered my mother. Yes, of course they did—very well indeed. She was one of the dear souls they could never forget. They had called her 'Pony.' Pony was not allowed to live in the slave quarters. She was too much beloved. She would have died for her masters. So she lived in the big house as a servant.

Mr. Hayes said that as he looked around about at the signs of poverty he could tell what trains of thought were running through the old man's mind—the contrast between the good old days when Pony had been a slave and they had been wealthy, and their present poverty with Pony's son, one of the world's most famous artists, standing before them. He wanted desperately to help them but wondered whether their Southern pride would permit it.

"So Pony's son has sung before the crowned heads of Europe," mused the old man. "Tell me what you sang for the King and Queen of England." "I sang a Negro spiritual entitled 'The Crucifixion'." "Why, that is the very song your grandfather sang the day he entered the ministry!" exclaimed the old man.

That was a story within a story. Roland Hayes' grandfather had been converted in Africa through the labors of a missionary. When he was sold into slavery in this country he brought his Christian faith with him. He did some preaching among his brethren. One text had made a very deep impression upon him—"Call no man master, for one is your master, even Christ." He was a hard worker and eventually rose to the position of manager. He could accept bondage, but he could call no man "master." This seemed to the proud owner to be stubbornness, and one day in a fit of anger he killed him.

Mr. Hayes knew that his mention of "The Crucifixion" had revived this ancient memory. Would this proud old couple accept his help? He used the familiar spiritual technique of the concert stage and obliterated himself. "Is there anything that Pony's son could do to help?" he finally asked. "Yes, I suppose so," came the answer. "I reached into my pocket then and pulled out a check," said Mr. Hayes. "It was a large one—the returns from several of my largest concerts. The old lady was ill and was lying on the bed. I walked over and laid it on her chest. She picked it up and saw the size of it and suddenly realized that it meant getting the old plantation back and ending their days in peace. She screamed and threw open her arms, into which the aged master and I both fell. In a moment we had our arms around each other and were crying like little babies—just three of God's children, with no dividing color line. The next day the old man walked five miles, hobbling along on his cane, to hear me sing 'The Crucifixion.' He sat on the front seat and let the tears course unhindered down his wrinkled cheeks."

As we rose to go I knew that this hour had been the most shining and radiant of my life because of this man's divine consciousness. We were too deeply moved to speak. In the silence and through our tears we merely gripped his hand. After a moment he asked, "And would you like me to sing 'The Crucifixion' tonight?" The people of Duluth will never forget it—the quiet announcement that two friends had requested it—and the spiritual power that surged through this little black man's soul, sweeping all before it. And as he sang and I recalled the thrilling story of what happened on that old Southern plantation, I saw anew, as in a flash of insight, the deeper meaning of the crucifixion itself and of the love that has emanated from it to save and heal and bless mankind. "And he that loseth his life for my sake shall find it."

Would you, my reader friend, care to tap the boundless spiritual energy that flows like a broad crystal stream from the heart of God? You can, you know. If a colored man, living in a white man's world with all of the attendant frustrations, can live and work happily, triumphantly and radiantly, so may you and I.

Resolve to pray without ceasing for a definite period of one week as a starter. As Muriel Lester advises, keep the conscious mind centered on God as Radiant Light, Shining Beauty, Contagious Joy, Creative Power, All-pervading Love, Perfect Understanding, Purity, Serenity. Soon these attributes of the character of God will come to permeate the deep or unconscious mind which must, in turn, bring these attributes into manifestation. Then indeed, "every valley shall be filled, and every mountain and hill shall be brought low; and the crooked shall be made straight, and the rough ways shall be made smooth."

At the end of one week you will most surely be eager to "pray without ceasing" for yet another week, and another, as you progress toward the achievement of that harmonious assembly of body, mind and spirit which transforms weakness into strength, despair into victory and darkness into the light that never was on land or sea.

March Toward Thine End

REVEREND ETIENNE ELIE LAURIOL

Pastor, the Church of the Oratoire du Louvre (Reformed), Paris, France

Pastor Lauriol is a successor to Wilfred Monod in the famous Church of the Oratoire, which was the royal chapel of Louis XIII, XIV, XVth, and which was given to the Protestant Reformed Church by Napoleon. It is one of the great pulpits of France; and his predecessor, Pastor Monod, was considered one of the greatest preachers of all European Protestantism. Pastor Lauriol's preaching is proving that he is a worthy successor of the great Monod.

He was in the war of 1914–1918, and from 1919–1930 was Pastor in Vezenobres-Ners (Gard), was President of the 7th region of Reformed Churches Union and Deputy Member of the National Committee. In 1930 he became Pastor at Nîmes (Gard), a historic center of Protestantism in France. In 1938 he was a member of the National Committee of the Reformed Church of France; reconstituted in its unity and was President of the 9th region of the Reformed Church of France from 1940–1946; 1945, President of the Social Christianism Federation; Member of the League of the Charter of Human Rights. In 1945 he became Pastor of the Church de l'Oratoire du Louvre in Paris, and a member of the Council of the Protestant Federation of France. He is a member of the Central Committee of the Judeo-Christian Friendship, and an important member of The World Council of Churches.

This Good Friday meditation has the full French flavor of his preaching, and has a distinct message for our own time. He discusses man and his fears, how man is harried by the world and how God tries to save him in spite of himself even when man flees from the God of Providence who would save him from himself and his sins.

Sermon Forty

TEXT: Blessed is he that waiteth, and cometh to the thousand three hundred and five and thirty days. But go to thine end till the end be: for thou shalt rest, and stand in thy lot at the end of the days.
DANIEL 12:12, 13

THE BELOVED PASTOR Emile Guiraud confided to us, one day, that whenever he was too harassed by worries, overcome by trials, or tormented by the wickedness of men, he would tell himself—thinking of the people of his own age—"In thirty years, at the most, we shall all be dead! That brings things back to their true proportions."

In fact, we ordinarily live with our joys and our sorrows, our passions and our disappointments, as though they were eternal and we were linked tightly to them and they forever riveted to us. From thence come our feelings of despair or frenzy, one as absurd as the other.

We live as though we did not know that if our lives are determined by the commandment, "March toward thine end!" nothing can be taken from us. But the fact is that though man wake or sleep, run or stand still, laugh or weep, does good or evil, becomes worked up and excited or remains morally as inert as an ox, he walks toward his end. He may accelerate his pace, but not hold it up. It is because we want to forget this fact that we conduct ourselves like fools. Happily, each end of the year which reflects the rhythmic functioning of the divine order comes to remind us that the only important thing in this world, among all those others to which we assign so much importance, is to be able at all times to account for ourselves to God. To render our accounts at once, because we know "neither the day nor the hour" of this end, which may be this very moment. I said "happily," for God, in doing this, has only our happiness in view. And that is why the Book of Daniel calls those who have understood this "intelligent."

This means that the primary question to resolve here below—in fact, the only question which is decisive since it is the only one which engages our eternal future—is that of our personal destiny.

For every man—even for the man who is the most harried with the preoccupations of this world—there comes a time when suddenly nothing exists except himself and God. When he is no longer concerned with the problem of his earthly success, or prestige, or economics, or social or international problems, or even the problem of the "Two World Powers," even that of taxes! Then there remains nothing but his personal problem, which isolates itself like a rock emerging from the flooding tide of death: "What shall become of me, when at last I confront God?"

This hour is the one which transmits to eternity our person, our personal records and our earthly balance-sheet. It is the hour of the tomb, when nothing which is not of ourselves descends with us; when the human being, divested of his possessions, his authority and his titles, has nothing left except what he has become.

In consequence, all our activities, our desires, our successes and our failures, our great achievements and our misery, our joys and our griefs must be appraised by their relation to this supreme hour of final reality. "Death," writes Vinet, "is the knot which draws together all morality." This is why the Book of Daniel calls those who have understood this "intelligent."

Notice that this command has a personal accent! It does not say, "You, men, march toward your ends!" but "Thou (Daniel, Victor, Helen or Joan), march toward thine end." And this emphasizes the importance of the person, the individual, before God. In our pagan-gregarian epoch we tend to overlook this.

In the face of the gigantic forces which today torture, buffet, and enflame the world, the human being is quite forgotten, quite overlooked. Amid such infernal thundering, who hears the sobbing of a mother? In the mist of such a holocaust, who sees the baby a bomb has torn apart? The official communiques sometimes remind us of this—with the terribly revealing phrase: "Insignificant losses."

But for God it is this "insignificance" which is everything. In a sense, we might say that God hears nothing else, sees nothing else. The vast geological upheavals of the tertiary epoch, which with colossal explosions and cataclysmic forces changed the aspect of this planet, were all insignificant for God alongside the appearance of the fragile, quartiarial animal called "man," because even in his poor fragile frame this creature was his son. Everything goes toward its destiny and knows it not, but this creature—his son—dare not be ignorant of it, because in all the world he is the only one who is able to turn his destiny toward salvation or perdition, to fall into darkness or to rise to the light!

The end of the world does not interest God. But he never ceases to emphasize the end of man, and to warn him to consider it. And yet man himself never ceases to refuse to think about it, a condition which perplexed a Pascal!

Man flees the God who seeks him in order to save him. He tries to escape him, with epileptic-like fits, with rapacious passions and orgies which greatly amuse the demons, with ambitions which are grotesque for the sons of God, in distractions unworthy of a thinking being, in pleasures which finish by becoming no more than stupefying opiates. He flees God also in his work, even in his art, and finally flees Him by losing himself in others, by hiding himself in the mass. Man, who is personally particularized by God, tries to escape Him by taking refuge in the anonymity of the crowd.

At first glance this is confusing, for, ultimately, it is not man's habit to want to pass unnoticed, to lose himself modestly in the crowd. On the contrary, he aspires to be noticed, to be singled out, acclaimed and decorated!

Then it is that man becomes afraid, afraid of death and afraid of God!

Haunted by the fear of death, he looks for some sort of permanence which shall save his person from total loss. He throws himself, especially when he is young, into the joyful abandon of the collective, but a pitiful postponement of famine for one who is famished of eternity! The more he matures, the more he finds this out. There is no way to play tricks with the command, "March toward thine end."

Moreover, because man is afraid of God, afraid of the judgment of God, he thinks that by thrusting himself into the mass, by dissimulating himself there, he can escape the judgment, because there is no last judgment for a crowd.

Man then turns to considering his nation, his race, his class, his party as one of those anonymous societies with limited responsibility, in which one might (if one were dishonest) engage in shady business, without too much personal risk, precisely because it was anonymous and of limited responsibility.

But the command of God destroys this illusion: "March thou toward thine end to give thy reckoning."

God will not judge the anonymous mass, but he will judge one by one the whole number of persons who comprise it.

Pastor Rauschenbush, to show the diabolic side of such institutions, said that "God himself would not be able to send to Hell a limited corporation." That is right. But he can, perfectly well, put its stockholders there. Not the group nor the party, but the group members and the participants!

No matter what society we may belong to, let us not imagine that at that time its responsibility will let us elude our responsibility for the mistakes or the crimes which we shall have been able to commit, thanks to it or even for it. Our personal responsibility remains. That's worth thinking about.

This is why the Book of Daniel calls those who have understood this "intelligent."

In this reflection resides, therefore, the well being of our present world. In fact, when men, as a whole, learn that it is this question: "Where do I stand with God?" which must come first, not after all others, when they realize that this question must be resolved first and command all our existence, instead of being relegated to troubled hours of anguish, then and then only will be solved (or set on the right road to solution) all those other terrestrial questions which harass and torture men and end by driving them to extinguishment in the terrible machines of mass destruction.

However, the majority of men do not realize this, and most Christians either; for though they may think them wrong in principle, they call them right in their practice, since they too put material questions foremost—and that is why everything goes so badly.

But the gospel persists in maintaining that these material, economic, social and international questions will remain insoluble as long as the personal question has not been solved. "Seek ye first, the Kingdom of Heaven and its righteousness and all other things will be given to you in addition."

Our terrible war, and this period following the war, which we might call the hell of unsolved questions, demonstrate to what degree men are mistaken —and the Gospel is not mistaken.

Thus in recognizing that on this earth we march toward our end, we see that God already wills even our terrestrial happiness. That is why the Book of Daniel calls those who have understood this "intelligent."

But beyond all this God wills above all our eternal happiness, which he has prepared for us from the beginning of the world!

Let us make no mistake about the meaning of the word "end." It signifies here, not cessation but goal. God does not make us march toward our end like cattle toward the stockyard.

It is not a question of annihilation but, on the contrary, of fulfillment, a realization, or better, a reintegration and restoration of ourselves, after the ruin of the fall from grace. After the plunge into darkness, the emerging into light.

Therefore, Christ defines the meaning of this Commandment: "The commandment which God has given to us, is Life Eternal." To march toward our end is thus to march toward Eternal Life.

There is no idea at all of growing old. What could growing old signify for one who is eternal? Who can state the present age of Moses, or John the Baptist, or St. Paul? One might as well speak of God's age! It has no more meaning.

Of course, our body, our worldly vestment wears out, which is normal. We await another which the Apostle calls (for lack of a better, since no term exists on this earth for what does not exist here) the "spiritual body," a body which is more than a body and a spirit which is not only a spirit.

But when a child prepares himself to receive a garment according to his new stature it is not a sign of decrepitude, it is an indication of growth, and to grow is the opposite of getting old.

Getting old is to descend toward death, and that is why, for a son of God, there can be no other way of getting old than that of separating oneself from his Father. But going toward Him who is life is to grow young.

March toward thy end, for that end is eternal youth. But take care that your earthly aims are in accord with it. In any case, do not disregard it.

Remember that your true destiny on this earth is not to enrich yourself, nor to make a name for yourself, nor even to instruct yourself, but to become capable, at any instant, of grasping Eternal Life.

Throw away, therefore, everything which prevents this!

Reject that burden of lusts which makes your path stony and deforms your soul and leaves your hands encumbered! Do you wish to arrive at the tomb a wretched, miserable being, lost and perverted simply for the pleasure of having cherished worldly goods only to have to leave them!

Reject this avarice which leads you astray—reject it!

Reject this hate which, like a horrible cancer, scars your face, eats away your eyes (so that you no longer know which way you may be going); tear it out! And these sins which dissolve your powers and corrupt your joy—reject them!

And these anxieties which sap the best of our lifeblood, and this fear which makes you quiver as if you went towards death, instead of holding out to life your free and eager hands—reject, reject them!

This end is not the end of the worker, but of the work, majestic but painful for which the worker will soon be paid.

When St. Paul felt himself about to leave this lodging, he did not write: "I am finished!" but "I have finished!" I have run my course and I await the crowning.

End of the race, the crown is near!

Brothers, may the end of this year tell us again that terrestrial existence is, in fact, the seeking of the crown, the crown of life amidst the flowers of farewell which nothing can cause to fade. Therefore, let us bless the Commandment which says to each one: "Thou, march toward thine end."

The Book of Daniel calls those who have understood this "intelligent." *Amen!*

THE HOLY SCRIPTURES (BIBLE)

The Bible in an Atomic Age

REVEREND CHARLES M. CROWE
Minister, Wilmette Parish Methodist Church, Wilmette, Illinois

Charles M. Crowe was born and raised in San Antonio, Texas, was educated at Southern Methodist University, Dallas, and Union Theological Seminary, New York City. He held pastorates in Texas, Louisiana and Missouri prior to coming to the Wilmette Methodist Church. He is a frequent contributor to religious journals, including "The Christian Century," "The Pulpit," and "The Christian Advocate."

He has written three books; Great Southern Preaching, Learning to Live, *and* Sermons for Special Days. *A book of his sermons is to be published in the Spring of 1952. He is the author of* The Sanctuary, *annual devotional booklet used by many denominations with a circulation of more than seven hundred thousand yearly. He writes and conducts a weekly radio program, "Day's End," on Chicago's NBC Station WMAQ, sponsored by the Church Federation of Greater Chicago. He is also a speaker on CBX "Church of the Air" and the Mutual Network program, "Faith in Our Time," sponsored by the Protestant Radio Commission.*

The value of Bible reading has been proved by millions who practice the daily habit of reading a chapter or a few verses of this great book every day. If men would read the Bible daily they would find greater enrichment for their lives than comes from newspaper reading or from other pastimes that return little for the time and trouble expended. Truly the word of God endures forever.

Sermon Forty-one

TEXT: The grass withereth, the flower fadeth, but the word of our God shall stand forever. ISAIAH 40:8

AT A MEETING of the United Bible Societies, Mr. Gilbert Darlington, treasurer of the American Bible Society, made an interesting revelation. He spoke of measures taken already to preserve the various translations of the Bible in the event of an atomic war. He said that copies of all important editions of the Scriptures have been deposited in three places in specially made vaults. These are placed at the University of Colorado in Boulder, at Dartmouth College, in Hanover, New Hampshire, and in New York City. If an atomic attack pulverizes one of these places, it is hoped that one or more of the others will be spared. Thus the Bible will be preserved for future generations, if any.

These precautions are probably important, unbelievable though they may seem. They raise the question of the place of the Bible in an atomic age. As important as it may be to preserve the actual Scriptures from destruction, it is clearly more important to preserve their meaning and spirit in the lives of men and nations in times like these.

Many people would say that the Bible hasn't a chance today. Strangely enough, however, there is a rising interest in the Bible even in this atomic age. Charles Laughton holds large audiences spellbound reading the Scriptures. He says his readings from the Bible are the most popular he gives. This popular actor has made $100,000 in fees just by reading the Bible! Raymond Massey recently did a series of sixty-five television programs of readings from the Bible for a commercial sponsor. Many colleges have had to enlarge their courses of Bible study. The New Jersey Supreme Court recently upheld Bible reading in the public schools of that state. The book *In Our Image,* Bible stories beautifully illustrated in modern form, sold one thousand copies at six dollars each at one bookstore in Dallas the first day of sale. In Formosa, Bible study classes are popular, with attendance up to two thousand. The American Bookseller's Association for the first time in history had an author on religious books for the speaker at a recent annual convention.

Why all this interest in the Bible in an atomic age? What chance has the Bible today? Is it out of date? What can it do for us and our world?

THE BIBLE IS STILL THE BEST GUIDEBOOK FOR VICTORIOUS LIVING. One night during World War I, General Edmund Allenby sat alone in his tent studying a map of Palestine. He was making plans for the capture of an enemy held village near by. He felt that there must be a weak point where the British could attack the Turks. He thought there was some key to the problem in the rough terrain but he could not seem to find it. The name of the town was Michmash. It had a familiar sound but he could not remember where he had heard it. At midnight the general went to his cot but he couldn't sleep. Surely somewhere he had heard that name, Michmash! Finally it flashed into his mind that he had read it in a book. He arose and took down the volume and scanned through the chapters quickly. After a time the name of the town, Michmash, was there on the page. He began feverishly to work with the book and the map. He worked all night. In the morning he unfolded the plan for the attack to his officers. They wanted to know where he had gotten his information. He read to them from the book and they said it would be a miracle if it worked.

In less than twelve hours the plan had worked. Allied troops had cut off the enemy and had taken the village. General Allenby had found a description of the land and a plan of strategy that had proven successful at that very spot more than three thousand years before. He found this key to the victory, of course, in the First Book of Samuel in the Bible.

To be sure, no one would think of the Bible as a book of tactics for modern warfare, any more than we think of it as a book of science or economics. But the story of General Allenby remains a striking parable. For all life is a warfare. The pressures of life are heavy; the enemies of life strong and pagan. If ever we needed defense for the spirit and ammunition for the soul it is now. And modern man finds in the pages of the Bible the strategy and plan of attack that brings victory. Across its pages there walk a gallant band. Men and women, warriors all. They were surrounded by enemies. They were without the trappings of modern civilization. They fought many a lonely fight against heavy odds. But they possessed a faith in God and in themselves and in the future that made them unconquerable.

Take Abraham. Faced with a civilization rotten to the core, he answered the vision of God and set his face toward the promised land. "He went out not knowing whither he went." Job, plagued with a thousand ills, cried out: "Though he slay me, yet will I trust him." St. Paul, beset with enemies of the flesh and the spirit, surrounded by an unfriendly and pagan world, dared to shout in triumph: "In all these things we are more than conquerors through him that loved us." Christ, under heavy pressure of evil temptations, suffering and poverty, with the shadow of the cross across his way, declared calmly: "Be of good cheer. I have overcome the world."

Whence come the victorious souls? Why have ordinary Christians through the centuries successfully defied dungeon, fire and sword? What are the weapons of their warfare? They are the secret weapons of the divine indwelling. They have strength from above. "They that wait on the Lord shall renew their strength; they shall mount up with wings as eagles; they shall run, and not be weary; and they shall walk and not faint." They are armed with the moral laws of God: "Thou shalt not steal. Thou shalt not commit adultery. Thou shalt not bear false witness. Honor thy father and thy mother." They are filled with the confidence of Christian discipleship: "I can do all things through Christ which strengtheneth me." They are sustained by an unshakable steadfastness: "Watch ye, stand fast in the faith, quit you like men, be strong." They possess a victorious faith: "If ye have faith as a grain of mustard seed nothing shall be impossible unto you." They know they are on the winning side: "Though Heaven and earth pass away, my word shall not pass away."

In other words, the Bible tells us our religion is valid and true for great living in time of war and trouble as well as peace and ease. Just because times are bad we cannot desert our faith. That is the time to use it!

THE BIBLE IS THE WORD OF GOD TO MEN. Here is assurance and reassurance. Here is power that nothing can stop. Here is the stuff that undergirds life and that thrusts it on from defeat to victory. Here is the heavenly leverage that makes the spirit of man forever triumphant. Here is the source of enthusiasm, confidence, and certainty that puts vigor and courage into living. Here is the message of the Bible to men in an atomic age or any age. Fearful to the point of hysteria, we need to partake of these everlasting fountains of faith. Obsessed with the question marks of a pagan and cynical world, we find here weapons of unfaltering hope and power. Downcast and downhearted, we can recover here a sense of the joy and glory that has characterized Christians through the centuries.

THE BIBLE IS STILL A BLUEPRINT FOR A BETTER WORLD. Dr. C. Darby

Fulton, of the Presbyterian church, was traveling on a train in Korea a few years ago. At a way station an old Korean man got on the train and sat facing him in the compartment. The man carried a large bundle wrapped in a white cloth. After a time the old Korean addressed Dr. Fulton in Korean which he did not understand. The American responded with the only Korean phrase he knew: "Chosen mal morogessimnidah." It means, "I do not understand Korean." The old man smiled. Later, however, he tried another sentence in Korean. Once again Dr. Fulton repeated glibly his one Korean phrase. Then the native tried still a third sentence. Just as the white man was about to give out with his one stock reply, it occurred to him that he detected a familiar word in the stream of Korean words. It was the word, Yesu. It was the word for Jesus. Dr. Fulton pointed to himself and nodded, saying, "Yesu." The old man did the same with a delighted smile.

Said Dr. Fulton: "No words can describe the sudden sense of fellowship that I felt with that old man. Here we were, traveling along together, two mutes, unable to reveal our thoughts to one another or to delve into the experience of the other. But we had one wonderful word in common, and that was enough. We were brothers."

Then the old man unwrapped his bundle. It was a large Bible in Korean. He turned the pages and pointed to a place he wanted Dr. Fulton to read. This, of course, was impossible. Then suddenly the American remembered that the make-up of the Bible is roughly the same in Korean and English, except that the Orientals begin at the back and read to the front. He took the Bible and counted the number of books and chapters from the back to the place the old man had pointed out. Consulting his own Bible he found it was the First Epistle of John, the third chapter and the fourteenth verse. "We know we have passed from death unto life because we love the brethren." Dr. Fulton then searched for a suitable reply. He counted it out and pointed it out in the Korean Bible, Psalm 133, verse 1. "Behold how good and how pleasant it is for brethren to dwell together in unity." The Korean read it with many noddings of the head and broad smiles. For the rest of the journey, these two men, ages apart in race, culture, and viewpoint, were brought together in a strange friendship as they pointed first to one verse and then another. Their separate Bibles had a common language of the spirit.

Yes, you see, it is not from books on economics or war or geography that men find the know-how that breaks the barriers and bridges the gaps between men, nations, and races. It is from the Bible, the source book of human brotherhood, that comes the only hope for re-uniting a broken and divided world. Books shape history. Karl Marx's *Das Kapital* and Hitler's *Mein Kampf* have succeeded in redrawing the map of the earth, even though millions of Russians and Germans never saw the books themselves. But the Christian Bible contributes more to human wellbeing, peace, prosperity, and progress than all other books put together. Wars may be necessary, but our problems of politics and peace will never be settled permanently until they are settled right. And they will never be settled right until they are settled on the conditions of the word of God. The causes of human dignity, freedom, decency and truth have their origin and primary support in this book. Only as we take these as a base and move out into economic and political areas do they have a chance of success.

We need to know again the word of the prophets: "Let justice roll down like waters and righteousness as a mighty stream." . . . "Remove violence and spoil, and execute judgment and justice, said the Lord God." "The work of righteousness quietness and assurance forever." Likewise we must hearken to the Master: "Whatsoever ye would that men should do unto you, do ye even so unto them." "Every kingdom divided against itself is brought to desolation." "The wages of sin is death." "I am the light of the world; he that followeth me shall not walk in darkness but shall have the light of life." "Thou shalt love the Lord thy God and thy neighbor as thyself."

Yes, across the ages the Bible is our guide, standard and hope. It tells us that this is God's world and that God has great things in store for His people. We would hear above the jungle cries of fear, hate and greed the strong clear voices of the prophets, apostles, and evangelists speaking a word of high challenge to the children of God even as Tennyson put it:

> . . . Come, my friends
> 'Tis not too late to seek a newer world.
> . . . My purpose holds
> To sail beyond the sunset, and the baths
> Of all the western stars, until I die. . . .
> Though much is taken, much abides; and though
> We are not now that strength which in old days
> Moved earth and heaven; that which we are, we are. . . .
> Made weak by time and fate, but strong in will
> To strive, to seek, to find, and not to yield.

THE BIBLE IS STILL THE PRESCRIPTION OF NEW LIFE FOR SICK MINDS AND SOULS. Millard F. Wright, age thirty-eight, underwent a delicate brain operation in Pittsburgh not long ago. It was known as a prefrontal lobotomy. The doctors hoped thereby to make Mr. Wright a better man. He was then serving a twelve-year term for burglary and had a long police record behind him. Unfortunately, the results were disappointing. Months later the clinic reported that Mr. Wright was still "unable to adjust himself to disappointments and frustrations." He was still "the evasive, crafty, cunning person" that he had been before the operation. There was no noticeable alteration of his ego-centricism. It certainly had not decreased.

Now, to be sure, there is a certain amount of anti-social behavior and of personality defects that probably can be helped by physical treatment. We wish that Mr. Wright had become a changed man. But it is not as easy as that. In fact, the case is a symbol of a modern tendency to blame our sins and our sickness of mind and soul on purely physical causes. Thus by finding a cure for syphilis, we hope to remove all stigma from adultery. By proper diet we expect to solve our personality complexes. By the use of barbituates we seek to dull our consciences and quiet our nerves. By controlling housing and income we want to control character and happiness. By doctoring the mind we would like to purify the spirit.

The popular literature of the day has encouraged this attitude. Much of the cynical, lustful, materialistic writing of the present completely ignores the capacity of the soul of man for godly living. Norman Cousins, of the *Saturday Review of Literature,* reflects this when he declares that modern writers have "been writing out of their egos instead of out of their consciences . . . been preoccupied with human neuroses to the virtual exclusion of human nobility."

No wonder, then, there are so many sick minds and unhappy hearts these days. They need the ministry of the Holy Bible. Here they will find that the human soul is at the governing center of life. They will find that the soul of man is inextricably related to God. They will find here that only as the soul of man is made clean and strong by the power of God can life be rich and good. We can never be happy outside until we are happy inside. Here is the prescription for true happiness. More than that, from its inspired pages breathes the divine holy spirit that makes life over. As Coleridge put it: "I know the Bible is inspired because it finds me at greater depths of my being than any other book." It is this, rather than any ecclesiastical magic, or literalistic interpretation, that makes it the sacred word of God to men.

Millions have found it so. Robert Ingersoll, the skeptic, urged General Lew Wallace to write a book exposing the falsity of Christianity. The general began to study the Bible in preparation for the book. So changed was he that the book he wrote, *Ben Hur,* became a classic about the beauty and divinity of the life of Jesus, and he himself became a devout Christian. . . . John Ruskin became disgusted with his own selfishness. In his Journal he wrote in 1852: "I resolved that I would act as if the Bible were true; that to disbelieve the Bible is quite as difficult as to believe it; that I should believe in Christ and take him as my master in whatever I did; (this was at two o'clock one morning) and when I arose in the morning I felt a peace and a spirit in me I had never known before." . . . A plain man of this country, Harry Doehla, was stricken with rheumatic fever, a hopeless cripple for years. He cursed his fate and God. Then he began to read the Bible. "Above all other things, as I dipped again and again into the Bible, a certain thought brought me up sharp: Of myself I am nothing; with God all things are possible." Thus began for him a new life. . . . Said a New York banker: "Reading the Bible is an intellectual and spiritual practice which gives me perception and insight."

Yes, here is the one unfailing prescription for human happiness and well being. Worried and unhappy? "The Lord is my strength and my shield; my heart trusteth in him, and I am helped: therefore my heart greatly rejoiceth; and with my song will I praise him." Heart hungry? "I am the bread of life: he that cometh to me shall never hunger." Dissatisfied? "Seek ye first the Kingdom of God and his righteousness and all things needful will be added unto you." Sad and discouraged? "Let not your heart be troubled. My peace I give unto you." Lonely and afraid? "Lo, I am with you always, even unto the end of the world."

It doesn't matter how literally you interpret the Bible. The compelling fact for us is that the Bible is a light and guide for modern man. With it we cannot lose the way and can never be defeated.

THE HOLY SCRIPTURES (BIBLE)

The Bible Book Series: Isaiah, Our Hopes

RABBI SOLOMON B. FREEHOF, PH.D.
Rabbi, Rodef Shalom Temple (Reform), Pittsburgh, Pennsylvania

Dr. Solomon B. Freehof was born in London, but came to the United States with his parents at the age of eleven. After graduating from the University of Cincinnati and Hebrew Union College, he served during World War I as a chaplain in the A.E.F. After the war he taught at Hebrew Union College for nine years, then became Rabbi of the Kehillath Anshe Maariv Temple in Chicago for ten years. He went to Rodef Shalom Temple in Pittsburgh in 1934 and has been Rabbi there ever since.

He has been the President of the Central Conference of American Rabbis (Reform), is a member of the Executive Board of the Union of American Hebrew Congregations, and is a member of the Division of Religious Activities of the National Jewish Welfare Board. Among his published works are a Commentary on the Book of Psalms, The Small Sanctuary, Modern Jewish Preaching, Reform Jewish Practice, Preface to Scripture. *His fame as a scholar and speaker is nationwide, and his "Book Review Sermons" are eagerly anticipated each year.*

This sermon illustrates what can be done with "Bible Book Review Sermons," using individual books of the Bible as the subject of occasional sermons to familiarize church people with the Bible itself. Or, one could follow Dr. Freehof's method and organize such sermons into a series of Bible book reviews at certain seasons of the year.

"Isaiah, Our Hopes," was delivered Sunday, March 19, 1950, at Rodef

Shalom Temple, Pittsburgh, where Rabbi Freehof's preaching regularly attracts large crowds. This is an excellent type of sermon for younger clergymen to study. With knowledge or a bit of research almost any clergyman could thus add a new type of sermon to his preaching and create a greater interest in the Bible itself. Other books beyond the Bible also afford opportunities to preach book review sermons in the Spring or Fall of each year. As Rabbi Israel Bettan commented, "This is a splendid example of an expository sermon. The method employed by Dr. Freehof, if applied to all the important books of the Bible, would make the Bible the Book of the People."

Sermon Forty-two

A BOOK, like a person, is subject to destiny. So said a Roman writer of books who, being a Roman, believed completely in fate or destiny. His idea was that just as it is foreordained that one person shall be obscure, another grow famous, one be effective in life and the other futile—as with a person, so with a book. A book seems destined either to have a happy career or a miserable one. "Habent sua fata libelli." Books have their destiny. To a considerable extent there is truth in this classic phrase. It can well be applied to the Biblical book of Isaiah.

Talent is fairly abundant. We all know a number of people who have considerable ability, enough to be described as talent. There are many who have fine skill in the arts or a clear foresight in business or a special deftness in political life and in government. Yes, talent is met with frequently. But genius is rare, rare indeed. How often do we meet anybody on earth of the caliber of Plato or Immanuel Kant or Einstein? The genius is a rare phenomenon, a mysterious phenomenon, so magical indeed, so inexplicable, so surprising that the mind of man is awed whenever a genius appears. No wonder that the Romans considered a genius a divine creature, born on earth, human only in form, but actually divine.

In speaking of the great authors of Scripture, we are accustomed to take on a mood of reverence. It is the proper thing to speak in awe of the authors of the Biblical books, though sometimes we may fall into the danger of overpraising them. There is a great temptation to consider all the Biblical authors geniuses. We should be careful not to use the word lightly. But the fact is that at least with regard to the great literary prophets there is much in their life and in the nature of their work that brings the word, "genius," to mind. In the first place, just as with genius, the great literary prophets appeared suddenly. It is true that they appeared against the background of a monotheistic religion; but they took such a tremendous leap upward in their concept of what religion means in human life that we must consider them originals. Like geniuses they were lonely men. They seemed to have no intellectual father or mother. Although there was a succession of prophets extending over a period of three hundred years, not one of the literary prophets belonged to any prophetic school. There is not much evidence that one knew the other. Each one was a lonely man, appearing self-taught, suddenly, and great. Their great thoughts were expressed with such magnificent eloquence that, when all the tides of ancient eloquence seemed to have receded forever, the torrent of prophetic eloquence still flowed like a perennial Gulf Stream, warming the coastlines of human life. There is much that tempts us to speak of the literary prophets as geniuses. But it is difficult to do so because there seems too much to say. For if a genius is rare, what shall we say of a galaxy of geniuses, a succession of great men who rose suddenly at a certain date and three centuries later disappeared, the like of them not to rise on earth since then, a galaxy of surpassing talent, perhaps a galaxy of genius, in the moral field!

Among those very great men, the literary prophets, Isaiah is definitely conspicuous. In the first place, there are certain superficial differences between him and the rest of that radiant company. They were, for the most part, humble folk who came from the lower social groupings and, therefore, were sensitive to the sorrows and the oppressions of life. He was an aristocrat. We know that from his at-homeness in the court of the kings and the way he speaks out to the nobles in Jerusalem. He was an aristocrat and yet deeply sensitive to the injustices and sorrows of life. It was a greater achievement for him. Most of the prophets were villagers like Amos and Micah, who lived close to the soil. He was an aristocrat of Jerusalem and lived close to the court of the king. That is why there is much more of politics, of statesmanship, of world problems in his prophecies than in any other. Also, he had a longer career than most of them. He preached and lived through the careers of four kings of Judea. In his long career, in his contact with the aristocracy and yet in his sympathy with the poor, he has given such a wide variety of ideas that one despairs of giving even a summary of them in one discussion. The difficulty is heightened by the fact that so great was Isaiah's name, so magnificently did it loom in the consciousness of his successors, that other writers, now anonymous to us, somehow contrived to have their writings appended to the book of Isaiah, as if to find shelter under his wing. In the present Biblical book of Isaiah, a book of over sixty chapters, there are certain chapters by other authors. But there is much in it authentically by the Isaiah of Jerusalem, who lived about the year seven hundred before the present era, the habitué both of the courts and the temple, the great eloquent voice of many ideals. We shall attempt, although perhaps superficially, to describe the main lines of his message.

All of Shakespeare's characters are robust with the notable exception of the vacillating Hamlet. All of them are willful. They know what they want, and nothing stands in the way of their taking what they want, by conspiracy, treason, or murder. They are typical, self-willed, Renaissance men. Nothing hinders Shakespeare's robust characters from grabbing what they want of life except one thing, and that is said by Hamlet, "Conscience does make cowards of us all." Another character of his, Richard III, says that conscience has a thousand voices and every one proclaims me for a villain. The word "conscience" appears very often in Shakespeare's writings, and the implication is that in a willful life, motivated by the satisfaction of selfish urges, the only real restraint is not the social world, which sometimes can be deceived, and not always the law, which sometimes can be successfully evaded. The only inescapable shackle to the evil intent is the conscience. So says Shakespeare, and we, of course, wish that it were truly so.

Sometimes we doubt the existence of the conscience, the moral guardian, within the soul of modern man. We know of the existence of biological urges, of hungers, of self-seeking, but after recent experiences in European history we tend to doubt that there exists in us something that makes us sensitive to other people's sorrows, genuinely concerned with other people's pain as much or almost as much as with our own. It would greatly advance our hopes for the human future if we could believe, and see evidences of that belief, that there exists in man, as Shakespeare believed, a guardian conscience. Were it true, then we would confidently declare that the hope for man's future exists within man himself.

The fact of the matter is that something like conscience does seem to exist in man. It is often rudimentary. Anthropologists who study savage man report with a fair degree of unanimity that they have not yet found a savage tribe anywhere in which there is not something like a sense of moral distinction. They say that they have not found people so primitive, so near the animal, that they do not recognize the difference between "I want" and "I ought." There is a moral sense. But the difficulty with this ancient and prevalent human moral sense is that it is generally merely conventional. It is static, it is a conscience that is easily soothed, as you can tell from the words we use to describe it. The word "ethics" is related to ethnology, to "ethos," "people." It simply means the custom of the people. And "morality" also, from "mores," means "custom." So while every primitive people has some understanding within it of the "I ought," of conscience, and understands that it differs from "I want," from mere willfulness, nevertheless their morality was simply an obedience to the customs of the tribe, the ethos, the mores. If you did what was customary to do, your conscience comforted you and fell asleep comfortably. We might say that the great problem in human moral development was how to *awaken* the somnolent conscience of ancient man and make it alert and sensitive. Or another way of saying it: the moral problem is how to convert the static ethics, the unchanging ethics of mankind, and make it a dynamic ethic; to teach human beings to grow *more* sensitive; to make their conscience *more* alert. If that were done and could be done, mankind would have real hope.

The awakening of complacent conscience was the chief accomplishment of the prophet Isaiah, although most commentators on his book do not quite put it that way. Let me call to mind his famous sixth chapter. That is said to be one of his earliest writings, his self-dedication to a noble yet bitter career of prophecy. It begins as a vision. Remember, he was a Jerusalem aristocrat, and so was frequently in the great Temple on Mount Zion. He says, "In the year that King Uzziah died, I saw the Lord in the Temple . . . Above him flew the seraphim and they cried 'Kodosh, kodosh, kodosh, Holy, holy, holy, is the Lord of hosts.' . . . Then said I: 'Woe is me! for I am undone; because I am a man of unclean lips, and I dwell in the midst of a people of unclean lips.' . . . Thereupon one of the angels picked up a coal from the altar and touched my lips and said, 'Thine iniquity is taken away.' And I heard the voice of the Lord, saying: 'Whom shall I send, and who will go for us?' Then I said: 'Here am I; send me.' "

This is a dedication address, a self-dedication address. There is something curious in it. This prophet, obviously a noble person, immediately says of himself, "I am lost, I am a man of unclean lips." How could he say that of himself? Because he saw a mental vision of the Lord of hosts, of the holy God, and, therefore, thinking of God of unimaginable perfection and turning again to himself as a human being, he realized by contrast how low were his standards, how weak his ethical intentions. What the prophet did was to bring the great concept of monotheism, of the great and only God (which in the book of Genesis is the basis of the law of the universe), and make it the law of conscience. Think of God in order that you should judge yourself. We see now the historic meaning of that great ethical use of monotheism. The ethics of all people up to that time was inescapably static. It had to be static and customary and conventional, because people were not better than their gods. If any Greek behaved like the god Zeus, he would clearly be a selfish person. When we consider the escapades of all the gods of all the nations, we could say "like god, like people; like people, like god." But the great moral value of a surpassing monotheism was clearly seen by the prophet. It represented a grandeur which we as human beings can never actually attain, and, therefore, it induced in us a constant ethical discontent.

As soon as Isaiah thought of the Eternal not merely as powerful but as holy, as ethically pure, then immediately monotheism became *ethical* monotheism and the moral life became an endless, dynamic, upward reach. God is the sunlight under whose beneficent rays the seed of moral discernment grows into the plant of conscience within the heart of man. This gives a new depth to religion. It means that hereafter men should not look upon their gods only as persons to propitiate; nor among monotheists should people only fear God the Almighty. They should also consider Him as the great sun of the soul towards whom we turn and we strive. We attribute man's moral coarseness to his distance from God. The more he is aware of "Kodosh, kodosh, kodosh," of the holy Eternal, the more humble is he and the more likely to try to attain whatever holiness is within his grasp. Thus what Isaiah says first of all is addressed to the conscience. "Do not be proud, do not be content; your very best goodness is as nothing when you compare it to the holy God." Should you be tempted to smugness and self-satisfaction, say, "I will never be content with my present moral state, because mine eyes have seen the glory of the holy God!"

Our country is now in danger. We are not accustomed to the idea that our United States should be in peril. It is something new to us. We had always been really isolated by the width of the Atlantic. In addition to that, we had become a powerful nation early. For those two reasons, our isolation and inherent strength, we were always safe; but now for the first time America as America feels insecure. In order to fight that insecurity we have developed a two-pronged foreign policy. One is to build up great military strength and the other is to build up diplomatic alliances. It is a foreign policy that seems so logical that it is bipartisan. Now if this is our procedure, when we, great as we are, feel insecure, imagine how a people would feel in a country which never was isolated, which was right on the border of belligerent neighbors all through its history, a country that never was large and strong, that always was small and weak. The land of Judea was adjacent on the south to Egypt, on the north to Assyria-Babylonia, the great rivals of the ancient world. Palestine was the bridge which both of these rivals constantly crossed. Ancient Palestine was a perpetual battle ground. Besides those two mighty world empires, there were smaller neighbors, always hostile. Thus the people of Judea and the people of the little northern Hebrew kingdom, Israel, lived in constant awareness of national danger. That was the permanent mood of every patriot. So they had their foreign policy, the same logical foreign policy which we now have: namely, to build up what strength they could in armies and to make alliances. But they were frantic alliances, with Assyria now when Egypt was the greater danger, and then with Egypt when Assyria became the danger. They were tempted always to make a confederation of the smaller states to fight against one or the other of the big ones, and yet also afraid to be in a confederation lest they be singled out as rebels against Assyria and Egypt. There was a constant succession of frantic, nervous, foreign policy proposals, military preparedness, new alliances.

Everybody agreed that it was the only thing a little state could do; everybody agreed—except Isaiah. He was a prophet who also had opinions about statecraft. He was close to the circles where policy was made. He saw that the people were nervous, pathetically persuading themselves that they had attained this security by this increase in the army, by this new alliance; but just when the people were beginning to feel secure, Isaiah rose before them in the Temple court and said, "Oh, sinful people. Oh, folk laden with iniquity. You have forsaken the Lord of Hosts, so your country will be desolate and your cities burned with fire. To what purpose is your coming to My house, saith the Lord of Hosts? I am fed full with the fatlings of rams, and when you lift up your hands to me in prayer I will turn my face from you; your hands are filled with blood. So wash ye, make yourselves clean, remove the iniquity from your heart and from my sight, and then perhaps a remnant will endure and Zion will be redeemed through righteousness." It is an eloquent but a sombre address. But the people must have wondered what it had to do with foreign policy. Yet it is a new idea in foreign policy. It amounts to a statement that military might and alliances may be important, but they ignore the greatest source of a nation's weakness or its strength. The greatest source of weakness or strength in a nation, says Isaiah (for the first time that it had been said), is in the character of its people. The real rampart is within. When a people is cowardly, when it can be said of it, as was said of the people of Judea by Isaiah, "the hearts of the people were shaken as the trees by the tempest," when a people is frantic because of its inner insecurity, no alliances will ever save it.

In this semi-religious age we have changed this moral explanation of Isaiah as to the strength of a nation. We have shifted it from ethics to psychology. Instead of calling morality the strength of a nation, we speak of "morale." But Isaiah spoke of moral stamina as the strength of any people. He is clearly right. National iniquity and hypocrisies are sources of inner corrosion. Nothing can save a nation as much as its inner strength. We Americans know it. The real strength of the American people today is not in its armies and its jet propelled missiles and its Atlantic alliances. None of these would serve to save us if we suddenly became a nerve-shaken people. If we lost our hope in the future, if we were not the confident, constructive Americans that we always were, all our diplomacy, our bipartisan foreign policy, would collapse, because the stones with which the ramparts are built would be rotten stone. The only real strength, the essential strength, is the moral strength.

But when Isaiah first said that to his people, it shocked them. They must have said, "And suppose we *are* righteous? Is that good politics? Cannot a righteous nation be destroyed too?" To that he says, "A remnant will be saved always." Thereby he expressed a true prophecy, because usually when a state is destroyed its civilization goes with it. The folkways of the people, their art, their language, their literature, go away into the dust that is carried by the wind of history. History is the graveyard of innumerable nations and civilizations. Isaiah said: You may not always save your government in a dangerous world but you can save your people by their inner stamina. Did he not prove it? Of all the people of the past who lost a state, we are the only people that remained as a cultural unit without armies and without navies and without geographical concentration. His teaching gave us immortality.

Hence he said, on another occasion, on the same theme, when the northern Hebrew kingdom had made an alliance with the Syrians of Damascus and came down to fight against Jerusalem, the prophet told the king himself, "Build up the character and the decency of the people of Judea and they will live, if only a remnant," and he ends with the great sentence, "Unless you have inner stability you will not endure."

Only with inner integrity does a people live. Thus Isaiah still speaks to the heads of our government these days. Indeed, build up the armies of America, make the alliances, but build up the American people. Do not induce panic in its heart, foster its normal confidence and its hope in the future. The strength of a people, the ultimate strength, is the strength within.

Sometimes we wonder how a person can endure the reading of history. So many noble experiments have vanished into dust, so many nations have disappeared, so many nobilities have been wiped out. How long can nations last? The prophet Isaiah must have confronted that fact. With his statesmanlike vision he saw that any average little state in that warlike world would not last very long. He gave the people the only advice that he could safely give: build up your inner worth and you will endure the changes of time. Will it always be necessary that states break down in this way, one after another? What of the future? Is that the perpetual future, an endless dying away of human governments and every state inevitably falling apart? So Isaiah says, it cannot be, not with the Just and not with a paternal Power governing the universe. Therefore, he gives a vision of the future, a vision which he uttered when the death of the state was staring him in the face. This vision, since he is an aristocrat, has to do naturally with the house of David. He foresees a noble child, born in the house of David, leading Judea and the world to a nobler day. In his description, scattered over two chapters, 9 and 11, is some of the most exalted writing that man has ever penned. He begins by saying that from the root of the tree of the house of Jesse, David's father—from the root of the family of Jesse will spring forth a twig. "Unto us a son will be born, unto us a child shall be given. And the government shall be upon his shoulders, and he shall be called wise councillor and prince of peace. And he will judge the fatherless in mercy and help the helpless with justice. Then men will learn to beat their swords into plowshares and their spears into pruning hooks, the lion shall lie down with the lamb, and a little child shall lead them. They shall not hurt nor destroy in all My holy mountain, for the earth shall be filled with the knowledge of God as the waters cover the sea."

Thus there was uttered, in a war-tossed, tiny land, the greatest vision of permanent peace ever spoken, a vision which did not die. It has beautified the picture of the future for noble people everywhere. A thousand Leagues of Nations may die, but Isaiah's vision will live, leading us with the confidence that in an earth God-governed, man, God's child, will yet learn how to beat swords into plowshares and to establish a world of permanent security.

The vision begins with this verse: "The people that walk in darkness will see a great light." Perhaps that is a sort of a description of genius, after all. Genius is a flash of a great light in darkness, a light, however, that endures. Men of talent beat with the sledge hammer of their ability upon the anvil of their opportunities. These men of talent strike off sparks that flame for an instant. But the sparks struck by genius are stars. They are fixed in the heavens of life to send radiance to generation after generation. Isaiah's words are stars. In almost every literature influenced by the Bible quotations from Isaiah are dominant. If, for example, you first study the book of Isaiah and know it well and then read the Gospels, you will understand one of the chief sources of the New Testament inspiration. All through the ages down even to the modern Hebrew novelist Mapu, who writes in Isaiah's style, the radiance of Isaiah's words has lived.

More important than his eloquence is his thought—the first to convert the cosmological, the world concept of God, into the ethical; to make us aware of God's surpassing holiness; and thus to awaken the sleeping conscience as a growing power in our hearts. He was the first to discover that the strength of a people is deeper than in military or in diplomatic devices, that a people ethically strong within is ultimately indestructible. He was the first to reverse the history of man. Where all mankind had hitherto considered the glory of humanity to be in the past, he showed us a vision of glory in the future, a grand hope which keeps us alive and our confidence strong. Isaiah's achievement, the awakening of man's conscience, the assurance of man's strength, and the radiance of man's hope, is more than enough to rank this man as one of the great moral geniuses of the human race.

Repentance and Resolve

REVEREND ALEXANDER WINSTON, PH.D.
Minister, First Parish Church (Unitarian), Portland, Maine

Alexander Winston was born in Seattle, Washington, in 1909 and grew up in that city. He attended the public schools and the University of Washington, being graduated from the latter cum laude *and Phi Beta Kappa in 1930. Later he received his doctorate in philosophy (1949) from the same university. While in college five of his one-act plays were produced, one was published. In 1933 he began studies at Meadville Theological School and the University of Chicago, receiving his Bachelor of Divinity degree in 1935. He was awarded the Cruft Fellowship and studied philosophy for a year at the University of Paris and at Marburg University.*

Since 1936 he has served churches in Seattle and Boston and since 1947 has been pastor of the First Parish in Portland. In 1944 he became a member of the faculty of Tufts College, teaching the philosophy of religion and ethics. He has been on the faculty of the Institute of Pastoral Care, Boston, and served on its Board of Governors. Two volumes of his sermons have been published: You Are the Key (1949) and I Leave You My Heart (1951). *In his college years and after, his summers were variously spent as a sailor on a freighter through the Orient, a cannery hand in Alaska, service station operator, radio announcer. Now he summers with his wife and three children on an ancient Maine farm in North Baldwin.*

Dr. Winston's sermons, printed in pamphlet form, have been widely circulated throughout the English-speaking world. He is a popular preacher and lecturer for college and club audiences.

"Repentance" and "Resolve" were preached as short sermons during Lent, 1950, as part of a series called "The Paths of Healing." Both are story sermons; each has a message for our times; the two are here presented as one to illustrate a type of preaching that has possibilities for other ministers.

Sermon Forty-three

THE OLD DEAN of the Cathedral had the face of a bulldog and the disposition of a chow. Let us say at the outset that he was not wanting in a certain solid wisdom. When our paths crossed we stalked each other like a pair of stiff-legged suspicious dogs, our theological fangs bared warningly. This most often happened in the undignified surroundings of a locker room, which can be more earthly than a clod turned by the plow and theologically as neutral as a concrete curbstone. He took his exercise wallowing philosophically up and down the swimming pool or sitting very still in a steam box. While I went in for athletics slightly more violent, we reached the locker room heated to about the same temperature. I was very young and he was—well, he *seemed*—very old.

"What can be done with this mess of warring nations?" I asked one day by way of opener, lest we find ourselves on a subject too amiable to be interesting. "What sort of a remedy can be found to still this immense distrust and warm men's hearts to each other?"

The Dean stopped mopping his neck long enough to measure me with one chilly glance. "The mourners' bench," he rumbled, "the mourners' bench," and resumed his mopping.

"Surely," I retorted quickly, "you don't think that a little repentance will shrivel the sinews of so much warlike preparation. Will you mend everything by setting Hitler or Stalin or Huey Long on the mourners' bench?"

"How beautifully you understand me," he answered; then added, after a moment, "If I could get them there a few of our troubles would be over at least. Oh, I don't expect to succeed, nor is it my business. I have a parish of one thousand souls. My job is to get *them* there. The mourners' bench. Repentance."

The Dean said no more for a while. He put on the most extraordinary length of heavy winter woolens that I have ever seen, wrapped a dry towel around his thickset neck and faced me with a thoughtful look. All these signals pointed to the fact that he was about to read me the lesson for the day. I braced myself and appeared not very interested.

"The mourners' bench," said the Dean softly. "What an unpopular idea! You and I love popularity. Preaching it to good-hearted middleclass Americans is like rubbing them with coarse sandpaper. Shrove Tuesday appeals to them more than Ash Wednesday. Life seems short and laughter shorter. Let us, they say, stretch the Mardi Gras beyond midnight's fatal toll. Forgive me, young man, if I quote, by way of illustration, from the ripe paganism of Omar:

Come, fill the Cup, and in the fire of Spring
The Winter-garment of Repentance fling
The Bird of Time has but a little way
To flutter—and the Bird is on the Wing.

That is how the natural, untutored, unawakened man sees this whole business of the contrite heart. It's unpopular."

The Dean stopped, looked thoughtfully at the floor as though he had forgotten me. "Is there a prophet who did not preach repentance?" he mused. "Not one. The whole glorious line of them, that astounding procession of God-endowed men, preached repentance. And they had to shout that they might be heard above the feet of those who danced before golden calves.

"You've read the Gospels, I suppose," he murmured, with the air of a man who believed otherwise. "What did John the Baptist say, that rugged, fiery-eyed searcher of men's souls? 'Repent for the Kingdom is at hand!' What did Jesus preach, as he came out of the desert's temptation to incarnate the Lord God himself in the streets and hillsides and lakes of Galilee? 'Repent, repent for the Kingdom is at hand!' Mark is the most reliable historian of Christ's life, the one who tells of him with the bare plainness of a man who was there. He takes Jesus to a populous city, sets him at meat and drink, tells how Jesus said: 'I came not to call the righteous, but sinners to repentance.' Oh, take that away from Jesus and you have left a good man, certainly, but no prophet, no Jesus, and above all no Christ. The Gospel of Matthew, more Jewish than the others, has it too, and the Gospel of Luke, more cosmopolitan than the others, has him say that there is a larger joy in heaven over one repentant sinner than there is over ninety-nine folks that don't need penitence, or don't believe they do. Even the disciples who thought they were voting with the right party and all their marks in heaven's book were writ in gold, were not exempt, let me tell you, young man. They gloated a bit because when Pilate had killed some Galileans they smugly concluded those men had died because of their sins. Jesus pulled them up short enough with a sobering remark: 'I tell you, Nay, but except *ye* repent; *ye* shall all likewise perish.' " The Dean rubbed his bald head and nodded to himself. "The mourners' bench," he said again. "Yes, Hitler and Stalin and Huey Long; all the German people, all the Russian people, all the American people, and especially you and I."

"That is all easy to say," I countered, "but how do you convince anyone of it?"

"Well," he answered, rising and beginning to dress with the automatic movements of long habit, "put it this way. Ask a man if he believes in God. Ninety-five percent of the American people will say 'yes.' Then ask him if he means a real God, not one who sits on a thundercloud and laughs ha! ha! but a God who labors with each portion of His creation to mold it nearer to His desire. If it is not the God of Epicurus, who lives between the worlds and cares not a whit what happens to us; if it is not the God of ancient Israel, who sides with one people against another, maiming and destroying so that a chosen few may be glorified and multitudes be cast down; if it is not a Vague Force—a sort of oblong blur—as many moderns think of Him; if, in brief, he believes in a Christian God who loves us all with a holy, fearful, just and righteous love, then He knows what we do and cares what we do and has a will for each of us. In our best moments, our cool and reverent moments, we have a mighty conviction that some things we do are right in His eyes and some things are very wrong. Should He have no will for us, it is true we can have no sin, but then we don't have a real God either. Because sin is simply this: to be aware of what God wants us to do but not to do it; to know what He doesn't want us to do and still to do it anyway."

While the Dean buttoned on his clerical vest, wrestling it around his portly figure, he paused and held his breath to make the operation easier—even possible at all. Then he suddenly turned.

"Are you," he said solemnly, "an angel, a beast or a man?"

"Why, a man," I stammered.

"Good," he said. "If you were an angel—something I have not had the pleasure of meeting—you would do God's will without fail and so not sin. If you were a beast you would not even know that God was in heaven or earth—and so not sin. But you are a man—as you admit and I assent—hence you live between the heaven of angels and mud of animality. You know God with your mind, but you are not near to Him in your heart. You know Him, but not well enough. You have an intellect of steel and a will of lead. There are two escapes from that predicament: either go up and fly with the angels or go down and run with the brutes. Which will you do?"

"Neither," I protested. "I can do neither."

"Then you are a mortal man, indeed," he resumed evenly, "and to you Jesus spoke when he said 'Repent, repent, for the Kingdom is at hand.' "

Taking his coat out of the locker, the Dean stood with it in his hand and told, as though reminiscing, a familiar story. "Two men went up into the church to pray, the one a business man and the other a racetrack bookie. The business man stood and prayed thus with himself, 'God, I thank thee that I am not as the rest of men, extortioners, unjust, adulterers, or even as this bookie. I go to church each week and pay my bills when they come due.' But the bookie dropped to his knees and bowed his head and wept and said, 'God be merciful to me a sinner.' I tell you, there was more hope for the bookie, for he knew that he was wrong and repented of it."

"That seems to me a bit unfair to the other man," I ventured.

"Perhaps," he said, raising his eyebrows and drawing down the corners of his mouth. "Perhaps. Still, what could be so unfair to God as to suppose that we—foolish, frail mortals—have arrived, and that, morally speaking, we are on a permanent high plateau. "You," he barked, waggling a finger at me, "will at least admit that in the secular pursuits of life the man who gets ahead is the one who never stops learning. A good engineer profits from his mistakes, sees them clearly and concedes them readily, then goes ahead to be sure that he never makes those mistakes again. A good salesman does the same. What housewife would be so stupid as the one who ruins one cake today and then ruins another tomorrow in the same way? Can you grow in a job if you never confess that you have done more poorly than your ideals aim at? Hardly. Well, how is the moral life different? The man who is still growing is the one who sees his mistakes, is sorry for them, and keeps on trying."

The Dean put on his hat and coat, thrust his hands in his pocket and planted himself for his peroration. "I have discovered this," he said, "that one of the keys to a man's character is the haste with which he defends himself. The man who feels sure of his place in life can accept the possibility that he is wrong on this point or that. He can relax and be impartial even in the face of his obvious errors. When a man does not take easily to repentance, and leaps hotly to argue his cause at the first faintest murmur of criticism, he has sand under his feet instead of rock. He needs a bit of humility. Give him a lesson in language: one Latin phrase—*mea culpa,* I am at fault; and one English phrase: *I am sorry*. From there on that man has set his feet in the paths of healing."

The Dean stopped. A crinkle began to appear at the corners of his eyes; the whole bleak steadfastness of the face softened with geniality; then he grinned a most lovable, friendly grin. I shook the outstretched hand, envying his communicants who could hear, each Sunday, such straightforward, unadorned wisdom. So we parted friends, he and I, for all our amiable insults, and I have never forgotten what he said about the "mourners' bench."

* * * * * * *

RESOLVE

EVERYONE loves a mystery story. Somewhere in the tangled skein of the story's plot—hidden, dissembling—the villain lurks. To pursue him, and pluck him forth unmasked constitutes the fascination of the mystery tale. Today I should like us to imagine just such a pursuit but with this difference: our setting consists of the splendid achievements of man rather than his sordid crimes, and our search is for the hero, not the culprit.

You may debate the ultimate moral significance of Oak Ridge, where atomic research in this country centers; however, the sheer audacity of mankind to put electrons to the yoke and move mountains with them, cannot but arouse our admiration. The day will come, and it will come soon, when coal and oil and gasoline to provide power will be as obsolete as the dodo. Oak Ridge symbolizes the revolution in man's productive means. Its enormous buildings, complex scientific equipment and immense staff of nuclear scientists is at once a gleaming hope and a dismal portent. In the long run, notwithstanding the short-term dangers to mankind resident in the Oak Ridge project, atomic power is going to completely revolutionize our society. Now comes the mystery: from what sprang so gigantic an achievement? Where shall we look to find the germ, the seed, the trigger, the initial word, the hero?

It helps us little enough to interview the contractors who built the vast structures, nor are we enlightened when we leaf through the legal papers or examine the war orders authorizing the whole project. We are still baffled, the mystery unsolved after tedious search; we are at the last chapter and the plot remains untangled. Our first promising clue takes us back to Germany, 1905, when an obscure mathematician published a theory about matter and energy. Early in the Second World War the same man, Albert Einstein, a refugee in America from the Nazi tyranny, proposed to President Roosevelt that his theory could be put to practical use. Until then nothing had happened; thereafter everything happened leading up to Hiroshima and the Japanese surrender. This meditative, doe-eyed refugee is the hero of the mystery of Oak Ridge: he is the germ, the seed, the trigger, the initial word. An idea, an act of will, was the beginning of the whole thing.

We can make our proposition broader than that and in doing so we state a Christian fundamental: an act of will is the beginning of all achievement. To see how sharply this idea is contested shows how fiercely Christianity must defend itself in the arena of world opinion. In the theory of Karl Marx, for example, economics decides everything, not only how goods are produced but also what we shall like and how we shall feel. Hitler used to say that the struggle of one race against another is what sets rolling the wheels of history. Politicians say it is politics, geographers say it is geography, some psychologists say it is our sub-conscious. Christian experience finds that these opinions always play around the edge of the matter and never touch its inner heart. We do not deny the influence of vast impersonal forces pressing in upon our lives; what we do continue to say is that we are free, that our destiny is in part our decision, and that an act of will is the beginning of every great achievement.

Generations of Christian experience prove it. That most eloquent of prophets, Isaiah, passed along the paths of healing through the stages we have described in earlier chapters. You will remember that, bedazzled by God's glory and overcome by his own littleness, his own wavering devotion, he repented and confessed his shortcomings aloud in the holy temple. Having thus prepared himself Isaiah received the forgiveness of God, for an angel of the Lord came bearing a live coal from off the sacred altar, and with it cleansed the lips of the prophet. This done, the voice of God came to him with an imperative question: "Whom shall I send, and who will go for me?" Isaiah entered into the fourth step along the path of healing by his answer: "Here am I, Lord, send me." An act of resolve, a decisive stroke of will, an emphatic "yes" to higher hopes and clearer insights—these are the beginning of achievement in your life.

Saul of Tarsus found the Christian groups harassed by the authorities and in danger of extinction, their fate in the balance, for division threatened them inwardly even as persecution threatened them outwardly. Saul might have made a career for himself, striding forward over the prostrate bodies of this new Messiah's helpless disciples. But when God said: "Whom shall I send to proclaim Jesus Christ; who will go for me?" Saul answered: "Here am I; send me." History is marked with such decisions, not so much the history of wars and dynasties as the history of man's slow march toward a Christian civilization. Francis of Assisi looked upon the corrupt magnificence of the church and organized the barefoot, begging friars pledged to poverty. "Here am I, Lord, send me." Columbus looked on the western ocean, Pasteur looked on contagious disease, Howard looked on prisons, Dickens looked on child labor. Lincoln looked on slavery, Joan of Arc looked on a dismembered France, Schweitzer looked on Africa; all heard a voice asking, "Whom shall I send?" and all replied, "Here am I; send me."

This kind of self-dedication is, strangely enough, our supreme act of freedom, bursting those very bonds which sociologists would say hold us to one place in society and one set of ideas. We expect a man of fortunate heritage to show it, and a man of unfortunate heritage to reveal his handicaps. These follow a natural law. We are filled with joy, then, when the chains of destiny are shattered and men rise up to a stature unbelievably above what nature gave them. Of Jesus the Fourth Gospel tells us that the eternal "yes" sounded in him. A great many people, however, were saying "no" to him. His family, we have reason to believe, was at first skeptical of his adopted life as a wandering religious teacher. His humble origin, far from the scholarly centers of rabbinical lore, said "no" to him. The priests resented his criticism, the Pharisees burned beneath his contempt; they said "no" to him. The Roman authorities said "no," as they did to anyone who threatened their peace. This ocean of negations Jesus met with his eternal "yes," his positive, steel-strong, settled, marrow-deep resolve. Ever since him, an act of will has been at the center of Christian achievement.

You may be thinking, however, of the resolutely evil men who have troubled their fellows by a determination just as evil as that of Jesus was good. I will agree. Look only at the dusky heroism of Milton's Satan in *Paradise Lost* to see that the devil can be as firm in his intention as are the angels of heaven. So dramatic is Milton's portrait of Satan that critics concede he is the epic poem's central figure. Hurled from the abode of song into Tartarus, drenched with the torment of a sulphurous lake, Satan addressed the defeated rebels:

> What though the field be lost
> All is not lost; the unconquerable will,
> And study of revenge, immortal hate,
> And courage never to submit or yield,
> And what is more, not to be overcome. . . ,

There you have a will to evil, a resolve to destroy man's happy state. Nor do we need turn to the Lucifers and Iagos of literature to find our examples. History is bloody with the record of determined men, whose very will-power constituted their threat—Nero and Cesare Borgia and Ivan the Terrible and Hitler and scores of others whose morality was the morality of the jungle, whose ethics the ethics of gangsterdom. What, then, is the difference between the resolve of Isaiah and the resolve of Napoleon? The difference lies in the presence or absence of the preceding steps along the path of healing. Isaiah worshipped God, repented, confessed, received forgiveness, said "Here am I; send me." Napoleon worshipped no God except that of personal prestige, repented no sins, asked no forgiveness nor received it, and decided his actions by the gross standards of power-politics. Christian resolve, by contrast, is fixed like a gem into a setting of Christian humility. Broad is the way that leadeth to destruction and many there are that travel it, but strait is the gate and narrow the way that leadeth to life. The kind of resolve that will bless men, not destroy them, comes through the strait gate of repentance, by the narrow way of God's forgiveness.

Having come that way, we are aware of an accession of power, a singular vitality which arms us against weariness, boredom or defeat. How many imaginary terrors vanish as soon as we confront them unafraid! You remember how frightened Dorothy was of the old Witch in that children's classic, *The Wizard of Oz?* When in a moment of revolt, the little girl doused the old witch with a bucket of water, she was dumbfounded to see her melt away, leaving only the magic cape behind. Half of our worries will melt, half of our little millstones will be cut loose from us at the moment of resolve, leaving behind only a laughable reminder of our foolishness. Most worries are as baseless as the shadowy coatrack we mistake for a midnight intruder; douse them with the cold water of your determination and they melt surprisingly away.

An act of will is the center of all achievement, great or small. "Impossible!" shouted Mirabeau, as the French Revolution raised the standard of freedom and equality against an entrenched nobility. "Impossible! Never utter to me that blockhead word!" How often has a desperate day been turned to victory by the man who refused to accept the impossible! Faith is an essential ingredient of every victory. Whole eras are eras of faith, of vitality, of controlled and decisive action. Personally, I do not believe that our civilization will be laid low by an A-bomb, H-bomb or any kind of bomb. Our threat is, finally, internal; it consists of a loss of nerve, an inability to arrive at vigorous and widespread decision as to courses of action. Arnold Toynbee's striking thesis, in his epochal histories, is that civilization survives as it successfully accepts a challenge and acts competently to cope with the challenge. Should this civilization close, it will happen only because of spiritual anemia, the relaxing of will, the loss of faith in what can be done if we but determine to do it.

After the 1922 attempt to scale Mt. Everest, the leader of the expedition gave his formula for a successful sortie against the world's highest known peak. One element was obvious enough, namely, perfect physical fitness. But both the other elements were so spiritual in character that the speaker might have been preaching a sermon on the Christian life. He said that whoever stands upon Everest's snowy crown must have first, singleness of aim, and second, unswerving faith. If we think of Christian democracy trying to surmount the dread heights of our present East-West conflict, it is easy to see what we need for the victory. To our armaments and our economic strategy—our physical fitness—we must add a single-minded determination to preserve democratic institutions, and an unswerving faith that we can do it.

Still, if such faith is in the end a global requirement, it must begin as an individual requirement. The Salvation Army has a motto: no man is a failure until he admits it. What can so devastate the forces of evil as a resolve that grows out of righteousness and flags not at the sorest set-back? The same hammer of adversity that shatters glass in one character forges steel in another. The origin of the word "tribulation" illustrates my point, for it comes from the Latin "tribulum," meaning a flail. Like a rod or whip, tribulation flails the human wheat. Under its blows which do you emerge: the whole grain or the chaff? It depends on the power of your resolve. John Wilkinson, the blind botanist who lost his sight at twenty-three years of age, learned to distinguish flowers by touching them with the tip of his tongue, and could instantly identify any of five thousand specimens. Emory Noyes, crippled with infantile paralysis, pitched for his high school team sitting down, and won seventeen games in a row. Remember too Washington undaunted at Valley Forge, remember deaf Beethoven writing his Ninth Symphony, Helen Keller graduating from Radcliffe with honors, and Walter Scott saying, "I am at sea in the dark and the vessel leaky into the bargain," then going on to his greatest romances. Remember Paul—stoned, whipped, mobbed, suspected by Christians and rejected by Jews—who could still dare the dangerous capital of the pagan world with the plain resolve: "Now I must also see Rome." These people were marked by a common characteristic: the power to will an act and perform it. Their habitual attitude was one of resolution. The word impossible, so quick upon the lips of others, was to them a blockhead word. They knew how to sustain an effort, keep it going, keep plugging. An arrow blown from its course goes forever amiss; but a determined man, buffeted by the cross-winds of circumstance, will come back straight for his target. If you were to peel the human character as one would an onion, you would strip off first the social conventions, then the more intimate habits, then the abstract reason, then the moral ideas, then the feelings—such as pleasure and pain—and finally at the center, first-sprung from the seed and last discarded, you would find the will. *As a man purposeth in his heart, so is he.* To change him, his will requires education, not his reason; to save him, his purpose must be pointed toward holy things, not his intellect.

You have walked the paths of healing to this point. You have seen God—somehow, somewhere,—no matter how misty your sight; you have perceived your errors and hoped to do better, and in that hope you have known God's forgiving love. Now He looks over your world, a world where war threatens, ambition sharpens its claws, weariness dulls the vital edge, confusion dissipates man's efforts. He says "Whom shall I send and who will go for me?" Don't wait a moment longer. Your best hour has arrived. Speak so that all men may know where you stand: "Here am I, Lord, send me."

THE SOCIAL GOSPEL

Building Tomorrow's World Today

Reverend William Orlando Carrington, D.D.
Minister, First A. M. E. Zion Church, Brooklyn, New York

Dr. Carrington was born in Georgetown, British Guiana, but came to the United States in 1905. He has held pastorates in Charlotte, North Carolina, Washington, D. C., New Rochelle, New York, and Hartford, Connecticut. He was dean of Hood Theological Seminary, Salisbury, North Carolina, from 1910 to 1920, instructor in the School of Religion, Howard University, Washington, D. C., from 1920 to 1924, and 1932 to 1936. He was editor of the "A. M. E. Zion Quarterly Review" *from 1924 to 1932. He has been pastor of the First African Methodist Episcopal Zion Church, Brooklyn, New York since 1936. In 1936 and 1937 he was a member of the National Preaching Mission, and was also a member of the National Christian Mission from 1940 to 1943. Livingstone College conferred the D.D. in recognition of his work for his people.*

In 1916 Dr. Carrington was the winner of the "Homiletic Review" *Sermon Contest and a contributor to* Prize Sermons *published by the Macmillan Company in 1932. He was the winner of the* Church Management *Sermon Contest in 1934. He is one of sixty-four representative American preachers whose sermons comprise* The American Pulpit Series *published by the Abingdon-Cokesbury Press, and the author of* Carry a Little Honey. *For the past ten years he has been instructor and lecturer at institutes for ministers. He is interested in literature, especially in Scott, Shakespeare and Dickens. His church membership is over three thousand, "good, bad and indifferent," and he has been gradually developing a seven-day church program. The church is meeting its challenging opportunity with educational, recreational and nursery work.*

As one of the great Negro leaders in the church today, Dr. Carrington has made a place of spiritual power for himself through his preaching and courageous social work. He is interested in building a better world for his people and for all people now, in making an ideal world in the midst of the real world, by creating better working and living conditions. He is anxious to see men of all races and all peoples have a chance for the good life in a world of science and faith. Through men like Dr. Carrington the Negro people will find their best and truest leadership as they look for a new day and a new chance in the spiritual and material world.

Sermon Forty-four

TEXT: Let us rise up and build. . . . NEHEMIAH 2:18.

MANKIND has always dreamed of a better order of things. That fact underlies the whole philosophy of progress, and it has not been more strikingly put than by the writer of the Epistle to the Hebrews when he says concerning Abraham: "Dwelling in tents . . . he looked for the city which hath the foundations, whose builder and maker is God." Such an attitude is indispensable to any worthwhile and noble achievement, and certainly there could be little hope of progress without it. George Herbert Wells spoke truly when he said: "The human mind has always accomplished progress by the construction of Utopias." But man's efforts have not been a sustained progress. The tragedy of his history is that he has so often destroyed the things he has builded at tremendous cost, that the painful labors and sacrifices of the centuries with their inestimable gains have been foolishly and ruthlessly laid waste. But as Dean Sperry of Harvard Divinity School has said: "It is because man's cities and societies are simply his life built into institutions, that when they are cast down, nothing matters so much to him as their recovery."

The seers and prophets have always envisioned a better order than the *status quo,* have looked beyond the shattered fortunes of state and nation, beyond the wreck of civilizations, and pointed, as did the Hebrew prophets, to some coming Golden Age when the nations shall walk in the light of the Lord, "and they shall beat their swords into plowshares, and their spears into pruning-hooks: nation shall not lift up the sword against nation, neither shall they learn war any more." So Jesus preached the Kingdom of God, and John on Patmos saw in vision the Holy City, new Jerusalem, coming down out of heaven from God. When Alaric and his Goths sacked Rome in A.D. 410, St. Augustine turned from a crumbling civilization and from the ruined city of men to the City of God, giving us, as Stringfellow Barr says in his recent book, *The Pilgrimage of Western Man,* "a vision of a Christian Society, a 'Pilgrim City,' moving through a Promised Land that would indeed be home. St. Thomas Aquinas had philosophized about that City; the Gothic cathedral had shown it forth in 'visible speech'; its greatest poet, Dante, had observed its stragglers in purgatory and its enemies in hell, and had ascended into heaven to seek the source of its hope, of its faith, and its charity." [1]

Thus age after age men have turned from some ruined City of Man, some crumbling civilization, some broken world to dream of—and sometimes to undertake to build—a new and better order.

To begin with, what kind of world do we want Tomorrow's World to be? Aldous Huxley has given us his *Brave New World,* a world cellophaned, polished, regulated. Communism has its own pattern of what it should be, and the democracies have theirs. Here in America we talk proudly and glibly, though somewhat vaguely, of "the American way of life" which we feel must be preserved. Ten years ago some eminent American scholars collaborated in producing *The City of Man,* a Declaration on World Democracy. It is good, even excellent, as far as it goes, but it doesn't go far enough. I have no blueprints of Tomorrow's World to offer. But of some things we may be sure, namely, that, whether we want it or not, Tomorrow's World will be a world of atom and hydrogen bombs, Bofors guns, jet and rocket planes, electronic marvels, and countless unimagined wonders of man's scientific ingenuity; and unless something happens to shake us out of them, it will be a world of increased cushioned comforts and soft satisfactions; and if we continue to build on the same foundations, it will still be a world of what Professor Sorokin calls "sensate culture," of clashing ideologies, power politics, rivalries, hatreds, fears, wars, haunting dissatisfactions, and persistent and inescapable frustrations.

But, if we would believe it, the ideal for Tomorrow's World was given nearly two thousand years ago by Jesus in His teaching about the kingdom of God. It has to do not so much with things as with people. It is concerned with building a Community of Love and Good Will, a society which, first and last, is based upon, and geared to, the Eternal Purpose—a society, therefore, which carries the promise and potency not only of the highest that man can hope or dream, but of the fulfilment of its own Divine destiny.

We know, however, that "the shape of things to come" will depend largely on the quality and temper of those who do the shaping. So that when we talk about "Building Tomorrow's World Today," if we are sincere and serious about it, we are confronted with the task of starting

[1] Used by permission of Harcourt, Brace & Co.

with ourselves. If we are going to build a better world we shall have to begin by building better men for, as Glenn Frank once said, "You cannot build a great civilization around sleazy individuals." Our Utopias have failed chiefly because we have neglected to build Utopians. Edwin Markham reminds us in his striking way of our mistake in this matter:

We all are blind until we see
  That in the human plan
Nothing is worth the making if
  It does not make the man.

Why build these cities glorious
  If man unbuilded goes?
In vain we build the world unless
  The builder also grows.[2]

But that isn't easy. It is so much easier to work with things than with ourselves, to improve something rather than to improve ourselves. Longfellow, in one of his poems, urges us

To build a new life on a ruined life;
To make the future fairer than the past,
And make the past appear a troubled dream.
E'en now, in passing through the garden walks,
Upon the ground I saw a fallen nest,
Ruined and full of rain; and over me
Beheld the uncomplaining birds already
Busy in building a new habitation.

But it isn't as easy as that. The birds meet no opposition or resistance from the straw or twigs or other materials with which they work. We who are to build Tomorrow's World will not work with inanimate things, with dull, insensate matter. We work with people, selfish, self-willed, cantankerous; with personalities, with clashing wills, sometimes unresponsive, uncoöperative, stubborn, recalcitrant, even as you and I. We are to be concerned more with the morals, than with the mechanics, of our situation. Jesus was always emphasizing the spiritual nature of the kingdom of God, and the fact that it is not simply something external and apart from us, but that it is within us, that it begins there, although it does not end there, for it transcends our little life and seeks its consummation in a perfect society.

There is another fact which accentuates the difficulty of our problem. I need only mention it. When we talk in terms of "Building Tomorrow's World Today" we should realize that it is later than we think. We are actually late beginning. We should have begun the day before yesterday, that is to say, this task should have been begun by our grandparents. Long ago when the Minister of Education for France spoke of improving the morals of the nation, Napoleon commented that if you would improve people you would

[2] "Man Making" by Edwin Markham from *The Gates of Paradise and Other Poems* (Doubleday & Co., Inc.). Used by permission of the publisher.

better begin with their grandmothers. Victor Hugo expressed a similar sentiment when he said, "If you would civilize a man, begin with his grandmother."

I am sure that capable students of religious education will not be disposed to question the truth of this view. In that way we should have been better trained for our task, and the task itself would not be so appalling. For if the grandparents of this generation had seriously and honestly faced some of the issues which confront us today we should have had a different world—such issues, for example, as these: shall we really accept Christ as Lord, or shall we continue the practice of race prejudice, for you can't have both? Shall we think and act in terms of race superiority and exploitation of weaker peoples, or in terms of respect for personality and mutual friendliness and service? Shall we regard war as inevitable, a sort of necessary evil, or as a deadly peril which should be denounced and renounced, outlawed once for all? Shall we build a new world-order along the lines of the kingdom of God, or shall we go on building on selfishness, hatred and injustice? If these people of day before yesterday, the grandparents, had seriously faced such issues and done something about them as they should, and the people of yesterday, the parents, had followed through, our own task would have been far less difficult. But even though it's later than we think, we should do the best we can, attacking our task with greater diligence and enthusiasm and a deeper consecration.

Let us go a step farther. We have said that Tomorrow's World will be a world of vast scientific discoveries and technological creations, hydrogen bombs, supersonic planes, and a thousand and one breath-taking inventions and devices. In II Peter 3:10, 11, it is said: "The heavens shall pass away with a great noise, and the elements shall be dissolved with fervent heat, and the earth and the works that are therein shall be burned up." And the writer raises the significant question; "Seeing that these things are thus all to be dissolved, what manner of persons ought ye to be in all holy living and godliness?" May we not just as pertinently ask the same question as we face Tomorrow's World?—Seeing that we are to live in such a world, what manner of persons ought we to be? It is a notorious fact that our scientific and technical progress has far outdistanced our moral and spiritual growth. As Professor Elton Trueblood has well said: "The awful truth is that our wisdom about ends does not match our ingenuity about means, and this situation, if it continues, may be sufficient to destroy us." A razor is a useful and quite essential instrument in capable hands, but in the hands of a little child it can work havoc with our over-stuffed furniture or the costliest library and, what is far worse, it can injure and destroy the child and others. An automobile in safe hands is a valuable, and sometimes a highly necessary, means of transportation, but in the hands of a "Hot-rod Happy" it may become an engine of death and destruction. A scalpel in the hands of an experienced surgeon may be used for beneficent ends, but in the hands of a gangster or maniac it may be made to serve deadly purposes.

The crux of our problem is moral and spiritual. We need to grow up to the point where we can be safely trusted to handle these vast powers, these amazing engineries and instruments wisely, beneficently, constructively. When we were still in yesterday's world Henri Bergson saw this and said: "The splendid powers which science has furnished are still in the hands of old sins, selfish ambition, cruelty. The innermost necessity of mankind is a spiritual life adequate to handle our new acquisitions. . . . We must have a new accession of moral vision and power or we are utterly undone." Of what use is our learning to split the atom, if we do not develop and master the powers of the soul which are mightier than the power of the atom and can control and direct it? No one is likely to dispute the point and pertinence of this statement by Lewis Mumford: "One may be able to chart the course of an electric charge, split an atom, imitate a thunderbolt, or send six hundred messages simultaneously through the same cable; but unless one knows in addition that good and evil are constant realities in human life, all other intellectual acquisitions are worthless. The consciousness of this truth turns the trifler into a responsible personality. Without it, the cancer of evil may work its way uncontrolled through human society until corruption and death are everywhere." [3]

Tomorrow's World ought to be, in the highest sense, one world. A few years ago the late Wendell Willkie captured our imagination with what has been called the "inspired title" of his book, *One World.* Our scientific and technological achievements have really made our world one world. Our wars have become global; any local incident may become a matter of planetary concern; disorders in any area, no matter how remote, may affect the entire organism; any sentiment expressed or deed accomplished in some obscure hamlet may be known to the ends of the earth almost in the twinkling of an eye; Emerson's lines about the embattled farmers who "fired the shot heard round the world" become more than poetry in today's world. We have been knit, as has been said, "into a perilous and contradictory unity." What should this mean to us? This, certainly: that we should learn to live as citizens of one world, should adjust our personal living and our social relations and, so far as we are able to influence it, the life of society, to this world-oneness. It should mean the getting rid of our insularities, our pride of race, our snobberies, bigotries, prejudices and all those divisive elements which set unnecessary boundaries to life. We need a sense of inclusiveness, the power to see life in this new perspective, and to envision the widening horizons of service for the common good.

But this is not as easy as blowing on your hands and getting it done. Most of us find it difficult, if not impossible, to live together in love and brotherhood, in peace and good will within our little communities, and much more within the extended area of natural life. For so many the democratic creed fails too often to express itself in the democratic deed, and the failure would be more pronounced if we substituted "Christian" for "democratic."

[3] *Faith for Living* (Harcourt, Brace & Co.). Used by permission.

The challenge of one world can only be met by dedicated persons of undiscouraged faith, who seek a creative fellowship and an adventurous coöperation. As Professor Nels F. S. Ferré points out in his recent book, *Christianity and Society,* "We need a whole new spiritual revolution and level of life to cope with our larger togetherness."

This is the task of the Christian Church. There have been eras when it has wrought gloriously and triumphantly at it, but there have been times throughout its history when it has failed ingloriously, when it has compromised itself and its cause and become a reproach, "an insult to its Lord and to its saints." Those who are familiar with the churches today cannot be ignorant of the fact that the religious life within them, speaking by and large, strikes such a poor average, is maintained at such a low level, as to render them practically impotent against the growing paganism of our time. That is our deadly peril.

A few years ago Bernard Iddings Bell gave us a book with the arresting title, *The Church in Disrepute.* Who is to blame if we have a Church in disrepute? How much influence for good can a Church in disrepute exert in the world? If we are to develop a really effective and influential Christian society, we must begin with individual Christians who will take their religion seriously enough to completely surrender themselves to Jesus Christ as Savior and Lord, and to dedicate themselves without reservation to the Eternal Purpose as the meaning and goal of all living. That, of course, will mean a minority, but it will be a creative and a growing minority, and such minorities, as history amply certifies, have been, time and time again, the saviors of society.

Build me a World, said God,
Out of man's fairest dreams;
Heaven must be its dome,
Lighted by prophet-gleams;
Justice shall be the stones
On which my World shall rise;
Truth and love its arches,
Gripping my ageless skies,
Out of dreams on the earthy sod,
Build me a World,
Said God.[4]

"Let us rise up and build." But surely we must know that men cannot build Tomorrow's World, or anything else that will endure, by themselves. They can only build it with God. Long ago a Hebrew singer reminded us in unforgettable words of the futility of all godless enterprises: "Except the Lord build the house, they labor in vain that build it: except the Lord keep the city, the watchman waketh but in vain." That does not mean that the builders and the watchman may sit down with folded hands, or go to sleep, saying, "Well, let God do it!" but rather that they do their work in the

[4] "Build Me a House," by Thomas Curtis Clark, from *Home Roads and Far Horizons* (Willett, Clark & Co.). Used by permission.

full confidence that they have the approval and help of God, and that apart from Him there can be no enduring achievement, no ultimate success. You recall that God said to Moses at the place of the burning bush: "I have surely seen the affliction of my people that are in Egypt, and have heard their cry by reason of their taskmasters; for I know their sorrows; and I am come down to deliver them . . . and to bring them . . . unto a good land and a large." But how does he deliver them? Listen!—"Come now therefore, and I will send thee unto Pharaoh, that thou mayest bring forth my people . . . out of Eqypt." And when Moses pleads his incompetence, God gives him the assurance: "Certainly I will be with thee." That is how deliverances are achieved and new worlds are built—God working with men and through men. Or take the battle cry of Gideon: "The sword of the Lord and of Gideon!" Not the sword of the Lord alone, nor yet the sword of Gideon alone, but "The sword of the Lord and of Gideon!" The secret of all triumphant effort, of all worthwhile and permanent achievement is to be found in this august copartnership, God and man linked together in coöperative enterprise. "We are God's fellow-workers."

And if, by the grace of God, a man have skill, or power, or the gift of genius, let him not in the spirit of pride, or vainglory, or arrogance forget God or try to discredit Him, or deem himself indispensable, as did William of Sens, the architect and builder of Canterbury Cathedral, in the play by Dorothy L. Sayers, *The Zeal of Thy House:*

> "This church is mine
> And none but I, not even God, can build it,
> . . . .
> He knows that I am indispensable
> To His work here; and for the work's sake, He,
> Cherishing, as good workmen do, His tools,
> Will keep me safe. When the last stone is laid
> Then may He use me as He will, I care not;
> The work is all; when that is done, good night—
> My life till then is paramount with God." [5]

But as he pulls himself up to the top of the scaffolding in a traveling cradle which he had contrived, a boy sees, as in a vision, an angel with a sword severing the rope which held the machine. Through what seems an accident, William falls to the ground and is left a hopeless invalid, and someone else has to finish the work he began. There is no need to point the moral.

A familiar story is told of a Negro who purchased a piece of land that was rather poor and unpromising and overrun with weeds. By patient toil and careful attention he transformed it into a fine garden of vegetables and flowers. One day his pastor, who was visiting him, observed: "Uncle Lige, you and the Lord have sure done yo-selves proud. This sure is a purty place."

Uncle Lige wrinkled his brow, scratched his head, and said: "Yes suh,

[5] Used by permission of Harcourt, Brace & Co.

preacher, we sho' has. But you' oughter seen this place when de Lawd was managin' hit by Hisself!"

Dr. Buttrick, in his book, *Christ and Man's Dilemma,* makes a brief reference to this story which he characterizes as poor and shallow, and makes the suggestive comment that when God had it alone, that acre "was a treasury of gifts—sun and shower, soil's mystery and seed's mystery, without which any man is helpless and which no man can create. Furthermore, that acre gave challenge and reward to the gardener. In short, if the gardener had 'had it alone,' he would have died before he could turn a spade."

If man cannot grow a little garden without God, how can he hope to build a better world without Him? Let us acknowledge that any enduring structure of life or of society must be the challenging adventure of the Divine and the human in mutual effort, must be the glorious result of fellowship of the Lord and His servants in common toil.

There is a tremendous implication here which we can ill afford to miss. If we are caught up in the Eternal Purpose and become fellow-laborers with God in working out that Purpose here in time, then these lives of ours, however little and insignificant they may seem, are really shot through with worth, dignity and meaning. We must hold to that faith, especially in a day when so many suffer themselves to be browbeaten by our modern astronomical science, or shamed by much of the new psychology and philosophy, into believing that we don't count and that all our efforts are doomed to frustration and failure. People who don't count can't build anything that counts. We must have great faith, courageous hope, and a high sense of our value before God, if we are to build a world worthwhile. Let us not take our cue from sources which attempt to belittle human life, but from God who made us and invested so much of Himself in us, and honors us by calling us to share with Him in His great creative venture.

In Hermann Hagedorn's striking poem, *The Bomb That Fell on America,* the Voice of God says to the man who had been fleeing Him across ten thousand miles of desert wastes, and finally fell with all his self-will broken:

"Give me your life, and day shall be like a new world.
The unclean shall be clean, the cowardice, courage, the weakness, power.
Give me your life and I will make it a spade to dig the foundations of a new world, a crowbar to pry loose the rocks, a hoe to mix sand and cement, a trowel to bind stone and make them a wall.
Man without God is a bubble in the sea, a single grain of sand on an infinite beach.
God without man is a mind without tongue or ears or eye or fingers or feet.
God and man together, We are such power as not all the atoms in all Creation can match!" [6]

And the man said: "I laid my hand there in the hand of God."

If we would build a new Life, or a new Church, a new Nation or a new World, we must come to that: "I laid my hand there in the hand of God."

[6] Used by permission of the Pacific Coast Publishing Co.

The House of Seven Pillars

REVEREND DANIEL L. MARSH, PH.D., D.D., LL.D., LITT.D., D.SC.
A Minister of the Methodist Church; Chancellor, Boston University, Boston, Massachusetts

Boston has great respect for Daniel L. Marsh as educator, friend and man of affairs. He was born in West Newton, Pennsylvania, in 1880, and studied at Northwestern University, Boston University, Garrett Biblical Institute, the University of Chicago, the University of Pittsburgh, the University of Geneva and Oxford University.

He began his career as a Methodist pastor in the Pittsburgh Conference, where he served from 1908 to 1913. Then he became general superintendent of the Methodist Church Union of Pittsburgh until 1926, when he became president of Boston University. There for twenty-five years, he has been a leader in the educational world, and fifteen important institutions have conferred the honorary doctorate upon him in recognition of his outstanding achievements. He became Chancellor of the University in 1951.

His interests are many and varied and he is prominent in numerous organizations. He is a director of the John Hancock Life Insurance Company, chairman of the Presbyterian Ministers Fund, president of the State Library, member of the American Council on Education, the Association of American Colleges, the American Association for the Advancement of Science and the American Academy of Arts and Sciences.

Dr. Marsh has a large vision and sticks to a project long enough to get big things done. His plan for a unified university on the bank of the Charles River, the great College of Business Administration, and his dream of education for the minds and hearts of young men and women mark him as one of America's truly great leaders. In large measure, the "new Boston University" is an enduring monument to Dr. Marsh himself.

Among his thirty-eight books are: The Challenge of Pittsburgh; The Faith of the People's Poet; Higher Education Plus the Highest Education; Bell, Benefactor of Mankind; The Patriotism of a Mature Mind; Freedom of Discussion Indispensable to Democracy; The American Canon *and* The Charm of the Chapel.

Daniel Marsh had the wholesome belief that a University President should give his students something to take away from the University with them, hence, for twenty-five years he delivered the baccalaureate sermon himself in the tradition followed by a few other great educators. In this sermon he outlines a series of simple rules and facts that will make life wholesome and worthwhile if followed by the graduates of Boston University and all the other universities in the world. It is refreshing to have a great University leader state his faith in God and Christ and the Christian way of life without fear or hesitation. The editor believes this sermon will inspire many others of a similar nature at graduation times all over America.

Sermon Forty-five

TEXT: Wisdom hath builded her house, she hath hewn out her seven pillars. PROVERBS 9:1

NATHANIEL HAWTHORNE wrote a novel about "The House of Seven Gables." It is an actual house, still standing in Salem; but Hawthorne filled it with fiction. "The House of Seven Pillars" is an imaginative house, but filled with wisdom. The Scriptural text which suggests the title is in the book of Proverbs, 9:1—"Wisdom hath builded her house, she hath hewn out her seven pillars." This text lends itself naturally to analogical treatment, but it is easy to overdo analogy. The frame of reference in the proverb is that of an oriental house of Biblical times. Wisdom is represented as building a house. Seven pillars have been hewn out and erected. They are the seven pillars which support the balcony around the banquet hall. The Scripture goes on to say that a banquet is being prepared by Wisdom.

It is the house of Wisdom, and Wisdom has hewn out seven pillars. Make note of that fact. These seven pillars have not been fashioned by somebody else, nor did they drop down ready-made from heaven. They have been hewn out by the builder of the house.

I am going to specify seven pillars for your house of wisdom; but you must hew them out for yourself. Every one of them is the result of attention to a preconceived plan, and of hard work in the plan's fulfillment.

The First Pillar is Knowledge. There is a tendency, in some quarters, to scoff at knowledge, at information, as though it were not a very important item in modern education. But let me assure you that your house of wisdom will collapse unless it has the support of knowledge.

Whatever else education may be or may not be, it certainly means an accumulation of factual knowledge. "Knowledge is power." The world is always ready to listen to the man who knows. Therefore, read, study, discuss, explore—and remember what you learn. There are no short cuts to learning. There is no excellence this side of drudgery. Genius is still one-fourth inspiration and three-fourths perspiration.

A large part of the responsibility of an institution of higher learning is the transmission of the accumulated knowledge of the ages. Curriculums have been devised with the expectation that students will learn something. Members of the staff of instruction are paid salaries to impart information and to stir up within the students an insatiable hunger and thirst for knowledge.

The money that a university spends for research is merely the quest of new knowledge—pushing back the horizons of knowledge in order to explore what lies beyond the horizon.

It is because this pillar of knowledge must be hewn out by the sweat of one's brain that I deplore the intrusion of lazy short-cuts which fool a student into thinking he is learning something when he is not. Francis Bacon's dictum is still true that "Reading maketh a full man; conference a ready man; and writing an exact man." Benjamin Franklin, probably through unconscious memory which he mistook for creative imagination, said nearly the same thing: "Reading makes a full man—meditation a profound man—discourse a clear man." Wise old Samuel Johnson declared that "Books have always a secret influence on the understanding; we cannot at pleasure obliterate ideas: he that reads books of science, though without any fixed desire of improvement, will grow more knowing; he that entertains himself with moral or religious treatises, will imperceptibly advance in goodness; the ideas which are often offered to the mind, will at last find a lucky moment when it is disposed to receive them."

The first pillar in the banquet hall of wisdom is knowledge.

The Second Pillar is Freedom. Freedom means to be untrammeled, free from restraint, from fetters. No house of wisdom can be built unless it be supported by freedom. True education is impossible apart from academic freedom, which means that each professor or student is free to seek the truth in his own way, to form his own opinions, to arrive at his own conclusions, and to announce his own convictions. He is not to be limited by patented dogma, faint-hearted consideration, inherited tradition or acquired prejudices. He does not need to bend the knee to error, nor to fawn before flattery, nor to cringe before denunciation, nor to yield to the lawless impulse of his own self.

Likewise, wisdom hews to the line of political freedom, which means that the people are free to govern themselves, free to live their lives in their own way, free to choose their own officers, free to hold those officers responsible for the conduct of government, and free to dismiss them if they fail to serve the best interests of the people. Freedom of enterprise is far more than a shibboleth.

For religious freedom, we plead also—freedom of conscience, freedom to worship God in one's own way. Freedom of faith means the right not only to worship undisturbed, but also to propagate one's religion. It means freedom from a dictator in any ecclesiastical system, freedom from burning of proscribed books, freedom of access for everybody to the open Bible, freedom from interference with the pronouncements of the pulpit.

And yet, we must never forget that freedom—religious, academic, political, or any other kind—is not absolute. Wisdom recognizes the limitations of liberty, the restraints of freedom,—not limitations or restraints imposed by some external authority, but by one's own sense of the responsibility of freedom. Freedom is never an end in itself, but a means. The end of academic freedom is the discovery of truth and beauty and goodness. The end of political freedom is the living of the good life. The end of religious freedom is the worship of God.

Freedom is independence from unjust restraints, not independence from properly constituted authority. A sense of responsibility is to freedom what a trellis is to a vine: The trellis restrains the vine in order that it may rise. Freedom means that a person is free to do what is right; but he is not to confuse liberty with license, nor freedom with irresponsibility, nor human rights with their perversion. He is free to the bondslave of Truth and Right.

Jesus gave the last word on freedom: "Ye shall know the truth and the truth shall make you free." Abraham Lincoln felt the burden of responsibility of freedom when he said: "I am not bound to win, but I am bound to be true. I am not bound to succeed, but I am bound to live up to the light I have."

The Third Pillar is Democracy. The best definition that was ever given of political democracy was uttered by Abraham Lincoln: "Government of the people, by the people, for the people." Democracy means that a just government is established with the consent of the governed, and is under their control. Government is the servant of the people and not their master. Ours is a government by law and not by men. Ours is a representative democracy, and hence is accurately called a Republic. Thomas Jefferson reminds us that "governments are republican only in proportion as they embody the will of the people, and execute it."

To Lincoln's compact definition of political democracy, we need to add the ethical conception of democracy as expressed by Theodore Parker, who said that "Democracy meant not 'I'm as good as you are,' but 'You're as good as I am.'" Democracy rests upon the dignity of the individual, the sacred worth of human personality. It implies equality before the law and also equality of opportunity. It expresses itself in brotherhood, especially brotherhood as interpreted by the Golden Rule.

Therefore, we crusade not only for political democracy, but also for the ethical conception and application of democracy. This includes democracy in education; for when a people undertake to be their own dictators, they assume the responsibilities as well as the privileges of the function. Therefore, education must be as widely diffused as the right of suffrage. When the people are ignorant, they become an easy prey to greed and corruption, and accomplish their own debasement and ruin.

We believe also in democracy in religion. That is the essential idea that underlies the doctrine of the priesthood of all believers,—that every man has access to the throne of God.

This pillar of democracy does not drop down from heaven ready-made. It can be achieved only if the great and small, here and there and everywhere, apply their hearts and minds in honest and passionate cooperation. The very essence of democracy is the consciousness of the individual's responsibility.

The Fourth Pillar is Solid Attainments of Character. Any so-called house of wisdom will collapse unless it is supported by solid attainments of character. In 1927, I chose for my Baccalaureate Sermon the theme "Higher Education Plus the Highest Education." In preparation of that address, I wrote to the fifteen thousand persons who were believed to be the most successful Americans, in the ordinary acceptation of the term "success." I told them that I was making a study of the development of character in college students, and desired their opinion on the following questions, namely:

1. What do you consider to be the essentials of sound character? In other words, if you were the head of a large University, what ideals of character would you seek to impress upon the students?
2. How can these best be developed in young people?
3. (a) Do you consider moral and religious training a necessary basis for the best development of character?
   (b) If so, should this training be given at home, in church, or at school?

I then submitted the same questions to the members of that year's graduating class at Boston University. Identical ideals, with slight variations in the order in which they were named, were listed by both groups, to-wit: *Honesty, Love, Reverence, Loyalty, Industry, Intelligence, A Moral Sense, Courage, Justice, Self-Control, and Patience.* In answer to the question how these ideals could best be developed in young people, both groups emphatically asserted that example and environment were most important, and that instruction and hero worship stood next. When asked whether they regarded moral and religious training as a necessary basis for the best development of character, two-thirds of both groups answered, "Yes, emphatically!"

There is a saying that is variously applied, such as: "You can always tell a school teacher, but you cannot tell her much," or "You can always tell a Boston University man, but you cannot tell him much," and so on *ad infinitum.* I would to God that the world could always tell a Boston University man—that is, could always distinguish or recognize him—by his possession and exemplification of these solid virtues. Conduct is character unfolding itself. Emerson reminds us that "character is higher than intellect. A great soul will be strong to live, as well as to think." Theodore Roosevelt unequivocally declared that "To educate a man in mind and not in morals is to educate a menace to society." I earnestly pray that every member of the Boston University family—trustees, administrative officers, professors, students and alumni—may choose the highway of virtue and piety as well as of learning. As John Oxenham indicates:

To every man there openeth
A way, and ways, and a way,
And the high soul climbs the highway,
   And the low soul gropes the low,
And in between, on the misty flats,
   The rest drift to and fro.
But to every man there openeth
   A high way, and a low,
And every man decideth
   The way his soul shall go.[1]

The Fifth Pillar is Understanding. The spirit of understanding is an essential element in any true education, and is indispensable to harmonious living. If you would build a durable house of wisdom, make sure to hew out a strong pillar of understanding. Aim to understand others, and aim to be understood. Woodrow Wilson in his notable address at Mobile, Alabama, in 1913—one of the great presidential utterances of our history—declared that "Comprehension must be the soil in which grow all the fruits of friendship." This is true of college friendships among individuals no less than international friendships among nations. The fruits of peace and love and friendships—personal friendships, business friendships and international friendships—grow out of the soil of understanding. Profoundly true is the old Arabic proverb: "Understanding is the wealth of wealth."

Understanding is reason expanded and glorified. Some persons never reason, but reflect only what others say and do. Some substitute feeling for reason. What they say stems from emotion, and what they do stems from passion. Others give a semblance of reasoning but lack either the knowledge or the common sense that makes their reasoning reliable.

Some there are who never read a book unless it agrees with their prejudice and bigotry. They participate in discussion only where their own point of view is accepted by others. To understand, we must see all sides of a question; read books that present various points of view; converse with persons of various and differing opinions. We must not parcel out to ourselves some little territory in the intellectual world which deprives us of the riches and resources of the great continent of human thought. We must not be content with traffic in some little brook through our parcel of land while the great ocean of truth remains unknown to us. It is said that the Mariana Islanders, before their islands were discovered by the white man, thought they were

[1] From "The Ways" by John Oxenham (Pilgrim Press). Used by permission.

the only people in existence. The fact that they thought they were the only persons in the world did not make them so. Likewise the bigotry and intolerance and prejudice that make one nation think it has all the virtues, or one ecclesiastical group think it has all the truth, or one clique think it has all the social graces is just as absurd as for the Mariana Islanders of old to think they were the only people in existence. The differences in men's understandings arise more from acquired habits than from natural faculties. We must learn to use the mind.

Understanding is needed in every area of tension today—in the home, between husband and wife and between parents and children; in industry, between labor and management; in fields of economic and political controversy, and even in the clash of religions which ought themselves to be the source and inspiration of understanding.

Especially should the international leaders in all the countries, including the United States and Russia, set themselves deliberately to understand each other's point of view. Our present-day international tension finds its explanation in the old Biblical proverb: "It is as sport to a fool to do mischief; but a man of understanding has wisdom" (Proverbs 10:23). The trouble is that too often present-day leadership implements only one-half of that proverb: "It is as sport to a fool to do mischief." We are living in a universe whose end is order and harmony. The fool does mischief by creating discord, by destroying, by tearing down, by disseminating lies, by stirring up fear and envy and jealousy. The other half of the proverb is the one that should be implemented, namely: "A man of understanding hath wisdom," and certainly if the destinies of the human race were in the hands of men of understanding and wisdom, they would find a way to solve international problems without resorting to the rude arbitrament of force and violence; for war is the summation of all villainies. It is the recrudesence of the jungle. It is organized insanity. As a matter of fact, war solves no problem. Why then do men resort to war? The answer is found in the lack of understanding, or the unwillingness to understand each other.

The Sixth Pillar is a Wholesome Philosophy of Life. It is an indispensable pillar in the banquet hall of wisdom's house.

"What is philosophy?" asks Epictetus, and then answers his own question by saying: "Does it not mean preparation to face the things which may come upon us?" Oliver S. Braston affirms that "Philosophy is commonsense in a dress suit," while Francis Bacon insists that "All good and moral philosophy is but a handmaid to religion. If you were to ask me for an outline of my philosophy of life, I could give it to you in five points of fellowship which a wise individual may have with abundant living.

The first point is expressed in a quotation from James Whitcomb Riley. In Riley's poetry, the vicissitudes of weather stand for the vicissitudes of life, fair weather representing the pleasant and prosperous things, and rainy weather the unpleasant and adverse things. In his quaint Hoosier dialect Riley makes the farmer say:

> It hain't no use to grumble and complane;
> It's jest as cheap and easy to rejoice.—
> When God sorts out the weather and sends rain,
> W'y, rain's my choice.

This is a philosophy of contentment, based upon a firm faith in the overruling Providence of a good God.

The second point is the dictum of Saint Paul: "We know that all things work together for good to them that love God." This means that I need to be sure of only one thing, and that is that I love God. To love God means more than mere conformity to some ritual or the repetition of some creed,—it means that I make my practices square with the will of God, and require myself to keep step with His Commandments against all the forces that oppose me. If I love God, then I can trust Him to make all things work together for good—work together as bitter and sweet work together for bodily health, as sunshine and shower work together for the harvest's fruition. I need not fill up my system with the poison of anxiety and worry. If I aim sincerely to love God, I can trust Him to make things work together for my good. That is, I must trust God as though everything depended upon Him, and I must behave myself as though everything depended upon me.

The third point is expressed in a bit of poetry—hardly more than doggerel—which I picked up in some fugitive way, but which I use almost every day:

> For every evil under the sun
> There is a remedy or there is none.
> If there is a remedy, find it.
> If there isn't, never mind it.

That saves the philosophy of contentment from becoming fatalism. Its propositions are self-evident: For every evil—personal, social, economic or international—for every evil under the sun, there is a remedy, or there is none. If there is a remedy, don't fret and fume about the evil: Find the remedy! If there is no remedy, then what is the sense of fretting and fuming? Never mind it!

The fourth point I do not express in a quotation, but in an aphorism of my own. It is this: *If you cannot realize your ideal, then idealize your real.* Henry Ford, when on the witness stand in his libel suit against the *Chicago Tribune,* said: "The idealist is a person who helps other people to be prosperous." J. C. Holland opined that "Ideals are the world's masters," and A. Bronson Alcott, in his *Table Talks,* declared that "Our Ideals are our better selves."

Have ideals! Carl Schurz reminds us that "Ideals are like stars. You will not succeed in touching them with your hands; but, like the seafaring man, you choose them as your guides, and, following them, you will reach your destiny." If you are not drawn upward and onward by ideals, I give you up now! But if you have an ideal whose realization is deferred, don't let it make your heart sick! Go to work, and idealize your real. Poet Walt Whitman says that

Through thy idea, lo, the immortal reality!
Through thy reality, lo, the immortal idea!

The fifth point of my philosophy of life is expressed in the immortal word of Jesus: "Whosoever would be great among you shall be your servant, and the greatest of all shall be the servant of all." That is, the true standard of real greatness is service to others. Humility is one of the most beautiful of all the virtues; but humility, rightly understood, is simply a disposition to serve. The person who thinks he is too big to serve others is really confessing that he is too little. Jesus performed the most lowly service that one person could render another, and His dignity suffered no abatement in the performance because He was essentially big! If you desire a radiant and wholesome philosophy of life, some rule by which to live that will make life meaningful to you, be sure to enthrone therein the idea and the ideal of service to others.

The Seventh Pillar is an Efficient and Rightly Integrated Personality. As you know, the word "education" is derived from two Latin words, *e,* meaning "out," plus *ducere,* "to lead,"—*to lead out.* It used to be thought that education consisted in leading out the different mental faculties, such as memory, imagination, reason. But if modern psychology has taught us anything, it is that we are unitary beings,—we are one self. Therefore, I am thinking of education as the leading out of the whole individual into a personality. Vivifying education does not come by knowledge alone, nor by experience alone, but by knowledge, observation, reflection *and* experience.

We have the Latin word for education used in a sentence written by the poet Vergil in his Fourth Book of Georgics (line 158), where, in discussing the division of labor in a hive of bees, he says that some of the bees are supplying food and are busy in the fields; some are constructing the honeycomb, and then he says: *"Aliae spem gentis adultos educunt fetus,"*—the translation of which is, "Others lead out the grown-up offspring, the hope of the nation." *Educunt* as here used means to lead out of doors, to teach to fly, to gather honey through the flowering countryside, to be at home in the world.

A Biblical text embodying this same idea is found in Deuteronomy (32:10–11), where God's dealings with the Israelitish nation, personalized as an individual, are described thus: "He found him in a desert land, and in the waste howling wilderness; he led him about, he instructed him, he kept him as the apple of his eye. As an eagle stirreth up her nest, fluttereth over her young, spreadeth abroad her wings, taketh them, beareth them on her wings: so the Lord alone did lead him."

This is education—*"educo"*—leading out the bees until they are at home in their environment; leading out the eaglet until it can soar against the sun and rest at ease on craggy heights.

The person who has an efficient and rightly integrated personality is at home in the physical world. He has a wise hearted grasp of terms. He is conscious of his dominion over this earth and the forces of nature.

He is at home also in the world of men, having an appreciative understanding of the Past of the Race from which the stream of history gathers momentum and direction. If he does not already know the story, he at least knows where and how to find out what is known concerning the path of human progress that begins in the state of tumulus ignorance and darkness that circumscribed Neolithic man, and that shineth more and more unto our own day. He knows the race through history as he knows the stars through the telescope, and thus is able to put each type and change of society into a general perspective. He marches in Caesar's conquering legions. He walks with Plato in his academic grove. He sings to the lilt of David's harp. The glory that was Greece, the power that was Rome, the irrepressible longing for spiritual reality that was Israel—all these are his.

He is equipped to earn his living, but he is concerned about far more than that. He has been led out from the egocentric love of a child into the altruistic love of an adult. He knows how to live happily with his fellows. He uses his mother tongue with fluency and precision, communicating his thoughts to others accurately and coherently, in words well chosen. He has magnified and multiplied the positive and attractive traits of character, and has minified and curtailed the negative and repellent traits.

He is comfortably at home with himself. He knows what to do with his leisure time. He knows that true happiness cannot be poured upon him from the outside like water on thirsty ground. It comes, rather, from within himself, welling up from his self-active creative life like a spring of living water. He is emotionally mature.

The person who has an efficient and rightly integrated personality is at home in the world of thought. He knows that thoughts are as real as things. He can now see the invisible, and feel the force of the intangible, and appreciate the weight of imponderables. He has maturity, experience, guiding ideals, adult capacities for reflective thinking. He is able to think independently and critically. He keeps an open mind, and listens to the man who knows enough to have a right to speak. He has within him the spirit of culture. He has been led out until he is at home in the magic world of thought—the lofty thoughts of the ages as expressed in poetry, art, music, philosophy.

Dallas Lore Sharp, of blessed memory, long time beloved and very distinguished professor in Boston University, was one autumn day returning from Boston University to his suburban home in Hingham, when he came upon one of his sons in the potato patch, digging potatoes. The boy did not note his father's approach, and just as Dallas Lore Sharp came up behind him, the boy caught sight of a wild duck sailing across the sky. He paused in his work, with his foot poised on the fork with which he was digging, and his hands grasping the handle of the fork, and so stood watching the duck fly clean across the sky until it was lost to sight in the dusk of the horizon. Then as he was about to resume his work, Dallas Lore Sharp spoke, and his manner of greeting was to quote a line from Bryant's "To a Waterfowl":

He who, from zone to zone,

and his son replied by completing the stanza:

Guides through the boundless sky thy certain flight,
In the long way that I must tread alone,
Will lead my steps aright.

And then he said to his father, "I am glad I know that poem." "Why?" "Because if I did not know it, I would have seen only a wild duck flying across the sky; but now I have seen both the duck and God!" What a beautiful picture of a father and son happily at home on the heights of thought as truly as on Mullen Hill!

This integrated personality is at home not only in the actual world in which he lives, but at home also with the ultimate spiritual forces that lie back of the visible phenomena of the physical universe. The living of the full life requires a complex equipment by which one makes adjustments in the spheres of science, the humanities, psychology, contemporary living, and all the rest. But to live the fullest life, it is essential that this complex equipment should be integrated into a personal religion.

This efficient and rightly integrated personality possesses knowledge which is balanced by character. He has entered into an offensive and defensive alliance with God, and God is forever on the side of goodness, freedom, love, spirit, person and progress.

The Best Things in the Worst Times

REVEREND JAMES A. PIKE, J.S.D.

Chaplain of Columbia University, New York and head of the Department of Religion; a Priest of the Protestant Episcopal Church; and Dean of the Cathedral of St. John the Divine, New York City

When Dr. Pike left the rectorship of Christ Church, Poughkeepsie, where he had also been in charge of Episcopal student work at Vassar College, (1947–49) to become chaplain of Columbia University, he entered upon the task of ministering to the spiritual needs of the students of one of the great universities of the modern world.

He studied law at the University of Southern California; was Sterling Fellow, Yale Law School, 1936–37 and took his J.S.D. at Yale University in 1938. He became a member of the California Bar, the Bar of the Supreme Court of the United States, and the Bar of the U. S. Court of Appeals for the District of Columbia, and was Attorney for the U. S. Securities and Exchange Commission, 1938–42.

He felt the call to preach, entered Union Theological Seminary in New York, and took his B.D. there. When World War II began he joined the Office of Naval Intelligence, then became Attorney for the U. S. Maritime Commission and War Shipping Administration. After the war he became curate of St. John's Church, Washington, D. C., following his ordination in 1944. He was on the faculty of the General Theological Seminary, New York City, as fellow and tutor, 1946–47. He is the author of books and articles in the field of federal judicial and administrative procedure, co-author of the forthcoming text on the Christian Faith being issued by the Episcopal Church; co-author of The Faith of the Church; *Fellow, National Council on Religion in Higher Education. He is also priest-in-charge, Summer Chapel of St. James the Fisherman, Wellfleet, Cape Cod. He has just become Dean of the Cathedral of St. John the Divine, New York City.*

This sermon was preached at the Baccalaureate Service in St. Paul's Chapel, Columbia University, on Sunday, June 3, 1951, and was much discussed at the time: here was a baccalaureate sermon that does not promise graduates that the world was theirs for the taking; rather, it shows the hard way in our new day.

Sermon Forty-six

BACCALAUREATE sermons are getting harder to preach every year. At a time in your lives when there is such a sense of achievement and fulfillment, it would be nice to say cheerful things. But these are not cheerful times. As a new phase of your lives begins it would be nice to hold forth predictions of security. But these are not secure times. Yet, impelled alike by time-honored precedent—and by the ever-present prophetic task that is the vocation of the preacher—I must produce some relevant advice today. What can I say that is honest as well as hopeful? The clue to what I would like to say comes from an obscure bit of English history I read a while back.

Nothing could have seemed more distressing than the tyranny of Cromwell and the Long Parliament—to certain groups of the populace, especially to those who loved the old Church and its ways of worship. Stained glass was broken in, altars were broken up, and conducting or participating in the worship of the Church of England rendered one a criminal. In the midst of these years a layman, Sir Robert Shirley, did an extraordinary thing: he built a church—one very well suited to the worship which had been forbidden and one which still stands in Leicestershire as a gem of architecture. For his pains he was summoned by Cromwell to London, was confined in the Tower and died. A full discussion of the issues need not detain us here; but I do believe we may gain a motif for our thoughts today from the inscription placed over the entrance of this old church:

IN THE YEARE: 1653
WHEN ALL THINGS SACRED WERE THROUGHOUT YE NATION
EITHER DEMOLLISHT OR PROFANED
SR ROBERT SHIRLEY, BARONET
FOUNDED THIS CHURCH
WHOSE SINGULAR PRAISE IT IS
TO HAVE DONE THE BEST THINGS IN YE WORST TIMES
AND
HOPED THEM IN THE MOST CALLAMITOUS

—*To have done the best things in the worst times and hoped them in the most calamitous.*

Thus my remarks today quite naturally fall into two parts: First, the worst times, and, Second, the best things.

As to the times, it would be rash for me to essay a blue-print of what lies ahead. But it is pretty clear that one of three things lies ahead:

First, there may be a war in which are employed by both sides all the devices which man in his scientific progress has designed, as a result of which our larger centers of population are leveled to the ground. (If this seems too theoretical I need only advert to the elaborate arrangements which have been made on the campus to meet the contingency of an air raid on Commencement Day; a mere reference to the "Shelter" signs all about the campus and the city would be unconvincing, since by now we take them as much for granted as traffic lights.)

Such an outcome would present our generations—yours and mine—with a major task indeed: if our way of life is to survive, a remnant must arise phoenix-like from the ashes. This will take more than trust in the works of our hands or than in the achievements of contemporary man. What raised up a new civilization out of the collapse of the old Graeco-Roman culture was the transcendent faith of men who knew all along—before, during and after the collapse—that "here we have no lasting city." And in their utter confidence in the heavenly city (which for them was no escape but rather power for action) they rebuilt the earthly city.

Second, we may have total war, but one in which we are successful in protecting our own cities and leveling off the other half of the world. This sounds better, but it isn't good news. It would present us with a grave task also—the task of feeding and ruling a world. All too quickly would an age-old hope of ambitious men and nations be fulfilled, namely, One World. Julius Caesar wanted one world; Adolf Hitler wanted one world; today the Communists want one world. The terms on which they want it would be indeed the terms on which we'd have it. For the utter ruin which would be the result, under modern scientific conditions, of beating our foes to their knees, would make impossible for many decades to come, any autonomous life or culture for them, nor soon again would we trust our enemies with the development of effective resources.

Now running a world represents grave dangers to the victors. Never has any nation been able to sustain this rôle without inner disintegration, for "power corrupts and absolute power corrupts absolutely." If we are to fulfil this task we will need that difficult spiritual resource of self-criticism. For those for whom the "American way of life" or any other form of the current mores is the ultimate dimension, this resource will be a scarce one. Again, a transcendent dimension is required—one in which there is a fruitful and continuing tension between the "ought" and the "is," a frame of reference in which we can be as dissatisfied with ourselves in our power as in our weakness.

Third, there may be no war. This sounds fine. But actually, under modern conditions, this result may be as spiritually devastating as total war. For this peace will have been bought at the expense of an "armed camp." With the threat of war ever present we will have been devoting month by month an increasing proportion of our manpower and other resources to the erection of a huge war-machine, and as there is increased attention to the problem of our national security there will be increasing nervousness over dissenting or critical voices, and we stand in danger of ending up with a police-state not unlike that of our enemies; and in a sense their cause—authoritarianism—would have won.

How can we have national security on the scale demanded today and at the same time maintain our freedoms as of old? Actually there are only two ways to keep a lot of people in line: they must be held together spiritually from the inside, or beaten together physically from the outside. Thus what is called for to save us from outward tyranny is the inner resource of self-discipline.

In short, in case of total war with destruction all 'round, what we most need is *vision* and *courage;* in case of total war with the subjugation of half the world, what we most need is *self-criticism;* in case of peace through an armed camp, what we most need is *self-discipline.* These then are the *best things.*

Thus for our national situation these qualities in the citizenry are basic requirements; but it so happens that they are equally requisite for satisfying personal life. For in every life there are crises which are microcosms of the world crisis we are in, and what minister or psychotherapist would deny that these qualities are fundamental to their solution? But neither for personal or common weal can we bootstrap-lift ourselves into courage, self-criticism, discipline. Unless there is a level of reality, a power, a norm beyond ourselves we have no purchase on ourselves. We have no scope for vision, no norm for self-judgment, no common spiritual source for collective self-discipline.

Hence the devastating effects to our intellectual equipment in times like these of several decades of the ascendancy of a purely descriptive sociology (which says we want "is's" not "oughts"), a deterministic psychology (which returns man once again to the rule of the Fates—though they now be recognized as the demons within by the depth psychologists), and a "scientism" (not to be confused with science) which explicitly or implicitly assumes that what can be weighed and measured is in the realm of truth and all else is in the realm of fancy. Equally deteriorating is an approach to history which assumes that ideas or doctrines are merely the ideological progeny of political or economic drives (*e.g.*, "what is important is not the soundness or unsoundness of Luther's views on justification by faith, but rather the rising nationalism in the German principalities") or an approach to personal counselling which would immediately assume that what the counsellee presents as a religious problem is really a manifestation of a psychological problem underneath, begging the question of which may be cause and which effect.

Granted there are no spiritual resources in this kind of thinking, may we not find a sufficient well-spring of what we need in our common humanity? I do not think so. First of all, it should be observed that, historically speaking, this humanism has arisen in a culture not yet bereft of power that for our grandfathers was based on deep religious conviction. No previous civilization seems to give any basis for the hope that altruism can survive for many generations the religion that gave it birth. Second, the norm "the good of humanity" leaves us with sheer relativism. As to "good," what is one man's meat is another man's poison (and this axiom is older than our modern knowledge of allergies), and as to "humanity," in case of conflict which men are to win the prize? Further, what is the good of man depends on the meaning of man, and in turn the question of the ultimate destiny of man—whether he lives forever or dies like a dog—determines the meaning of life now.

Some of you may have read about the dreadful event which occurred in Boston one day last week. A young soldier, disappointed in love, poised himself on an eighteen-inch ledge outside the tenth floor window of a hotel facing the Boston Common. Intending suicide, he nevertheless hesitated. Meanwhile, first a young woman and then a priest volunteered to go to the room and try to persuade him to come back in. A huge crowd had gathered during the ninety minutes he was on the ledge. Members of the crowd cried, "Jump! Jump!" repeating the words in the form of a regular chant, as one would at a sports event. The young man was half-way toward trust of those speaking from the room, but with tears in his eyes said, "I've got to jump now; down there they want me to jump." But at length he decided the voices of the two from the room, promising new possibilities for life, were the more reliable, and he left the ledge.

The story can serve as a parable of the situation we are in, in times like these. We are, as it were, on a ledge. Those that believe and teach that "this is all there is" (almost literally what *secularism*—"this-age-ism"—means) invite us to jump, and settle for what they would call the realism of positivism and relativism. The motives of the many who are sincere in this position can perhaps be compared to certain of those in that Boston crowd who yelled "Jump! Jump!" because apparently they thought it was a planned spectacle.

But in our time there are also voices who invite us to leave the ledge into a room whence one may go forward to new creative possibilities in the here and now in the confidence of an eternal meaning and security. I bid you listen to those voices.

In assessing the four years you have just spent, I hope you will value whatever has been honest description, whatever has delineated the material factors in any situation. You should value, too, any information about human behavior or mores which this short time has permitted you to accumulate. Even more, you should value those universal and transcendent meanings which are more primary than any physical forces or psychological compulsion or social conditioning. But most of all, you should value any mature religious understanding of reality which sets in an abiding frame of reference all of these other things so that you have, "oughts" as well as "is's," and the conviction that in the realm of the spirit the total of life can be greater than the sum of the parts, "that so among the sundry and manifold changes of the world," in the words of an old prayer, your "hearts may surely there be fixed where true joys are to be found." It is when life has such a dimension that there is the possibility of vision which transcends the lineaments of any given culture and thus provides the courage ever to build anew; it is when life has such a dimension that there is the possibility of self-criticism even in the midst of power and success; it is when life has such a dimension that there is the possibility of the reconciliation of discipline with freedom. It is one who all his days addressed God in common prayer as Him "whose service is perfect freedom," whose deistic notions did not deflect him from saying, "To avoid tyrants obey God"—here, of course, I refer to Thomas Jefferson. In short, it is when life has such a dimension there is the possibility of the courage of vision, the humility of self-criticism, and the freedom based on self-discipline. These are the best things in the worst times. If we are earth-bound, these things cannot long abide.

If by chance the studies of some of you have been so directed that you have not been able to see the forest for the trees and have reached this day without a mature and vivid spiritual understanding of reality, then make this the first concern in your lives ahead, for the sake of your own personal serenity and effectiveness, for your potentialities in building sound family life, and for your rôle as leaven within our common life.

MEMBERS OF THE CLASS OF 1951:

You go forth from this University into a world that is in great travail. You go forth as responsible citizens of a nation to which are committed tasks fraught with material and spiritual dangers. Though now we all "see through a glass darkly," so discharge the stewardship, in the years ahead, of all that has been committed to you by Almighty God, by the love and concern of parents, and by the dedicated scholarship of teachers in school and college, that at the end the Judge Eternal may find that you "did the best things in the worst times and hoped them in the most calamitous."

On Growing Up

REVEREND LEONARD HODGSON, D.D., S.T.D., HON. D.C.L.
Regius Professor of Moral and Pastoral Theology
Canon of Christ Church (Church of England), Oxford, England

Canon Hodgson is a popular preacher at Harvard and in Oxford, where he is Regius Professor of Moral and Pastoral Theology, and Canon of Christ Church, Oxford. Born in London in 1889, he was educated at St. Paul's School; Hertford College, Oxford; and St. Michael's College, Llandaff. He was Curate of St. Mark's Church, Portsmouth, 1913–14; Vice-Principal of S. Edmund Hall, Oxford, 1914–19; Official Fellow and Dean of Divinity, Magdalen College, Oxford, 1919–25; Professor of Christian Apologetics, The General Theological Seminary, New York 1925–31; Residentiary Canon of Winchester Cathedral, 1931–38; Examining Chaplain to the Bishop of Lichfield, 1917–25.

His writings include The Place of Reason in Christian Apologetic, *1925;* And Was Made Man (1928); Essays in Christian Philosophy (1930); Eugenics (1933); The Lord's Prayer (1934); Democracy and Dictatorship in the Light of Christian Faith (1935); The Grace of God in Faith and Philosophy (1936); This War and the Christian (1939); The Christian Idea of Liberty (1941); *Joint Editor G. R. Driver of* Nestorius: The Bazaar of Heracleides (1925). *Recreations: reading, walking, motoring. Since 1938 he has been Regius Professor of Moral and Pastoral Theology and Canon of Christ Church, Oxford.*

This sermon was preached in 1950 at Columbia University chapel and in the chapel at Harvard University for the students and guests. His discussion of the need of growing up and the place of the Church will bear consideration by students and adults alike.

Sermon Forty-seven

TEXT: When I was a child, I spake as a child, I felt as a child, I thought as a child: now that I am become a man, I have put away childish things. I CORINTHIANS 13:11

THERE IS A RHYTHM in the life of man which is indicative of a wide range of God's dealings with us human beings both in creation and in redemption. When we are first born, little new-born babies, our individuality as persons is at a minimum if, indeed, it can be said as yet to exist at all. We wriggle, kick, turn our eyes to light, and our mouths make noises and take food from breast or bottle. But in none of these activities are we giving effect to consciously thought-out purposes of our own. Each one of us is moved by racial habit working in and through this or that particular physical organism.

The months and years go by. It is not long before the growing infant begins to express a mind and will of its own. Children are seen and heard to quarrel for the possession of the same toy, to disagree about what game to play next, to assert themselves in opposition to their parents' wishes. These are symptoms of developing selfhood, and for many years wise parents and teachers will foster this development, giving freedom to make mistakes and be naughty and so become persons, responsible for running their own lives.

Thus we pass through a phase in which self-assertion is the right and proper activity of growing manhood. But there comes a time when we have to learn that this is not the whole of the story. What have we become selves for? The answer is to be found in one of the characteristics of human beings which distinguishes them from animals. Instances of reasoning, observed in animals, are concerned with finding ways and means to satisfy their bodily desires; man knows what it is to recognize that the claims of truth, of beauty, of goodness demand his devotion, the surrender of himself. The phase of self-assertion through which he has become himself must, as it were, go into reverse if it is to be justified. "He that saveth his life shall lose it; he that loseth his life for My sake shall gain it unto life eternal." We have to beware of the danger of arrested development, of what psychologists call "fixation," of failure to go forward, and this in both phases. In the first, if we fail to become fully mature responsible selves we go through life carried hither and thither at the mercy of demagogues and advertisers; if we stick at the achievement of self-assertive selfhood we become those "economic men" who achieve eminence by preying upon their fellows:

"getting and spending we lay waste our lives." We escape the danger of being tied to our mother's apron strings only to become the self-made man who worships his maker.

What to begin with is a two-phase rhythm, self-gaining for the purpose of self-giving, becomes a continuing harmony of alternating phases. "Now that I am become a man, I have put away childish things." Here we must discriminate. The things I am to put away are the irresponsibility of the infant and that exaggerated self-assertiveness of the growing child through which I have become my adult self, not the readiness to take charge of and be responsible for running my own life. I must continue to be myself, and to grow and develop in myself, in order to continue to have a self to give.

In individual men and women this two-phase rhythm is worked out within the span of a single lifetime. One sees the same rhythm pervading the history of man's political development, both within nations and internationally. Here a much longer time is needed for the development of the theme.

Within a nation an age of liberalism, when the cause of liberty for the individual subject or citizen runs like wine to the head and courses like fire through the veins of a people—such an age is comparable to the phase when the growing child wakes to the realization that he cannot be content to be governed by race-habit or by his elders, but must become the free and independent arbiter of his own fate. Here, too, the age of liberalism needs to be succeeded by the realization that the new-won freedom must not be used for the selfish advantage of individual, of party, or of class, but for coöperative devotion to the common good. And just as for the adult individual, life's problem is the problem of harmoniously interweaving the alternating phases of self-being and self-giving, so the political problem for mankind is the coördination of respect for liberty with devotion to the common good. Until that problem is solved we can expect nothing but recurrent swings of the pendulum between revolution and totalitarian tyranny. And in what nation can the solution yet be said to have been found?

Similarly between nations. The history of the world since the middle ages has been the history of the birth and growing childhood of nations, of nations finding their selfhood through the assertion of independent national sovereignty. Some of the western nations have gone far enough in this to be thinking that the time has come for the second phase, for the sovereign states to pass from competition to coöperation. This transition, difficult enough in itself in all conscience, is complicated by the fact that nations to the East are at earlier stages in the younger phase and, as Samuel Pepys would have put it, what will be the end of it, God knows.

In this rhythm, pervading the lives of individuals and of states, we see revealed God's method with us in creation. Too often our religion suffers from the fact that we fail to realize that God's method with us in redemption is similar. Let us now consider the importance of this, both for the individual Christian and for the Church.

"As my Father hath sent me, even so send I you." That is our Lord's commission to His disciples. "Ye are the body of Christ, and members in particular." That is how St. Paul addresses Christians. Think what these words mean. They mean that to be a full-grown, adult Christian is to be one through whom the love and power of God are flowing out into the world around for the overcoming and casting out of all that is ugly and base and evil, for the discovery, encouragement and building up of all that is true and good. But we men are not by nature fit for this work. Our cowardice, our selfishness, our lusts and ambitions all get in the way. We need ourselves to be cleansed from our sins before we can be used as channels for His power and love. We have to be born again and become penitent and forgiven selves before we can give ourselves to our Lord for Him to use.

But it is for that use that He calls and cleanses us, and here, too, there is danger of fixation, of arrested development, of resting in the thought that our calling and cleansing is simply for our own perfecting and eternal bliss. This danger is all the greater because of the widespread and deeply rooted notion in men's minds, a notion too often encouraged by misguided evangelistic preaching, that the purpose of religion is to secure one's own eternal bliss. But in the healthy religious life, when it comes to maturity, there will be that alternating rhythm between the pursuit of holiness and the forgetfulness of self in the service of God. Indeed, to redress the balance of misguided notions it would not be too much to say that the Christian life is at its best and healthiest when a man thinks of his own salvation as a by-product of his forgetting himself in his devotion of himself to his calling. "He that loseth his life shall save it." When I was a child in the faith it was right that I should speak and feel and think as a child and be so concerned about my own soul; now that I am become a man in Christ I must put away childish things.

So I come to the real point and purpose of this sermon, which is to speak about the Church. During this last summer I have spent a good deal of time as a representative of our Church of England at meetings of the Faith and Order Commission of the World Council of Churches, and in conversations with representatives of the Free Churches in England. From all this there has been borne in upon me the conviction that what is true of individual men and of states in creation is also true both of individual Christians and of the Church in redemption, and that we shall not make progress towards restoring the broken unity of Christendom until the Church, too, becoming a full-grown man, puts away its childish things.

When the Christian Church began its work, its first task was to make men Christians. Our Lord Himself had had to begin His ministry with the cry "Repent ye"—it was only to penitent and cleansed disciples that He could say, "As my Father hath sent me, even so send I you." So the risen Lord set His Church to work with the command, "Go ye into all the world, make disciples, baptize."

The Church went out into a world where Jew and gentile alike thought that the whole purpose of religion was to secure to the devotee salvation and eternal bliss. It naturally followed that they thought of the Church as existing to provide this security; to be baptized into the Church was to be rescued from among the perishing multitudes into the blessed company of the saved; the Church fulfilled its function in being the ark within which believers could sail safely through the stormy waters of this troubled world with a guaranteed safe passage to the eternal heavenly shores.

So deeply ingrained was this notion of religion in the human mind that two thousand years have not been enough for the Spirit of Christ to eradicate it from our thought. Here note the similarity of God's method in creation and redemption. The individual man must grow through self-gaining to self-giving in the course of a single lifetime; it takes longer for states to grow through the assertion of independent national sovereignty to effective membership in a world-wide community of nations. So, too, in one lifetime the Christian must learn that, being saved, he is saved to serve. The Church has needed a longer adolescence to realize its selfhood as the fellowship of forgiven sinners; if it now knows itself to be the redeemed community it is time for it to realize that it is redeemed in order that it may give itself to the service of mankind in the name of God.

Do we realise this? Do we think of the Church—of our baptism, our confirmation, our Bible reading, our worship, our communions, as all existing in order that through us God may be at work in His world, working for its redemption and perfection? If I were the vicar of a parish instead of a professor in a university, my congregation and I would think of ourselves as bound together to care in God's name for the welfare of the community in which we were set—interested not in ourselves, but in sharing God's interests in the world around? Or should we fall into the temptation of thinking of the world as existing for the Church, instead of the Church for the world, and only being interested in our neighbors in so far as they could be used for building up the life of the Church? When the priest allows himself to slip into that attitude, or when priest or layman allows himself to think that the Church has fulfilled its function if it has ministered to the salvation of his own soul, in such individuals the Church is suffering from arrested development, from failure to grow up.

So long as we continue to think of the separated churches in our divided Christendom as rival purveyors of guaranteed communion with God and safe passage to a blessed eternity, so long shall we seek in vain for light upon the path to unity. But suppose we think of the Church as the body through which our risen Lord is seeking to rescue His world from all the evil with which it is infected, in which artists may be cleansed and inspired for the service of beauty, scholars for the pursuit of truth, politicians and industrialists, merchants and laborers, doctors and lawyers, for the better ordering of our common life—how different then the prospect becomes! We may be—we still are—far from the goal of one worldwide united Christian Church. Only those inside the movements working towards that end know how apparently irreconcilable are the sincerely held convictions that need to be reconciled. But the whole prospect changes when we view those from whom we differ and are divided, not as rivals in a common market, but as allies in a common cause. But this can only be if Christian people learn the lesson that God is seeking to teach us in this twentieth century of ours, that for nations as well as individuals in the order of creation, for the Church as well as for its members in the order of redemption, the self-gaining which is the task of childhood and adolescence must bear fruit in the self-giving which is the glory of maturity.

VISION

The Tragedy of Gehazi

Reverend Clayton E. Williams, D.D.
Minister, The American Church in Paris, France, and a Minister of the Presbyterian Church

The full French flavor of this sermon, preached in the American Church in Paris, makes it fruitful for men who want to know what is being preached by Americans in foreign lands today, for Clayton Williams is an American at home in Paris. Since 1926 he has been helping Americans to find themselves in Paris and has had a brilliant part in keeping faith alive for diplomats and G.I.'s, for artists and teachers, for visitors, students and members of his regular congregation.

Born in Peoria, Illinois, he is the son of a Presbyterian minister and studied at Butler College and the University of Pittsburgh. He went into war service in France as a Y.M.C.A. secretary in 1917–18, and became an officer in the air service of the United States in 1918–19. He studied at the University of Paris in 1919, did social work at Château-Thierry in 1921, and later in that year came to the United States as assistant at the First Presbyterian Church in Indianapolis. From 1921 until 1925 he attended Western Theological Seminary and had the highest standing in his class. During 1925–26 he was assistant pastor at the First Presbyterian Church in Poughkeepsie, New York.

Upon the termination of that year of service he was asked to go to Paris to join Dr. Cochran at the famous American Church as the assistant minister in charge of religious education and young people's work. This position he held until 1933, when Dr. Cochran resigned and the full charge of the church went to Dr. Williams. The American Church in Paris was founded by Americans and is largely supported by Americans in the United States and in Paris. It is the oldest American church outside of the United States.

When World War II came and the evacuation of Paris by most Americans took place, the church committee urged Dr. Williams to take his wife and children to America, so that they would be out of the danger zone. This he did, returning at once to France in the hope that he might continue the work in Paris, if it proved possible. However, it did not. In France and in the south of Spain and Portugal, he assisted in many ways with relief and rescue work.

He finally had to return to the United States, and while here, he served as pastor of the Seventh Presbyterian Church, Cincinnati, Ohio, from December, 1941, to May, 1945. In June of 1945 it was possible for Dr. Williams to secure transportation to France. He immediately resumed his pastorate in Paris. The church was the one American church in the war zone which was able to continue to hold services in English all through the conflict. In 1937 the French government made him a Knight of the Legion of Honor for his work for Franco-American relations, and in 1948 he was made an officer of the Legion.

Clayton Williams makes faith a glowing and glorious matter, shows the futility of pessimism and cynicism. His use of history, his knowledge of man and the application of his message to the problems of daily living add a touch of genius to his preaching.

Using the story of Elisha and Gehazi, Dr. Williams points man's faith upward to powers higher than his own in the daily struggle between materialism and vision and faith. This sermon represents Clayton Williams' preaching in Paris, where a minister must preach "up" to his congregation, never "down" as some men in smaller places in the United States feel is in order.

Sermon Forty-eight

SOMETIMES an interesting insight is given into a situation by the accidental grouping of a few phrases in such a way that their sequence suggests the true significance of the occasion. And though the narrator probably never intended it, this passage from the story of Gehazi has such a sequence and it is very suggestive.

"All is well: . . . Give me, I pray Thee, a talent of silver and two changes of garments" (II Kings 5:22).

These words are the key to the tragedy of Gehazi.

Let me remind you of the occasion on which they were uttered, and something of the man who said them. They are the words of Gehazi, the servant of Elisha the prophet of Israel. In many ways Gehazi had not been a bad servant. He had risen to a place as the intimate assistant to the prophet and throughout the stories of Elisha's activities, he bears an almost unbroken record for faithfulness and obedience.

However, this passage tells the story of his downfall and reveals his fatal weakness.

As the result of a suggestion made by his little Hebrew maid, Naaman, the Syrian prince had come to the prophet to be healed of the disease which harassed him. Despite his reluctance to fulfill the prophet's conditions, his mission had been successful and his gratitude and joy over his recovery, both in body and in spirit, prompted him to offer his benefactor rich gifts of silver and festal garments, all of which were immediately refused.

Stern and uncompromising Elisha did not want it thought that he sold his powers for personal gain, as the sorcerers did their magic charms. Perhaps he overdid the matter and was rude about it, failing to crown a generous deed with a gracious acceptance but, be that as it may, Gehazi thought it all a mistake not to profit by the opportunity. After all, Naaman was a foreign prince from a wealthy state, and he could easily have given them enough to leave them comfortable for the rest of their lives, and he personally considered it very poor judgment on the part of his master not to have taken advantage of this opportune occasion.

And the longer he thought on it, the more he deplored it. At last he set out to take things into his own hands, and running after Naaman, he overtook his chariots.

On seeing him, Naaman was concerned lest he might be the bearer of evil tidings. Gehazi, however, assured him that all was well, but that two impoverished theological students had arrived unexpectedly, and Elisha had sent him for a talent of silver and two changes of raiment. Naaman was delighted to show his gratitude and pressed upon him double that amount, much to Gehazi's satisfaction.

Elisha, in the meanwhile, missing his servant, had guessed the reason for his absence, so he sent word that upon his return Gehazi should appear before him. And there he stood, condemned for his covetousness with a condemnation that left a mark upon him forever after, for he had contracted the leprosy of Naaman. But the leprosy of his body was of little import; he had already fallen a victim to leprosy of the soul.

He had been a good servant, far better than most. He had had a rare experience, living daily in the presence of a great prophet, sharing in his work and witnessing his power, but that did not save him from a great failure! Indeed, it helped in his case to bring it about, for the secret of his trouble lay in the fact that Gehazi was ruined by "familiarity with sacred things."

Day after day he had gone out with the prophet and had seen his power flow out into life, changing it, restoring it, renewing it. I do not doubt that in the beginning he had been much impressed by the wonders that he saw and that the things he felt had moved his heart. No doubt he had followed Elisha with genuine admiration and had attached himself to him from the purest of motives. His very obedience was evidence of his loyalty. But little by little—and here lies the tragedy—the unusual became commonplace, the wonderful became ordinary and the lustre that creative contact with God can give to a life was dulled. He lost a sense of their supreme significance. Life had been drained of its wonder by repetition and familiarity.

He had witnessed so many wonderful things, he had lived so long and in time, so carelessly, in their presence, that what should have always been sacred experiences became just ordinary occurrences. And when fine things are reduced to the ordinary, when the vision dims, and the burning bush, at first a miracle before which we pause and kneel and take off our shoes, becomes at length only a curious phenomenon and at last a commonplace, that means spiritual disaster.

There is a very subtle danger here. It lies in the fact that those things which should always be sacred, tend to become commonplace. It ought to be the other way around: the commonplace should become sacred. Like Brother Lawrence, we should worship God in the common task and find a divine glory in everyday living. But that takes the inbreathing of a spirit and a presence. And, if that is lacking, even divine things will turn cold and lifeless with continued repetition.

Take our worship, for example. We come here Sunday after Sunday for prayer and praise and worship—and that is good. But those things can become mere formalities, worn dull by constant repetition, and our worship can fail miserably, unless we meet God here. And that won't be because of fine music or a good sermon, but because of the eagerness of our desire and the depth of our loyalty and our sincerity and willingness to wait upon God.

Don't misunderstand me. I'm not suggesting that we should not come to worship if we do not feel like it. Even religion grown formal and mechanical can have a disciplinary function in life and there is always the chance that we may catch a new vision.

Sometimes we hear people say that they don't go to church because they had too much of religion in their youth. They are "fed up!" No, they are not "fed up" with too much religion, they are "fed up" with too much parental authority. A vital experience of Christ doesn't leave a person "fed up."

Like Gehazi, one can live so continuously and so carelessly in the presence of fine things that they lose their significance. But mark you, it isn't because we live so continuously in the presence of religion that it means nothing. It is because we live so continuously and so carelessly in its presence. So carelessly. That was the trouble with Gehazi. He had lost his sense of faith's reality in his own life.

God's power demonstrated in another's life meant little to him because he himself did not feel dependent upon God's power for his own life. Religion had become a formal affair of fulfilling certain duties which life imposed upon him. And since his own life knew no deep need of the miracle of grace, he had, in consequence, no real appreciation of Naaman's need and, therefore, no realization of what had been done for him.

If we do not feel the urgency of our fellows' need of God, isn't it often because we, ourselves, no longer feel God is indispensable, no longer know the radiance of His presence, are no longer sustained by His love?

Let me speak frankly for a few minutes. We no longer sing songs like "Rescue the Perishing." Now granted that our forefathers made a travesty of damnation and hell, and almost as much of a burlesque of salvation, until it has become an outmoded term, nevertheless, the fact remains that we are not greatly moved by mankind's desperate need of God. We think man needs a great many things: the solution of his economic and his international and his personal problems, but I'm sure we're not concerned about his need of God, or we would do something about it.

But isn't it the plain implication that if a man isn't overconcerned with making it possible for others to know God, it's because God doesn't mean very much to him? I say, isn't it?

Gehazi didn't count spiritual health worth very much because it didn't mean much to him. He was much more interested in how much Naaman appreciated his restoration to physical health. All he saw was a man healed of a disagreeable disease, for which he thought he ought to compensate the one who had cured him. He saw the process and its results, but not the hand behind it; the miraculous was not a window through which to catch a glimpse of something wonderful beyond—an earnest of the power or willingness of God to break through the limitations that life seemed to impose—the source of a new hope, a new faith, a new vision. It was something to be exploited.

You see, Gehazi had lost the capacity to wonder, to marvel, the most precious power we have, for there is a time in everyone's life when the world is full of wonder, and to be a child is to believe in love, to believe in loveliness and to believe in belief, to see a miracle in every common process; but as time goes on we become accustomed to miracles and take them for granted, and wonder is supplanted by sophistication, and we awake one day to realize that, as Wordsworth said, "There has passed away a glory from the earth!"

Our smattering of science, which is all the best of us have, convinces us that everything can be explained. We are tempted to think, since we see how God works, that we can dispense with religion. We have seen behind the scenes, like the country yokel who went to the theater for the first time in his life and was wonderfully impressed by the storm that was staged during the play, but whose wonder suffered a collapse when they took him backstage and he saw that the thunder was made by rolling heavy balls about in a box, the rain by quivering of great sheets of metal, and the wind by whirling fans, and he remarked: "Well, I thought it was something marvelous, but I see it wasn't anything after all!"

I fear that is the way we often feel about God's wonders. Like the small boy who took apart the watch to find what made it run, we have taken our world apart. We have found the wheels that make the ticks and discovered what makes the hands move. But for all that, we have missed the Mind that produced it and the Spirit that sustains it. We live in a day of miracles, miracles that should touch our spirits with awe and wonder at the wealth of our universe. Today, we switch on the electric light or tune in on television or fly through the air with scarcely a moment's thought of the wonder of it all, because these things have been so quickly turned into accessories to our comfort or made adornments for life. We fail to see the miracle.

Miracles are doorways through which to catch faith-creating glimpses of another world and if they can break our earthborn bonds and give birth to a spirit of divine expectancy, they can renew our courage and inspire our hearts.

But too often, like Gehazi, we have thought of them merely as a means to serve our temporal ends. Gehazi had ceased to wonder at the miracle and had turned his attention to the contribution it could make to his physical comfort and pleasure. His interest in the miraculous had shifted from the point where it brought a vision of God's presence and power to the point where it represented a way to get something he wanted. He thought of God's wonders as something to be exploited for personal gain and advantage.

As Dr. Kelman once said: "He had just seen the most wonderful thing in all the world—the dawn of faith in a human soul—and the only impression which was left with him is the fancied vision of himself clad in coloured silk."

Now, Jesus was keenly aware of that tragic pragmatic tendency of the human spirit. He was constantly having his own power misconstrued and misappropriated. You remember how, when the crowd sought him on the lakeside where he had recently fed the multitude with the loaves and fishes, he frankly rebuked them saying: "Verily, verily I say unto you, ye seek me not because ye saw the miracles, but because ye did eat of the loaves and were filled." You didn't see the miracle. It meant nothing. You were only interested in what you could get out of it. That, my friends, is the spirit of opportunism that prostitutes the spiritual. It penetrates even into religion.

There are some people who look upon faith as the touchstone to their heart's desire. "Faith works miracles," they say to themselves. "If only I can have faith enough, I'll have no troubles." They think of religion as a way to get what they want in life, and so we have any number of faith cults.

There are Christians who are Christians because they think it pays to be Christians. They go to church because it puts them with the right group and gives them the best contacts, and there are Christians who look upon miracles as mere proof points to confirm some doctrine or corroborate some statement. When a man begins to use the best things in life for what he can get out of them, they begin to lose their spiritual significance, and he begins to lose his soul.

"But surely," says some very modern person, "in the face of modern science and universal law, you don't believe in miracles." "Indeed, I do—more than ever." After all, what is a miracle? Is it some strange phenomenon that violates all our expectations, without rhyme or reason, injected into life just to prove that there is a God or that He is on one side or the other? Or is it something that the creator and sustainer of all that is good and beautiful and wonderful, who holds the forces of the universe in His hands, uncovers in life to meet our deepest heart's desire, a gracious act of God that sets aglow the fires of expectancy and faith and renews our confidence in His presence and grace in the heart of life itself?

The secret of the miraculous does not lie in having something so unusual happen that it bludgeons our poor dull senses into seeing it. The secret of the miraculous lies in the delicate sensitiveness of a soul so responsive as to see God's hand at every turning and to stop and marvel at the grace of it.

In a wonderful world like ours—that is to say in a world full of wonder—everything is a miracle revealing the creator's grace; from the song of the lark to the sheen of a butterfly's wing, from the tiny handclasp of a trusting newborn babe to the heights of heroism and sacrifice to which men rise in times like these. If it's miracles you want, the world is full of them.

Our lives are hemmed in by God's mercies, and lightened by His miracles of grace and beauty. He touches our lives at a thousand points, and there are a thousand things to move and thrill our spirits if only we will pierce the veil of casual indifference that hides them from our eyes.

There is the "Sacrament of Spring," as Dr. Gossip calls it. There is the song of the wood thrush, and there is a Beethoven sonata. Those are miracles of God's grace, both of them. There is the glory of a gorgeous sunset and the misty gray beauty of dusk along the river, and there are the marvelous wonders from which science has torn the shroud of ignorance. Those are miracles.

And then there are the miracles of the spirit. The wondrous freshness of a child's unfolding life, the peace and contentment written on an old man's face, the redemption of some derelict spirit long given up as hopeless, the birth of faith in a man's logic-bound heart, the wholehearted dedication of those who go out and give themselves in service for the common good, and the epic of heroism and sacrifice written in your neighbor's life. Miracles all. Only most of us haven't the eyes to see them or the heart to wonder at them. The pursuit of the commonplace obscures them—two changes of raiment and a talent of silver. Mind you, it isn't the commonplace itself. We all live in the commonplace, and there must be changes of raiment and bank-notes to purchase them.

It isn't that we live in a humdrum world, but that we live in it and seek no more than that in it, and so we see no more in it, missing the best for which life exists. Seeing miracles in life is seeing God in life, and missing them is missing Him. And everything is an opportunity. A cake of soap can be a miracle, or a spool of thread or a needle. When I returned from France in the summer of 1941, I used to tell about the things I had seen in Western Europe, how people were living without soap and towels and paper, many of them, and where common cotton thread was worth its weight in gold. And once when I had finished, a man came up to me and said: "I have heard you speak three times and each time I have been struck by your account of the shortage of soap in France. And do you know, I never hold a cake of soap now between my hands but I give thanks for it." That's what makes a miracle: the sensitive soul and the grateful heart.

Take that matter of saying grace at table. It has almost disappeared or lingers only as a perfunctory gesture. I wonder if that doesn't indicate that, after all, we think of our bread as something we alone have produced rather than a token of God's gracious mercy. There is a miracle in every loaf of bread, a miracle of growing life and sunshine and rain and all the hidden forces that make it possible—all conceived by God's love and sustained by His power.

There is something significant in the fact that it was in the breaking of bread that Christ was made known to the disciples at Emmaus, something in the way in which He accepted God's mercies that made them know that it was He.

We have lost something from our life with all our cleverness and all our conveniences and comfort, something very precious: seeing the hand of God and marvelling. Dr. Gossip points this out. "Brother Lawrence," he says, "was a raw lout of a soldier lad. But one day, limping, foot-sore, among his fellows, long past whistling, stolidly holding out, he saw beside the road the first tree bursting into leaf and all at once his heart stood still. 'If God can bring that wonder of fresh life out of a thing so dead, what can He not do to my soul,' he thought, and there on the dusty road it all began. Linnaeus saw a bank of broom trumpeting its riot of glory to the heavens and sank upon his knees, because God in His majesty was passing very near."

The tragedy of our civilization lies in our not seeing the world with all its beauty, its glory, its miraculous possibilities, its abundant opportunities as the gift of God, a revelation of God's love, a miracle of grace, but seeing it only as something to exploit, to use for our own selfish purposes. That's the tragedy—to face the world in the moment of revelation, when God's love manifest in Christ lingers by us and to say, "All is well, give me I pray Thee, two changes of raiment and a talent of silver." "All is well: all I need is pleasure, adornment and security." That is the tragedy. Gehazi thought that was what he needed when what he really needed was a miracle in his own life. He needed a miracle of grace to open his eyes, to make him see and appreciate God's love and power in His world. And that is what we all need, a new vision of God's love and power to redeem life, a new experience of God's grace.

The mystery of the cross has become an old story. We have heard it times without end. It does not mean much to us now but it ought to hold us bound by loyalty to Christ, it ought to keep us ever marvelling at his love, it ought to make love overflow from our lives. Only a realization of the grace of God manifest in Christ can save us Christians from the tragedy of Gehazi.

Stars Above the Freight Trains

Bishop Donald Harvey Tippett, D.D., LL.D., L.H.D., Litt.D.
Resident Bishop of the San Francisco Area, The Methodist Church, San Francisco, California

All his life Bishop Tippett has been a courageous fighter and leader. When he gets an idea he holds to it as firmly as any man can. If the community needs cleaning up or the children need a new playground or the Church needs renewed spiritual life, Donald Tippett starts things in motion to accomplish the required result.

He was born in Central City, Colorado, in 1896, graduated from the University of Colorado, took his B.D. at Iliff Graduate School of Theology, followed by graduate work at New York University. His honorary degrees include the D.D. from the University of Colorado, an L.H.D. from the University of Southern California, LL.D. from the College of Surgeons and Physicians, and the Litt.D. from Samuel Houston College.

In 1919 he entered the Methodist ministry and served at Longmont, Colorado; Johnstown, Colorado; then to Christ Church, Denver, 1922–25; Gunnison, Colorado, 1926–27; the Church of All Nations, New York, 1928–31, Bexley Church, Columbus, Ohio, 1931–40, and First Methodist

Church, Los Angeles, California, 1940–48. In 1948 he was elected a Bishop of the Methodist Church and has been resident Bishop of San Francisco ever since.

He had his hand at teaching at the University of Colorado, Western State College of Colorado, Teachers College, Columbia University, and was lecturer in homiletics at the University of Southern California. He is the author of Voices of Yesterday Speak to Today, The Desires of a Godly Man, The Desires of a Religious Man, *and chapters and articles in various publications.*

"Stars Above the Freight Trains" illustrates the Bishop's characteristic method of applying a strong Bible text to modern conditions in an attempt to help right contemporary wrongs through spiritual powers.

Sermon Forty-nine

> TEXT: Woe unto them that go down unto Egypt for help, and rely on horses, and trust in chariots because they are many, and in horsemen because they are very strong, but they look not unto the Holy One of Israel, neither seek Jehovah. ISAIAH 31:1

ERNEST POOLE in his novel, *The Harbor,* describes a lad in Brooklyn Heights whose habit it was to wander out to a secluded spot which gave him a glorious view of the harbor. Almost at his feet, however, were nests of railroad tracks which carried hundreds of freight trains laden with precious cargoes from all over the world. But these trains frequently hid his view of the harbor. Then one night he discovered that by lifting his eyes ever so slightly he "could see the stars above the freight trains."

We are so situated in history and by geography that it is not at all difficult for us to see the freight trains. That we see the stars above the trains is not so certain. The freight trains burdened with the material things of life are always before our eyes; so near indeed, that they tend to obscure our vision of the stars—and beyond. The failure of our generation to see the stars has given secularism a terrible power over our lives and over our nation.

The age in which Isaiah lived was not unlike our own. He saw his friends obsessed by things and his government dominated by materialism. The moral failures of selfish men had brought his nation near to destruction. His contemporaries were putting their trust in objects. They behaved as if there were no God, no spiritual resources and no moral law. His age, like ours, had failed to see the stars above the freight trains.

Isaiah—the greatest prophet of his age—saw, as no one else in his day did, that the exigencies of his time required not more power and not more things, but a greater emphasis on spiritual values. And he cried out against the materialism which he saw engulfing his nation: "Woe unto them that go down unto Egypt for help, and rely on horses, and trust in chariots because they are many, and in horsemen because they are very strong, but they look not unto the Holy One of Israel, neither seek Jehovah."

His warning to his people is filled with meaning for our time. The chariots and the horses are obsolete, but the underlying sin is still the same. Secularism is still secularism whether it hides in the cargo of a transcontinental train or in that of a camel caravan. We, like Isaiah's generation, have lost ourselves in the meshwork of things. Assuming that the freight must be moved, we have given ourselves over with self-abandon to the task of perfecting the trains to move it. Of course we have to run the trains—or else we starve. But when the freight demands so much of our attention that we never see the stars—then—we have become secularist.

When spiritual values are no longer important to a nation the misuse of power by its rulers becomes a terrible danger. Isaiah saw this when tragedy struck a man whom he greatly admired. Uzziah the King had been a good ruler and worthy of Isaiah's admiration. Successful in war and peace, he had brought great prosperity to his nation. But a revealing vignette is preserved for us in "The Chronicles of the Kings of Judah" where we read: "God helped him and made him to prosper and his name was spread far abroad, and he was marvelously helped till he was strong. But when he was strong his heart was lifted up to his destruction; and he trespassed against Jehovah: for he went into the temple of Jehovah to burn incense upon the altar of incense. . . . Then the Lord smote the King so that he was a leper unto the day of his death."

I used to think that God was a little petty in this episode. To send leprosy upon a man for so trivial a thing as burning incense on the altar! But I have lived long enough now to see other rulers arrogate unto themselves prerogatives which did not belong to them, and the terrible tragedy that resulted from such behavior.

I saw, for example, Adolf Hitler and his cohorts usurp one power after another—powers that were rightfully vested in others—until a totalitarian state emerged, diabolical and cruel. Fiendishly a handful of men under Hitler sought to dominate the entire life of all the citizens of Germany, and eventually the lives of all men and women everywhere in the world. In Nazi Germany we saw the fruits of secularism and its attendant evils. But make no mistake, such secularism is not confined to Nationalist Socialist Germany. There are abundant evidences of it in Soviet Russia too. But why go so far away in search of it? Our own land has enough of it to keep us busy for our lifetime.

Our current variety of secularism started with Niccolò Machiavelli a little more than four hundred years ago when he dared to propound, for the benefit of Cesare Borgia, the principle or theory that the state is beyond morality. The state, according to Machiavelli, is a law unto itself, is responsible unto no one but itself. The state which so considers itself feels free to assume control of the whole life of a people. It does not hesitate to seize power regardless of means. Amenable to no law, since it is a law unto itself, it employs force, deceit, slander and even murder to achieve its goals. Usurping civil control it arrogates unto itself the prerogatives of the Church; indeed, it dismisses God Himself and then takes over complete control on the assumption that there is no God. Many leaders in a totalitarian state would not bluntly assert that there is no God; yet, in practice, following the Machiavellian dictum that the state is responsible to no one but itself, they behave as if the state were an end in itself and as though God did not exist. Indeed, in the alleged or assumed absence of God the state becomes a god!

In the totalitarian secularist state the government assumes the right to tell the universities what to teach, the scientific researchers what to find and what not to find, and the preachers what to preach and what not to preach. If Machiavelli is correct and there is no law higher than its own to which the state is accountable, then it may be just as well that the pulpit preaches what the state dictates. If, however, the state is answerable to the laws of God, then the pulpits had better be kept free to make articulate the voice of God and the voice of conscience. The modern state is in desperate need of a pulpit that can serve as its true conscience. No pulpit or church can be an accurate conscience if it is compelled to mouth over prefabricated platitudes authored by the very men who need most to hear some voice other than their own speaking with an authority that emanates from a power greater than their own. For the health of a nation nothing can be more important than to have a conscience untrammeled and uncontrolled, which freely, courageously and honestly makes known to a nation and to a people the errors of their ways and their sins. The Church, unfettered by the state, can be of inestimable service by being a barometer, registering faithfully the moral and spiritual climate of a nation. The state which usurps the prerogatives of the Church deprives itself of an unbiased diagnosis of ills and, not knowing precisely what ails it, a nation cannot expect to find its way back to health. Maybe God was not so petty, after all, in his dealings with a king who took it upon himself to assume the office of a priest.

The health of our nation is dependent, more than many people realize, upon a free pulpit speaking to a free people encouraging them in their well-doing, facing them with their sins, and warning them of impending evil. There are those who would silence the voice of the pulpit in these days when it too pointedly warns of secularism taking over. Frequently a pulpit voice has been accused of being unpatriotic or downright disloyal because it dared to sound a warning of imminent danger. Far more patriotic is that courageous voice from the pulpit, which sounds such a warning and quickens the conscience, than all the strident voices of those who insist that all is well, when in reality we are merrily on our way to hell.

Thank God, warnings are being sounded all about us. The Kefauver exposé, revealing corruption and graft in high places as well as low, was a testimony, too eloquent to be pleasing, to the fact that secularism is on the march. The recent investigations into the narcotics racket have left us with a sick feeling. That men, for money, would prey upon children and young people in such a despicable traffic is hard to believe. Such men are beneath contempt, but their foul practices are but further evidences of the growing menace of materialism.

There are other danger signals we should heed. Un-American Activities Committees can in themselves be a threat to democracy unless directed with great care. Such committees under Congressman Dies in Washington or Senator Jack Tenney in California have gotten so badly out of hand as to give real concern to all who love the liberty and justice we speak of when we pledge allegiance to the Flag and to the Republic for which it stands. The intended purpose of these committees was to protect our liberties from all subversive agencies and agents. But what happened in practice was something else. Too often these committees violated the very American way of life they were claiming to protect. Most of the people investigated were never given an opportunity to explain; often they were denied legal counsel; never were they permitted cross-examination. Proof rested almost entirely upon the associations of the accused. Anyone who joined in a protest against the Dies Committee or the Tenney Committee was branded a Communist. Hiding behind their legislator's immunity some committee members have blackened reputations and been guilty of character assassinations. Star Chamber proceedings like these were as bad and as Un-American as the Gestapo, the Ogpu, and the secret police of any police state, in principle. That such committees could do as much damage and operate as long as they did before they were repudiated is but another warning that such things can happen here.

Our forefathers were wise in establishing three divisions of government, each to serve as a check upon the other two. These divisions were (1) the legislative or law-making body; (2) the administrative or executive whose duty it is to enforce the laws; (3) the judicial whose function is to interpret the law. This division of governmental responsibilities was a wise one. I am convinced that it is a serious mistake for any of these branches of government to infringe upon the prerogatives and responsibilities of any of the other two. Let the Uzziahs of today take warning: let them never assume the powers belonging to another.

Just as Uzziah had no business doing the priest's job, so the law-makers commit a sin against the Constitution when they arrogate unto themselves the role of judge, jury, prosecuting attorney and policeman. Law-enforcers must not, on the other hand, preempt the work of the legislators. No traffic officer has the right to make the law. He can only enforce it. And law-interpreters do the state a great disservice if they usurp the duties of either the legislators or the law-enforcement officers. So long as we keep these branches of government separate and so long as each serves as an honest check against the other two, we shall not be in too great danger from the inroads of totalitarianism. Uzziah's offense was not a petty one.

Uzziah's state had been a thoroughly secularized one. It is true the priests were going through the motions of religion but Uzziah demonstrated how little he depended on them. Even the altar was no longer sacred to him. And Isaiah, admiring his king, and taking part in the life of his nation had become as secularized as his contemporaries. And it was not until the hour of the leper king's death that Isaiah's eyes were opened. For in that tragic moment he saw the Lord. He came to realize that he had been so occupied with matter that he had failed to get acquainted with spirit.

Isaiah had been a victim of the secularism of his day but in the darkest hour he saw God. Having seen both ways of life he had no doubts about the one to choose and to recommend to his native land. Above the dust of horse and chariot he saw the stars. He became an uncompromising foe of secularism. And even now after twenty-seven hundred years, Isaiah's battle has not been won. Although the locale is different today, Isaiah, who raised his voice for the supremacy of the spirit, has a message for our secularized age which rings with a strange modernity and urgency.

"'Woe unto them that go down to Egypt for help and rely on horses, and trust in chariots because they are many, and in horsemen because they are strong, but look not unto the Holy One of Israel, neither seek Jehovah."

VISION

Three Doors and a Dream

Reverend Rolland David Snuffer
One of the Ministers of Montview Boulevard Presbyterian Church, Denver, Colorado

This unusual sermon created a sensation when first preached at Montview Church, Denver. Possessing vision, faith, courage, the sermon was written all in one sitting, one morning when the lamp of inspiration was burning. It is given here as an example of what can be done on occasion when the spirit and the mind are one in Christ.

Mr. Snuffer's life is mixed with the military and the radio. He grew up in Kansas City, Missouri, in Westport Avenue Presbyterian Church, attended Missouri Valley College, McCormick Seminary, Chicago; did post-graduate work in drama and speech at Northwestern University and in radio at the University of Denver.

In college, being an ex-service "man" from World War I he joined the National Guard, then went into the reserve chaplaincy. Always fascinated by drama, in Wichita he worked with his first professional radio writer, collaborating on a thirteen-week, half-hour dramatic "show" called "Religion in Action." He was approached on the problem of broadcasting religious services. After working out a procedure and schedule, he suggested a dramatic program on the work of the chaplain, which continued once a week for three years, "The Diary of a Chaplain's Clerk," which he wrote, produced, acted in, directed, and ran the sound track.

He was pastor of Presbyterian Churches in Mt. Vernon, Iowa; Iola, Kansas; Grace Presbyterian, Wichita, Kansas; and is now Associate Pastor with Dr. Arthur Miller of the Montview Presbyterian Church, Denver. He has frequent opportunity to use his radio experience for the Church, and the Council of Churches. He preaches frequently, still keeps in touch with the military as chaplain for the Colorado Wing of the Civil Air Patrol, in which capacity he accompanied a group of Civil Air Patrol Cadets to Switzerland last August.

Montview Church is large, it has 3,415 members. His favorite ministers are Dr. Arthur L. Miller and Dr. Orlo Choguill of First Presbyterian Church, Topeka, Kansas.

Sermon Fifty

Scripture—MATTHEW 17:1–8

THERE IS a fascination in—Doors.
They stir curiosity and imagination.
Whether we stand within or without, we wonder what lies beyond—a door.

There is a significance to Doors.
For whom are they opened? To whom are they closed?
Who is welcome? Who is refused admittance?
Who comes out, and how, and for what purpose?
When and how and for whom and why
A Door is opened, or remains fast barred and closed,
Determines the destiny of him who is charged with the keeping of the Door.

There is a strange fascination
There is a peculiar significance,
There is an inescapable destiny,
in—Doors.

This morning—we'll think—of *Three* Doors—and a Dream.
Or perhaps it was not a Dream, perhaps it really happened.
Or better still, perhaps by the grace of God, 'twill happen yet.
As I speak, you may have the uneasy feeling
that I am speaking—not of myself, but of you.
You may be right.
For I shall recite the story of something which has happened to
countless thousands. It may well have happened—to you.
If it has not, I will join you in a prayer that some day,
perhaps even this day,
it shall.
This strange dream of three doors.

I will pray that you may have this same experience
I relate to you now.
An experience in which fantasy and reality
are so closely interwoven,
are so perfect in pattern of similarity and contrast,
that I still cannot tell for sure
which is a part of my living,
and which is a part of my high dreams.
I cannot tell—nor do I care.

I will pray that you may stand, as I stood,
behind that first fast barred door.
That you may hear, as I heard,
that firm insistent rap upon the panel of my door.
You will hesitate, if this experience should be yours,
You will hesitate as I hesitated.
Now that I know, I wonder why I did not rush to the door
And fling it open—with joyous welcome.
But I did not.
Even though I heard the voice saying,
"Behold I stand at the door and knock.
If you will open, I will come in. . . ."
Still I hesitated, and looked around.

Looked into the eyes of long-cherished companions
Who were in the room with me.
Looked to see what they would say.
There was Pride, with her tilted chin. Pride lifted her haughty head, still higher,
and shook it emphatically—"No!"

Cadaverous Envy was with me—Envy, with his close companions,
the twins, Covetousness and Gluttony.
Deep-eyed Lust looked at me, and Indolent Sloth.
And each, with an all knowing smile,
Shook his head—"No! Do not open the door."

Finally one of them, Anger,
stepped toward me threateningly.
"You know better than to open that door, you fool.
You know what will happen.
And besides—we promise you,
If you let Him in, we go.
We've meant much to you—we've MADE you. Now—
Admit Him at your own risk."

They had meant much to me.
They had satisfied desires,
They had brought happiness—
Though many times with heartaches afterwards.
They had given me satisfaction
and power,
and some riches,
and independence.
(At least I thought, independence.
Later I realized they were not my boon companions,
but my masters—and I was their slave.
But I did not know then.)
How could I let them go?

And I knew what Anger meant when he said,
"You fool, you know what will happen!"

I had long been fearful of that huge vat of scarlet vitriol
Which stood almost in the center of my room.
Once the door of my soul was opened,
I knew this bubbling, burning cauldron of sin,
to which I had been adding, each day of life,
would topple and flow out through the door,
out into the streets,
Searing and scarring helpless children, innocent friends.
And worst of all,
Twisting and distorting souls
of those whom I loved.
My children,
My family,
Friends who trusted and respected me
for all I knew I was not.
Should I open that door, many of them would be disfigured
beyond recognition.
Sick beyond the power of cure.

Yet—I was faced with a dilemma.
I knew the door must open—some time—
I knew all these lives must be burned and crippled—sometime.
What was I to do?

Others were with me, too, in this soul house of mine.
There was a Secret Sorrow I needed so much to share
With someone who would understand.
I knew that He who knocked—would understand.

There was a Great Longing,
A yearning which had been with me for years.
Somehow I had never quite had the strength,
the courage,
to move across the invisible line
which separates dreaming from acting,
which makes longing become reality.
He who was knocking now could give me that strength,
could give me that courage.
It was in the record. He had given it to others.

And there was a Craving for Fellowship.
A craving for wholesome, heartening, inspiring Fellowship.
So often these others had steered me on false paths,
Introduced me to unfaithful companions.
Yes—I needed Him—

With the knock—Others, whose presence I had almost forgotten
Made me aware that they were still with me.

Beyond that vile vat, three who one time had meant the most to me,
Stirred—and their eyes met mine.
Dear companions once, but now
Abused by the evil company I had admitted,
Stifled by the gaseous stench of the cauldron of sin,
On which the others had seemed to thrive,
These three were sick almost to death.

Faith looked at me—how precious He had been once.
Now Faith raised to his elbow on his cot.

And Love—ah, Love, once so beautiful—now weak, and emaciated.
Love sat on the edge of her bed, leaning forward eagerly.

Hope looked at me.
Hope had been the farthest gone of all.
Now he had risen to his feet
With a strength he had not shown
or known
for years.
And with a voice that stirred all the secret longings of my life,
Hope was saying:
"Open, can't you see!
It's our *chance*.
It's *your* chance.
Perhaps—your last.
"Open—in the name of all that's—Holy."

And then I did.
In spite of the angry cries,
The warning screams,
The violent protests
Of all the others.

I unbolted the door
And flung it wide!

O, why had I waited so long
He had been there before—many times.
Now I realized how much I had been denying myself
by refusing Him.

What happened?
I will tell you,
Perhaps you would like it to happen to you. It can.

When He stepped across my threshold,
He did not point to the parasites who had been feeding off my life.
He did not even scourge the money-changers
from the temple of my soul.
He simply—Smiled.
And the curtains darkening the windows of my soul
were pushed back.
And the windows of my soul were opened wide.
And the clean air
and the sunshine
swept through that room which so long
had been dark and foul.
Faith and Hope and Love stood resurrected and strong
My smiling friends, whom never again
would I allow to be overcome by evil.
Pride, Envy, Anger, Covetousness, Gluttony, and Lust and Sloth.
They were gone.
They could not abide His presence.
In their places came Prudence, and Justice,
Came Fortitude and Temperance.
Fit companions for those who had stayed with me,
For Faith, and Hope, and Love.

To my surprise, I was glad those former companions,
Whom I had considered almost indispensable, were gone.
I knew now that all the time they had been with me,
I had felt a sense of guilt.
Now I felt, clean—and cool.
I hoped they would never return.

But greatest miracle of all—
The stinking, nauseating cauldron of sin,
of which I had been so fearful,
Stood,
Not a boiling mass of vitriolic horror—
But a crystal clear vial of beauty,
From which the sweet perfume of rightness
Scented the air.
Causing those who passed

To turn their eyes toward my home,
and my life,
coveting my acquaintance with Him
who had entered and transformed.
They need not have envied me.
He would gladly—be their guest.
Too.

While I was thrilling to the sense of satisfaction
and contentment,
and the joy of having opened unto Him,
He spoke to me.
"I rejoice that you let me enter," He said,
"But I must remind you of another door
within your house which too long
you have neglected."

I was taken aback for a moment.
Surely opening the door of my heart to Him
should have been sufficient.
I went so far as to question Him.
"A door in my OWN house?" I asked.
"Yes," He replied.
"A door within your own house.
A door which has been here always.
Once you used it often, but now
It is almost forgotten.
Its lock corroded,
Its hinges rusted.
But it still can be used—Come, I will show you the way."

I followed Him through familiar halls,
We traveled to a remote section of my soul house
Into which, it was true,
Of recent years I had seldom gone.

As we came nearer I knew where we were going.
We turned into a room,
A quiet room,
Which once I had been careful to keep in order.
It was unkempt, disorderly, dusty now.
Several times in my mature life
I had visited the room.
Each time I had hurriedly straightened it,
and put it in order,
as I started to do now.
For the door at the end of the room could not be opened
As long as the room was in disarray.

"Do you remember that door?"
My companion was speaking to me.
"In your youth you used it often, almost daily."

"But I have used it since," I protested,
"It has always been one of the favorite doors in my house."
But my voice did not carry conviction.
I knew that what I said
Was not entirely true.

I *had* been to the door once in a while.
I had used it once when our little baby
Was ill almost to death.
What a comfort when it had swung open.
I had sought this room frantically when war's alarms had shrieked
A call to arms!
When I had been forced to part, possibly forever, from those I loved.
And fear had sent a devastating chill over my life.
I sought it, I remembered,
The time I had heard someone whisper
It might be but a matter of moments
Until I should slip from this world—into the next.
At this door I had found calmness—and consolation.
Again when dear ones had walked the valley of the shadow,
I had heard His voice come through the door—
"I am the resurrection and the life—Peace I leave with you—
Let not your heart be troubled, neither let it be afraid."

These were not the only times.
I came once to the door after a great sermon had shaken my soul.
And again when a splendid sunset made me keenly aware
Of the goodness and greatness of God.
Once when I had heard a mighty oratorio,
Once when I had seen a little child crying,
Once when I had heard children's voices singing
of a child in Bethlehem.
Once I came when I had seen a man, ragged and hungry himself,
Drop a coin into the hand
Of one more ragged—and more hungry.

"Do you know this door?"
The voice of my companion repeated the question
Bringing me from my reveries.
"Yes," I said, "I know it well,
It is the door to the Heart of God."

His voice went on with hardly a pause.
"Ask, and ye shall receive.
Seek and ye shall find.
Knock and it shall be opened unto you."

"I know—I know—"

I would have dropped to my knees in sheer humility,
But I felt His hand beneath my elbow—holding me erect.

I knew that I had not come here often enough.
I knew that each time I had come,
In dire distress—or translated by joy—
I had found the door more difficult to open.
I knew that should it open now, there would pour into my life
Such deep abiding joys as I could find nowhere else.
Joys I had sought in such foolish places.
I knew I could cause the door to open
and remain open always.
As long as this Companion was with me.
I knew He wanted me to kneel—And so I tried again—
I tried again—to kneel.

But no—His firm hand still kept me to my feet.
And His voice, at once insistent and stern,
Yet withal kind and understanding, said,
"No—not now—in a little while perhaps—but—
There is yet another door."

"I would open this door," I protested. "I have neglected it so long."

"You shall, perhaps—but not now—come—"

So I walked with Him again,
Not following now,
For His hand was upon my arm—He was guiding me.

We were out-of-doors—following footpaths deeply worn.
I had the feeling we were walking a way
Which countless thousands had walked.
We were going to a place where peoples of every nation under heaven
had gone, and were going.

We walked through the warm sunshine of summer,
And the bright glory of autumn;
We walked the brittle whiteness of winter
And into the fresh newness of spring.

His hand always upon my arm—guiding—guiding—
Through a lifetime,
And yet through the swift seconds
of but a single moment.
Then—we were there.
Ah, well I knew the place.
The door was of massive oak, set in stone.
Ponderous—almost foreboding,
yet protective—and promising. I knew it well.

In eagerness now I pulled myself loose from the guiding hand of
my companion.
I rushed forward, throwing myself against its oaken massiveness.

To my surprise it did not give.
I stepped back, and then again
I put my shoulder to its unyielding panels.
I was indignant!
This door should be always open!
I raised my fist and pounded!
I raised my voice and called!

Then His hand upon my shoulder reminded me that I was not alone.
And His voice said, "Yes, it should be open.
  But look. You are not on the outside—
  You are on the inside. . . ."

He was right. I was standing within the walls, inside the door.
I saw it clearly now—
And outside, where but a moment ago I had been,
I heard the shuffle of many feet,
The testing push of many hands.
Looking through the window beside the door,
I saw men, and women—
  A countless number it seemed,
  A countless number of eager faces—
    All turned toward the door.
  Many of them pushing through the crowd to get nearer the door.

I saw friends, close friends,
Who answered my greeting as I waved to them.
Answered—and then pointed toward the door.
I saw many familiar faces.
Folks to whom I had nodded, and smiled,
  with whom I had a passing acquaintance.
They, too, pointed impatiently toward the door.

I saw strangers—complete strangers—
Surging in the tide that eddied and swirled around
  that oaken portal.
Some looked at me with anxiety,
  Some with pleading in their eyes,
    Some with anger and indignation,
As if they had a right to demand that the door be open.

I turned to my companion.
I spoke to Him, and I am afraid my voice was harsh,
  and impatient.
"But why don't they come in?" I cried.
"They have a right in here—
  This is the door of my church!"

He smiled in benevolent understanding,
And He answered my brash interrogation like a tolerant Father
Speaking to a child who insists on overlooking
  the very obvious solution of a problem
  that is frustrating Him.

"Of course they have a right in here—
I am so happy to hear you say that,
It *is* the door of your church,
But *you* must open the door."

Of course I must open the door.
I smiled my thanks and hurried to do what I should have done
without His saying.

I had known—how foolish of me.
I had opened the door before.
I would show Him there was one thing I knew how to do.
I would fling open the portals!
Fling open the portals to let them in!
And there would be joy in heaven!
There would be stars in my crown!
And all these people, friends and strangers alike, would thank me.
I would fling open the portals!
And I did!

O what a happy time that was,
As the happy crowds surged around me—and past me.
I clasped hands and voiced welcomes
Until my hands ached and my voice was hoarse.
O what a rejoicing that day.

We sang songs—
We gathered in groups for discussions,
We talked of civic affairs and social problems.
We debated world governments
And we told of the missions in Alaska, and the hospitals in Iran.

Once or twice I looked around to find how my companion was faring.
But I did not see Him.
He was probably meeting the folks.
Somebody else was introducing Him to somebody else. Probably.
I would find Him later.

We played games.
We pointed with pride to our building and our organizations.
With provisions for everyone from tiny tots in cribs
to those who could hardly see or hardly walk.

We talked of our growth, and we talked of our great plans.
We had the most wonderful dinners—and the finest programs
with the greatest speakers.
I wore myself out rushing hither and thither.
What an opportunity for our Church! I must make the best of it.

But almost from the start,
I noticed that our guests seemed always to be looking for something
else, or someone else.
What could it be?

We had provided for everyone—we had thought of everything.
They were having a wonderful time—enjoying themselves.
They said so—time and time again.
But always—that wistful look!
I even asked some of them about it.
But they just smiled a bit—gazed over my shoulder,
And then said, "No—everything is fine!"

Then—they started to leave.
Just one or two at a time—
But the party had begun to pall.
They were losing their enthusiasm.
  I tried to be twice as helpful—twice as cheerful—
  But still they wandered out the door—and were gone.
I wanted them to stay—I told them so.
I even begged them, promised them more—but still—they went.

Even as they shook my hand and smiled farewell,
I saw that mysterious, longing look.
As they walked away, many of them turned and looked back,
Hesitating, as though they still hoped for something,
  They didn't know themselves just what—
  But something they greatly wanted.
They had been disappointed.
I knew I would not see them again.
My soul was sick.

When the last guest stood lingering on the threshold,
I pleaded with him to stay longer.
"No," he said, "I—there was someone I had hoped to meet.
"I thought He would be here.
  But I did not see Him."
    Then he turned—and was gone—
And I was left alone.

But I was not alone.
I turned, with my heavy heart—and saw my Companion there.
Standing where He had been when I had opened the door to my Church.
I looked at Him with all the righteous indignation I could muster.
He would see—He would understand.
  "Look," I said,
  "I opened the door.
  "They came—but they would not stay.
  "What more could I do? What more could they want?"

"Perhaps," He said,
And I saw a sadness in His eyes,
"Perhaps they wanted to meet—Me."

And then I knew.
My guilty heart could hardly bear to face Him.
"Come," I said. "Come—please."
"I'll go," He answered, "I'll go,
For I know where you are going."

Then back we went together—
Back to the little room
Wherein was the Door to the Heart of God.
And there, with His hand of blessing upon my head,
I kneeled in humble confession and repentance.

"O God," I cried, "be merciful to me, a sinner.
May I never again forget.
Though I must open the doors to my Church,
To others it is not my Church,
But Thy House.
Wherein they have a right to meet Thy Son.
May I never forget, as I bid them welcome,
That my first duty is to let them know
Thy Son is here.
That He is anxious to knock upon the door of their heart
And enter—
Even as He knocked—
And entered into mine."

* * * * * *

Three doors—and a Dream.
Or perhaps it was not a dream—
Perhaps it really happened.
Or better still—
Perhaps by the Grace of God
'Twill happen yet—
  to me—
    to you.

Let us pray:
"Oh Thou from whom can come the Dream,
And the courage to make that dream—Reality.
Hear now the prayer
  Born of our High Hope
    And Deep Desire.
The prayer that we may hear, even this hour,
Thy knock upon the door of our hearts.
That we may hear—and open—
Open unto Thee—
O Jesus Christ Our Lord—*Amen.*"

What Happens in Worship

Reverend James S. Stewart, D.D.
Professor of New Testament, New College, Edinburgh;
formerly Minister, North Morningside Church (Church of Scotland),
Edinburgh, Scotland

James Stewart's great preaching gift is one of the spiritual treasures of Scotland today. Men and women find faith in Christ at the altar when Stewart preaches. "Stewart of Morningside" is able to preach and also to teach other ministers how to preach with more effectiveness.

In his church in Edinburgh he was so successful that men thronged to church even on Sunday nights. It was worth a pilgrimage to his city to hear him. As he enlarged his work at New College to impart his preaching secrets and plans to others, he joined a long line of great men who have made New College trained men world famous. His preaching is popular in America as well as in England and Scotland and he visits America every few years for a series of preaching and speaking engagements.

His books have enriched our contemporary religious literature: A Man in Christ (*the vital elements of St. Paul's religion*), The Life and Teaching of Jesus Christ, The Gates of New Life, The Strong Name. *His latest book,* Heralds of God, *is a study of preaching and practical advice to preachers. Those who read the book will find part of the secret of his great success in preaching.*

There is something of the profound, a deep spiritual insight, and a persuasiveness in all his sermons. The sermon given here was chosen for its inspiring interpretation of the meaning of Christ's Resurrection. All of James Stewart's preaching brings the Gospel near men's hearts and shows God's redemptive love and Christ as the hope of the whole earth.

The place of the church and worship, eternal life, faith, and consecration have a large share in his thinking. This is an excellent modern textual exposition and application, both Scotch and American in its idiom. Dr. Stewart's explanation and interpretation of the spiritual meaning of worship and what takes place within the worshipper's own heart and mind is an important contribution to current religious thinking.

Sermon Fifty-one

TEXT: Ye are come unto mount Sion, and unto the city of the living God, the heavenly Jerusalem, and to an innumerable company of angels, to the general assembly and church of the firstborn, which are written in heaven, and to God the Judge of all, and to the spirits of just men made perfect, and to Jesus the mediator of the new covenant, and to the blood of sprinkling, that speaketh better things than that of Abel. See that ye refuse not him that speaketh.

HEBREWS 12:22–25

WHAT THIS writer is doing is to describe what it felt like in those early generations to belong to the Christian community—and what it ought to feel like to belong to that community still today. He is trying to give some idea of the amazingly rich inheritance which is ours in the Church, the Body of Christ. "When you meet in your places of worship," he says in effect, "with whom are you meeting?" And then he proceeds to tell them. He tells them first negatively, casting his mind back to the old days before ever there was a Christ at all. "You are not come unto the mount that might be touched, and that burned with fire, nor unto blackness, and darkness, and tempest": you have not come, that is, to Sinai, to the voice of thunder that might freeze the blood and terrify the soul; not come to the old rigid system of exclusion, where only Moses could see the face of God, and common sinners had to keep their distance and tremble in their God-forsakenness—you have not come to that! All that, having served its day, is past and finished. And then he goes on to tell them positively: "You are come to mount Zion, to myriads of angels in festal array, to the Church of the firstborn written in heaven, to the God of all who is Judge, to the spirits of the blessed departed, and to Jesus and His saving blood." This is your Christian heritage, declares this man to his readers, and says it still to you and me; this is the truth about the Church at worship. And notice precisely how he puts it. He does not say, "You shall come," forecasting something dim and far away, in another world beyond the grave. He says, "You are come": meaning, "This is what actually happens every time you meet for worship. This is the fellowship into which you enter." If only we could realise the riches of our heritage!

It is an amazing wealth of suggestion that this writer has piled up here in disorderly profusion. Can we get some order out of it? I think we can. He is saying five things about our fellowship of Christian worship in the Church.

He begins with this: it is *a spiritual fellowship.* "Ye are come unto mount Zion, the city of the living God, the heavenly Jerusalem." "You Christians," he means, "have direct touch with that invisible spiritual world which is the only ultimate reality. You are not prisoners behind the bars of a narrow earthbound existence, where men push and jostle one another for their tawdry, perishable prizes, and breathe the suffocating, poisonous air of a materialist philosophy. You are done with that! You have had your horizons stretched immeasurably. You are breathing the ampler air of spiritual truth. You have come unto mount Zion." For the bedrock reality of the universe is spiritual.

But we all know how this highway of the spirit tends to get blocked. The great Indian mystic, Rabindranath Tagore, once wrote a poem in which he compared our daily life to a narrow lane overhung with high buildings, between which there could be seen above a single strip of blue sky torn out of space. The lane, seeing the sun only for a few minutes at midday, asks herself—is it real? Feeling some wayward breeze of spring wafted in from far-off fields, she asks—Is it real? But the dust and the rubbish never rouse her to question. The noise of traffic, the jolting carts, the refuse, the smoke—these she accepts, these she concludes are clearly the real and actual things of life; and as for that strange strip of blue above, she soon ceases even to wonder about it, for so manifestly it is only a fancy, nothing real. This, says Tagore in effect, is precisely the truth about our ordinary mundane existence. The near things, the tangible, material things—these we accept, these we say are obviously the things that matter, they are solid, substantial fact: not recognising that it is that streak of blue above, that far glimpse of the spiritual, which is the essential reality for which every soul of man is made.

Yes, we forget it; and yet sometimes, thank God, it comes back to haunt us. Sometimes, as Browning knew, "a sunset touch, a fancy from a flower-bell, someone's death" will disturb our unspiritual security. And especially in the worship of the Church, and above all through the mystery of Holy Communion, the world unseen may break right in upon us. Eternity then stirs within our hearts, and we can doubt the spiritual realities no longer, and we know we are going to be restless until we rest in God. God grant that this may happen every time we come up into His courts. "Ye are come unto mount Zion, the city of the living God, the heavenly Jerusalem."

I pass to the second fact our text underlines concerning the fellowship of Christian worship: it is *a universal fellowship*. "Ye are come to the Church of the firstborn, who are written in heaven." The writer is thinking there not of the Church in glory but of the actual Christian society. In other words, this is the magnificent idea of the beloved community. You who belong to Christ, he declares, are no longer isolated and alone. You are members of the greatest fellowship on earth, the Church universal.

Now, of course, it is a great thing to be loyal to one's own congregation, to be able to say that the very stones of the building where you worship week by week are dear to your heart. That is splendid. But do let us remember the greater heritage and the wider horizon! Think of belonging to a fellowship which from its small beginnings in an upper room has grown and extended until today it is as wide as the world. Dead must he be of soul who does not feel the thrill of belonging to a fellowship like that!

Faber in one of his hymns complains—"We make God's love too narrow by false limits of our own." He might have said the very same thing about the Church. We make God's Church too narrow by false limits of our own. Too often men have built again with mistaken zeal the very barriers Christ came to level to the dust, racial barriers, class barriers, barriers of sect, denomination and government, of taste and temperament, until the whole fellowship ideal has been pitifully impoverished and restricted. But in the providence of God, such has been the innate vitality of this ideal that not all the excesses of militant nationalism, not all the outbursts of racial bitterness, not all the mad follies of the sectarian spirit, not all the pathetic spiritual exclusiveness which has claimed a monopoly of the grace of God, have ever been quite able to destroy it. It is a tremendous fact, the universal fellowship. Surely it does mean something to us all that the hymns we sing here in church are being sung in Africa, in India, in China, in the islands of the furthest seas. It does mean something that a Kagawa in Japan, a Schweitzer in Africa, a Berggrav in Europe, can feel themselves to be blood-brothers in Christ. It does mean something to me to receive, as I did not long ago, a letter from a Scots missionary in the city of Sian in the province of Shensi, telling me of a great open-air gathering of young Chinese students and others who had come together to listen to half-a-dozen of their own number speaking of—what do you think? Communism? Nonsense! "What Jesus Christ means to me." And it does matter mightily that you and I within this church today should know that we are not isolated units brought together by the fact of meeting beneath the same roof or belonging to one congregation, but that outside these walls there is a great unnumbered host worshipping the same Lord Christ, feeling towards Jesus just as we feel towards Him, thanking God for Jesus just as we thank God for Him, pledged to stand up and fight for Jesus just as we are pledged to stand up and fight for Him.

There is no other fellowship—social, political or international—which can compare for a moment with this. Some of you may remember those dreadful Nazi demonstrations which were broadcast occasionally before the War, thousands of voices shouting fanatically "Sieg Heil!"—yelling it in menacing unison, "Sieg Heil!" There is a louder chorus than that across the world today, if only we were able to hear it. And if I could stand in this pulpit and shout the words "Christ is risen" across the thousands of miles to Calabar, the Gold Coast, Uganda, Bengal, Manchuria, Jamaica, back would come the cry, drowning all other sounds, "Yes, He is risen indeed! Hallelujah!"

It is an incomparable thing, this universal fellowship. And every time you come to church, says this writer to the Hebrews, this is the fellowship you enter. "Ye are come to the Church of the firstborn, who are written in heaven."

I pass to the third description he gives of our fellowship in Christian worship. It is *an immortal fellowship*. "Ye are come to myriads of angels in festal array, and to the spirits of just men made perfect." He is away now, you see, across the river in the Church invisible and triumphant. He is away in that other world where all the celestial hosts sing to the Lord, ten thousand times ten thousand uniting in the great Te Deum of heaven: "Worthy is the Lamb that was slain to receive power, and riches, and wisdom, and strength, and honour, and glory, and blessing." And what he says is immensely important for all of us who here on earth have had to face the sorrow of separation, all for whom the poet in his bereavement spoke:

Thy voice is on the rolling air;
I hear thee where the waters run;
Thou standest in the rising sun,
And in the setting thou art fair.

For what this writer says to us is this: "When you Christians are at worship, bowing in prayer before the throne on high, then your loved ones on the other side are very near to you, and the cloud of witnesses is all round about you. In coming to commune with the world unseen, you are come to the spirits of just men made perfect—the immortal fellowship."

There is a lovely old Greek play, the *Alcestis* of Euripides, which tells how the hero Heracles, the Samson of the Greeks, once met and conquered death. There was a day when Heracles on a journey came to the palace of King Admetus; and there he found everyone desolate with grief because Death, that bitter tyrant, had carried off the fair young Queen Alcestis. Whereupon Heracles, who in his day had fought and tamed many wild beasts and dangerous monsters, the lion and the bull, the Hydra and Cerberus, offered to go out to the grave and face this last grim enemy and rob him of his prize. Away out to the lonely tomb he went; and there he met the monster Death, and grappled with him and vanquished him, and set his victim free. And the most beautiful scene in the drama is that in which Heracles comes leading by the hand someone completely covered with a white veil, and stands before the heartbroken King, and cries:

Look on her, if in aught she seems to thee
Like to thy wife. Step forth from grief to bliss.

And he lifts the veil, and there is Alcestis, alive and fair and smiling as of old. "See, O King, I give her back to thee." Of course the Greeks knew that that was myth and fable. But come across into Christianity, and it is no myth nor fable now. A greater than Heracles is here. Our Master has met the last enemy and vanquished it forever: so that today from across the river there comes, blown back to you, the sound of singing:

> Ev'n as a bird
> Out of the fowler's snare
> Escapes away,
> So is our soul set free:
> Broke are their nets,
> And thus escapèd we.

"See," says Christ to you whom death has robbed, "I give him back—I give her back—to you!" And with that, you know that your loved one is alive forever and intimately near you still.

No, they are not far away, the spirits of just men made perfect; and those who love God never meet for the last time. They are still watching over us who are left journeying and battling here, still rallying our faint hearts to cheer us on our way. It is said that when Napoleon with his Grand Army was crossing the Alps the troops at one point began to mutiny, refusing to march further, giving up the whole adventure in despair, beaten by the cold and the toilsomeness of that frightful journey. But someone suddenly had an idea: why should the band not play the Marseillaise? As soon as the notes of that thrilling, jubilant melody were heard, there in the wilds of the Alps, the light flashed back into listless eyes, and the strength returned to weary limbs, and they went on and breasted the summits and turned defeat to victory. So our dear ones, from where they dwell with Christ in glory, still cheer and urge us on; and every echo of the new song they sing is a thing to bring the light to our eyes and strength to our fainting hearts. And remember, says this writer to the Hebrews, when you come together for worship they are very near you then. It is just as if they were holding Christ's right hand, and you His left. They are as near you as that. "You are come to the spirits of just men made perfect"—the immortal fellowship.

Fourth, it is *a divine fellowship*. "Ye are come to the God of all who is Judge, and to Jesus the Mediator of the new covenant." In your worship, he tells them, reaching now the very heart of the matter, you have come to God as revealed in Jesus. And, indeed, without this all the other great things he has spoken of would be insufficient and unavailing. But beyond all these, he now says, deeper than all these, you have come right through to Christ.

Now never a congregation meets for worship but some are present perplexed and baffled and oppressed by the difficult problems of life; and the inarticulate longing of their hearts is—"Sir, we would see Jesus." Are there, indeed, any of us exempt from this experience? Somewhere, we feel sure, there must be a solution to this vexing enigma of life, somewhere a transfiguring light across all its sorrows and frustrations, somewhere an interpretation of all the discipline the human spirit has to bear. And no mere philosophy of life is adequate, no Stoic abstract view can finally avail. It is a divine heart we want to speak to, a Friend who will understand, a living love on which we may confidently lean. "Sir, we would see Jesus."

There is a characteristic little story which is told about Dr. John Duncan—"Rabbi" Duncan, as he was familiarly known. He was scholar, mystic, theologian. He knew the Hebrew language like his mother tongue. It was rumored among his students that when Rabbi Duncan said his prayers at night he prayed to God in Hebrew. One day two of them resolved to listen outside his door at the time of prayer: they would hear those flights of mysticism and theology going up to God in the Hebrew tongue. They listened, and this is what they heard:

> Gentle Jesus, meek and mild,
> Look upon a little child,
> Pity my simplicity,
> Suffer me to come to Thee.

There is nothing deeper than that. We would see Jesus. We are all one in this critical, final need. Suffer me to come to Thee!

And those happy souls who do in fact discover Christ—how different life looks to them! You know the kind of testimony they can bear. "I was battered by trouble, and Christ was my strength and stay. I was lost and groping, and Christ was my guiding light. I have trembled in the fierce fury of the storm, and Christ came to me over the waves and held me up, delivering my soul from death and turning the tempest into calm." This is the testimony.

When the earthquake broke on Philippi, and the prison walls were shaken and the doors burst open, and the gaoler rushed in on Paul and Silas, crying "What shall I do?"—I do not read that the apostles answered, "Run, man, for your life! Quick, before the walls crush you!" What they did say—and the words ring like a trumpet—was, "Believe on the Lord Jesus Christ and thou shalt be saved." This is our deepest need. "If there be one of you," cried a Covenanting preacher to his flock dispersing across the moors when the approach of the enemy was signalled, "if there be one of you, He will be the second. If there be two of you, He will be the third. You will never, never lack for company." And every time you come to worship, says this writer to the Hebrews, you can be quite certain you are coming to Jesus. Do you think any of us could stay away from church if we realized that?

One other fact about our fellowship in worship he adds, and so makes an end: it is *a redeeming fellowship*. "Ye are come to the blood of sprinkling, that speaketh better things than that of Abel." For when all is said and done, it is sin that is the trouble. It is these weak-willed, wayward hearts of ours that are so desperately undependable. It is not only the stubborn sway of the sins of the flesh: it is even more the satanic subtlety of the sins of the spirit, which makes us amid all our religious profession so terribly unlike Jesus—the pride of intellect patronising simpler folk, the pride of judgment imputing motives, the pride of religious life making us feel that we are doing rather creditably and certainly a good deal better than many others. It is these things that make our very virtues shabby and our righteousnesses like filthy rags; it is these which, when we come alive to them, may well reduce us to despair, as we realise the devastating truth that even at our best we are bankrupt utterly, and that all we have ever done or can do is nothing.

But, says this man, "Ye are come to the blood of sprinkling." Where should any of us be if that were not true? If Christianity were not above everything else a religion of redemption?

And, he goes on, it "speaketh better things than that of Abel," or as it might more accurately be translated, "it cries louder than that of Abel." "The voice of thy brother's blood," said God to Cain in the old story, "crieth unto Me from the ground": our secret sins, our frequent defeats, go crying up to God in heaven. If this were all, hope would be gone, and we could never lift our heads again. But here enters the Gospel. For when our sins cry out to God for punishment and vengeance, something else also happens—the blood of Christ cries louder, overbears and silences the very crying of our sins, and God for Christ's sake forgives.

There is a moving scene in Ian Maclaren's *Bonnie Brier Bush* where the old country doctor MacLure, who for more than forty years had been a familiar and well-loved figure in the glen where he had gone his rounds, has come himself at last to the end of the day. Beside him in the gloaming there sits his friend, the farmer Drumsheugh. The dying man asks him to read aloud to him out of the old Book; and Drumsheugh opens the Bible at the fourteenth of St. John. "Ma mither aye wantit this read tae her when she wes weak." But MacLure stops him. "It's a bonnie word, but it's no for the like o' me. It's ower gude; a' daurna tak it. Shut the buik an' let it open itsel." The farmer obeys. He shuts the book, and it opens of itself at a much thumbed page—the story of the penitent sinner in the temple who "would not lift up so much as his eyes to heaven"—and through his tears he reads these words, "God be merciful to me a sinner."

And whoever we are, and whatever we are—minister, elder, deacon, church member, social worker—we all come to that at last. God be merciful to me a sinner. And the only question that ultimately matters is, Can God be merciful to such a one as I am, who must have wearied His mercy so often? Can His patience indeed hold out to the end? Let me answer it, not in my own words, but in words that have gone on singing their way down three hundred years from the Elizabethan age, since first they were written by John Donne, Dean of St. Paul's, in his *Hymn to God the Father:*

Wilt Thou forgive that sin where I begun,
  Which was my sin, though it were done before?
Wilt Thou forgive that sin, through which I run,
  And do run still, though still I do deplore?
    When Thou hast done, Thou hast not done,
      For I have more.

Wilt Thou forgive that sin which I have won
  Others to sin, and made my sin their door?
Wilt Thou forgive that sin which I did shun
  A year or two, but wallowed in, a score?
    When Thou hast done, Thou hast not done,
      For I have more.

I have a sin of fear, that when I have spun
  My last thread, I shall perish on the shore;
But swear by Thyself, that at my death Thy Son
  Shall shine as He shines now, and heretofore;
    And having done that, Thou hast done,
      I fear no more.

And every time you come to church, says this man, you are coming to the blood of sprinkling, that forgives—everything. Everything!

"See that ye refuse not Him that speaketh."

YOM KIPPUR

And Tomorrow Will Come

RABBI CHARLES E. SHULMAN
Riverdale Temple (Reform), New York

An outstanding rabbi and a navy war veteran, Charles E. Shulman is the Rabbi of Riverdale Temple in the Bronx section of New York City, where he has undertaken the pioneer experiment of building a new congregation along the lines of liberal Judaism. To do this he gave up his position as Rabbi of the influential North Shore Congregation in Chicago, Illinois, a post he had held for sixteen years.

A graduate of Hebrew Union College, Cincinnati, Ohio; Ohio Northern University law school, and the University of Chicago, he has held pulpits in Wheeling, West Virginia, and Chicago prior to his present position. He was formerly President of the Chicago Rabbinical Association as well as Vice-President of the National Association of Jewish Chaplains. He is

Chairman of the Urban League Board of the Bronx and a member of the Greater New York Urban League Board. He is also a member of the Executive Board of the New York Board of Rabbis and a member of the Synagogue Commission of the Union of American Hebrew Congregations.

During World War II he served as a navy chaplain, attaining the rank of Lieutenant-Commander. His war services included active duty with Naval Bases in New England, and, later, as a staff member attached to Admiral Thomas C. Kinkaid, Commander of the 7th Fleet and Commander of the Allied Naval Forces of the South Pacific. His next war duty saw him attached to Admiral Kauffmann, Commander of the Philippine Sea Frontier. He was the only Rabbi among two hundred twenty-five chaplains attached to the 7th Fleet, and traveled constantly between Australia and the Philippines, with the greater part of the Southwest Pacific as his parish. Because of his outstanding service he was presented with the Navy Commendation Award.

He is the author of Problems of the Jews in the Contemporary World *and* Europe's Conscience in Decline. *He has traveled widely and is well acquainted with foreign problems as they affect Jewish life today. He has been in Europe and Palestine on several occasions and recently returned from an extended stay in Israel where he gained fresh insight into the problems affecting the Jewish people there in relationship to the Arabs.*

"And Tomorrow Will Come" shows his style of preaching, talks about the Jew of the ages, his contribution and message. It is a fine Yom Kippur sermon.

Sermon Fifty-two

TONIGHT THE JEW of the ages would converse with you. He stands before you and yet he is unseen. His voice is very clear, but it seems to come from afar. For he is of you and in your heart, yet he seems beyond your reach. He is in the synagogue which you have created. He is in the prayer which you read, in the thought which you silently voice. And his tone is not commanding. For he feels as you feel this night. It is the great eve of Yom Kippur, the Sabbath of Sabbaths, and he has come, as you have come, impelled by the spirit of this occasion, wondering, groping, yearning, caught somewhere in that vast realm of infinity bordered by yesterday and tomorrow. And as he speaks, the pages of history appear before your eyes, and you behold the world when time was yet young and growing with pangs of pain. The dead are not yet dead, and Israel is not yet ancient, but as if he were just born. You hear his familiar invocation—it seems but a brief moment since your grandfathers heard it—*solachti kidvo recho*—I have forgiven according to thy word. There is vibrant power in his utterance. And as he speaks you recognize yourself, driven to great heights at this moment. The hour is sacred. You are sacred. Your life thoughts and deeds are sacred. You listen, so still you can hear your own heart beat, and the words which sound in your ear tell of the forgotten things:

"Yesterday you gave no thought to this hour. You were engrossed in your affairs of business and pleasure. One day was as another to you. Nothing stirred you. Nothing awakened you to the consciousness that you were a Jew. And what has made time different? Why are you suddenly solemn, thoughtful, forgetful of your daily occupation and pleasure? Is it because you know that in every synagogue in the world, be it New York or Chicago, San Francisco or London, Bombay or Tel Aviv, or even behind the iron curtain in Moscow, your people are congregated at this hour to meditate and pray? Or can it be that something of the divine has touched you and made you great because it has made you think of your fellow beings looking for comfort and goodness and hope in the face of the evils that nature commits against you and that human beings add to in their thoughtless cruelty? Or can it be that you in this tiny world, lost among the gigantic universes in the galaxies of the stars, are for this one instant face to face with life, and you realize with the tenderness that comes only of deep understanding, what glorious achievement can come to the weakling called man?

"Perhaps you have not thought of it before. Perhaps you took it for granted that you are here today and gone tomorrow and that it makes little difference to the generations engulfed in eternity, since they, too, go to the land where dead dreams go. But let me tell you of life as I have seen it and as I have lived it. It is like a musical instrument. God himself touches that instrument and melody comes forth. There comes a time when the instrument, that has felt the divine touch of the Master's hand, crumbles and falls into broken ruins. But the melody has permeated to the hearts of others. It has become part of the traditions which people treasure as a rich heritage. And so the song of life does not die, but remains unbroken throughout the generations even though the instrument is clay and its hours on earth are limited. Just think! They cannot find the grave of your father Abraham. But you do not need his grave. You who are four thousand years away from him can hear the song of his life. You treasure the early childhood pictures of this patriarch who was upright in an age of evil. You cannot even imagine how they looked—those prophets standing among the hostile crowds in ancient Palestine, but their words ring in our ears and touch your hearts and make you, who are of the people of Israel, sensitive to wrong-doing, sympathetic to those who are oppressed. Those who wrote your laws cannot be identified among the ruins of the orient, but the truths they declared are imperishable and appear as if written this very day by one of your number—

Justice, justice, shalt thou pursue.
Thou shalt not go up and down as a talebearer among thy people.
Thou shalt love the Lord thy God with all thy heart.
Thou shalt love thy neighbor as thyself.
Honor thy father and thy mother that thy days on earth be prolonged.
Wash you, make you clean, put away the evil of your doings before
Mine eyes, seek justice, relieve the oppressed, aid the widow and
the fatherless.

"Those ancient Rabbis of yours with such strange names as Akiba and Hillel and Resh Lekish and Judah Hanasi and Yochanan HaSandler and Joshua ben Chananya are apparently lost in the great folios that are known as the Talmud. And yet their wisdom and their sanity and their code of holiness are found in your home life, in your thirst for education, in your desire to rise above the clods of the earth.

"You are the musical instrument of today! In you is the stored melody of these thousands of years your people have lived on earth. And tonight, at this moment, God has touched the strings of your soul and opened up the floodgates of your memory. You belong to Israel! You have a history! You occupy a place in the world! From your midst, O children of Abraham, there rose men and women who loved peace and pursued peace, and saw truth and fought for truth. In your midst the Lord God Jehovah rose from a crude fighter of battles, from a victor in gory combat to an exalted ideal of righteousness and beauty among men. In your blood there is the pride and courage of the Maccabee, the humility of the Psalmist, the wisdom of the sage, the passion of the prophet, the vision of the dreamer and the energy of the world builder. You were never lacking in warriors to fight the battles of the underdog in the struggle for existence. One age it was Amos and Jeremiah, another age it was Hillel and Shammai. In another age it was Samuel HaNagid and Spinoza. In still another age it was Benjamin Disraeli and Max Nordau. And in latter days it was Louis D. Brandeis and David Ben Gurion. You are not a narrow-minded provincial, but a world citizen. You have mingled with the nations and you have seen them in the heyday of their growth. You remember the Greece of Pericles, the Rome of Virgil and Horace, the Carthage of Hannibal, the Arabia of Al Ghazali and Avicenna, the France of Charlemagne, the Germany of Frederick the Great and the Russia of Peter and Catherine. Tomorrow I shall not remind you of these things. But tonight you would sit silent and hear. You feel near to your people. You are in the mood to reflect and to wonder.

"Aye, for one moment you shall open with me the pages of your lore and literature. You shall see yourself in every synagogue in the world. You shall extend your hand to the ages which preceded you and you shall search, as they searched, for your brothers on earth. Your understanding shall be the prayer that you weave in tribute to your God and to humanity:

"The Great Sanhedrin is in session in Jerusalem. It is forty years before the time of Jesus of Nazareth. You listen as Hillel the elder explains the meaning of Judaism in the brief moment that a man can stand on one foot—'Whatever is hateful to thee do not unto another. This is the whole law of Judaism. The rest is mere commentary.'

"You listen to the golden words of wisdom flowing from the lips of scholars who taught Saul of Tarsus his ethics and his religion before he became the Apostle Paul:

If the young tell thee, build! And the old tell thee, destroy!—follow the counsel of the elders; for often the destruction of the elders is construction, and the construction of the young is destruction.

As the ocean never freezes, so the gate of repentance is never closed.

The best preacher is the heart, the best teacher, time. The best book, the world, the best friend, God.

In Palestine it is considered a sign of descent from a good family, if any one first breaks off in a quarrel. The greatest of heroes is he who turneth an enemy into a friend.

Iron breaks the stone, fire melts iron, water extinguishes fire, the clouds drink up the water, a storm drives away the clouds, man withstands the storm, fear unmans man, wine dispels fear, sleep drives away wine, and death sweeps all away—even sleep. But Solomon the Wise says: Charity delivereth from death.

"A Rabbi is standing in the midst of a semi-circle of scholars in ancient Babylon and commenting upon individual responsibility in the Jewish life of the sixth century. That responsibility is for the moment yours also. He is saying:

He who forgets his teachings and his upbringings, causes his children to be driven in anguish from their dwelling and from their patrimony.

"A scholar is comforting his brethren. It is during the dark ages, the year twelve hundred, when the Jewish people are at the mercy of treacherous associates and ribald peasants. You hear his words and understand how much in need of comfort and hope are his people:

If the means of thy support in life be measured out scantily to thee, remember that thou hast to be thankful and grateful even for the mere privilege to breathe, and that thou must take up that suffering as a test of thy piety and a preparation for better things.

But if worldly wealth be lent to thee, exalt not thyself above thy brother, for both of ye came naked into the world, and both of ye will surely have to sleep at last together in the same dust.

"A prayer is being uttered in the fifteenth century in Spain. Perhaps it is offered in an underground chamber concealed from the fanatical agents of the Inquisition hunting for heretics. Perhaps the contrite individual has been forced to accept Christianity and is now, on the eve of his Yom Kippur, pouring out his heart to his God. You hear and you recognize the universality of human emotions expressive of sorrow and hope. You, too, have felt the beautiful words he is uttering. They were expressed originally by the exalted Hebrew poet, Ibn Gabirolf, and they are always recited on this holy night:

My God, I know my sins are numberless,
More than I can recall to memory,
Or tell their tale; yet some will I confess,
Even a few, though as a drop it be
    In all the sea.

I will declare my trespasses and sin
And peradventure silence then may fall
Upon their waves' and billows' raging din,
And Thou wilt hear from heaven when I call
    And pardon all.

"A leader is exhorting his people to leave their lives in the valley and ascend to the mountain top. He is teaching that the true holy man is the incarnation of righteousness in the service of humanity. He is laying the foundation for a new world building—the foundation made of the delicate stuff we know as dreams and hopes. He is telling of great workers engaged in forwarding the happiness of the world, great thinkers, writers, poets, visionaries. And Israel has had its share of them! He is speaking in the twentieth century—this very night—in some synagogue in America. And his thought was born in Judea thousands of years ago, and has reached him through God his Father. You know now how close you are to your people. You feel the poignantly beautiful words of the great modern Hebrew poet Tchernichovsky as he beholds an old, old synagogue which time has not destroyed. It is, he writes, as if a maternal voice whispered to him:

Peace unto thee, my son! Hast thou, too, come back?
During these hundreds of years I have seen many
Captives . . . men of renown . . . those who escaped baptism,
Some withered, others sprouted again, like grasses.
From the North and the West, they join together in
    "The Lord is One!"

"And tomorrow will come! And you will forget this brief hour when your thoughts have traveled to realms vaster than your little sphere. You will tire of hearing of your people and their history. You will wonder why Jews still recite the glories of their past. For your fathers, you will say, have lived their day, and you desire to live your own precious hours as you see fit. You will call yourselves modern and think yourselves very wise because you are informed about the meaningless jumble of present-day politics and economics. You will be impatient with the things that can bring rest to your wearied spirit, when all the latest mechanistic philosophies have led you farther and farther from peace of mind. You will become merchants again, professional men and housewives preoccupied with your distractions.

The world will narrow itself once more to fit the limits of your little city with its bridled imagination. You will demand practical evidence of the worth of the dream of Israel which you have witnessed and believed at this moment. You will content yourselves with passive contacts with your institutions and consider these your bonds that link you with your great people.

"Aye, tomorrow will come! I know that you will forget! I have seen your brothers in other cities as they came to their synagogues. They as well as you could carry this hour into a lifetime and make that lifetime beautiful. But they are lost in the valleys of grosser things. They are proud of accomplishment that is not accomplishment. They consider themselves men of the world when they are not even men of their own community. They think themselves wise, but they have not comprehended the stream of history and the quest of mankind.

"And you, who are weary of preaching in this advanced age of freedom, will travel your life paths serenely confident in your self-reliance, mistaking a wonderful experience of thousands of years for a complicated theology, denying your own existence in the espousal of an attitude which is indefinite and uncertain as it is unsatisfying to your inspired living.

"Thank your God, then, for this wonderful hour which has brought you face to face with your long life on earth! Thank your people Israel which has given you this atmosphere where you can rise to the dignity of thought! For this holy night shall remain to you as a memorial. It shall consecrate you and your labors and teach you that not until you have grown to understand your people and your place among those people can you attain to greatness. Pray that the Divine shall touch you, that from you shall sound the melody of life ennobling and sweet to the humanity near you!"

INDEX

This index has been prepared for clergy and laity alike, but particularly with the needs of the busy pastor-preacher in mind. It has been designed to be suggestive, helpful, and convenient for the man in search of sermon ideas and illustrations. To make the book his own, the reader should add other words and pages as he finds them useful.

"And as ye go, preach."